THE RISE AND FALL OF THE THIRD CHIMPANZEE

Jared Diamond is among America's most remark-able scholars. While, by appointment, a Professor of Physiology at the University of California Medical School, he is equally celebrated for his brilliant contributions to ecology and evolutionary biology, and for his explorations of remote parts of New Guinea. *The Rise and Fall of the Third Chimpanzee* won the Rhône-Poulenc Science Book Prize in 1992, as did *Guns, Germs and Steel* in 1998.

JARED DIAMOND

The Rise and Fall of the Third Chimpanzee

How Our Animal Heritage Affects the Way We Live

VINTAGE BOOKS
London

Published by Vintage 2002

34

Copyright © Jared Diamond 1991

Jared Diamond has asserted his right under the Copyright,
Designs and Patents Act, 1988 to be identified as the author
of this work

First published in Great Britain in 1991 by
Radius

Vintage
Random House, 20 Vauxhall Bridge Road,
London SW1V 2SA

www.randomhouse.co.uk

Addresses for companies within The Random House Group Limited
can be found at: www.randomhouse.co.uk/offices.htm

The Random House Group Limited Reg. No. 954009

A CIP catalogue record for this book
is available from the British Library

ISBN 9780099913801

Penguin Random House is committed to a sustainable future for
our business, our readers and our planet. This book is made from
Forest Stewardship Council® certified paper.

Printed and bound in Great Britain by Clays Ltd, St Ives plc

Dedicated to my sons
Max and Joshua,
to help them understand
where we came from
and where we may be heading

THEME

How the human species changed, within a short time,
from just another species of big mammal
to a world conqueror;
and how we acquired the capacity
to reverse all that progress overnight

CONTENTS

Maps

Figures

Illustration

PROLOGUE

It is obvious that humans are unlike all animals. It is also obvious that we are a species of big mammal, down to the minutest details of our anatomy and our molecules. That contradiction is the most fascinating feature of the human species. It is familiar, but we still have difficulty grasping how it came to be and what it means.

On the one hand, between ourselves and all other species lies a seemingly unbridgeable gulf that we acknowledge by defining a category called 'animals'. It implies that we consider centipedes, chimpanzees, and clams to share decisive features with each other but not with us, and to lack features restricted to us. Among these characteristics unique to us are the abilities to talk, write, and build complex machines. We depend completely on tools, not just on our bare hands, to make a living. Most of us wear clothes and enjoy art, and many of us believe in a religion. We are distributed over the whole Earth, command much of its energy and production, and are beginning to expand into the ocean depths and into space. We are also unique in darker attributes, including genocide, delight in torture, addictions to toxic drugs, and extermination of other species by the thousands. While a few animal species have one or two of these attributes in rudimentary form (like tool use), we still far eclipse animals even in those respects.

Thus, for practical and legal purposes, humans are not animals. When Darwin intimated in 1859 that we had evolved from apes, it is no wonder that most people initially regarded his theory as absurd and continued to insist that we had been separately created by God. Many people, including a quarter of all American college graduates, still hold to that belief today.

On the other hand, we obviously are animals, with the usual animal body parts, molecules, and genes. It is even clear what particular type of animal we are. Externally, we are so similar to chimpanzees that eighteenth-century anatomists who believed in divine creation could already recognize our affinities. Just imagine taking some normal people, stripping off their clothes, taking away all their other possessions, depriving them of the power of speech, and reducing them to grunting, without changing their anatomy at all. Put them in a cage in the zoo next to the chimp cages, and let the rest of us clothed and talking people visit the zoo. Those speechless caged people would be seen for what we all really are: a chimp that has little hair and walks upright. A zoologist from outer space would immediately classify us as just a third species of chimpanzee, along with the pygmy chimp of Zaire and the common chimp of the rest of tropical Africa.

Molecular genetic studies over the last half-a-dozen years have shown that we continue to share over ninety-eight per cent of our genes with the other two chimps. The overall genetic distance between us and chimps is even smaller than the distance between such closely related bird species as red-eyed and white-eyed vireos, or willow warblers and chiffchaffs. So we still carry most of our old biological baggage with us. Since Darwin's time, fossilized bones of hundreds of creatures variously intermediate between apes and modern humans have been discovered, making it impossible for a reasonable person to deny the overwhelming evidence. What once seemed absurd – our evolution from apes – actually happened.

Yet the discoveries of many missing links have only made the problem more fascinating, without fully solving it. The few bits of new baggage we acquired – the two per cent of our genes that differ from those of chimps – must have been responsible for all of our seemingly unique properties. We underwent some small changes with big consequences rather quickly and recently in our evolutionary history. In fact, as recently as a hundred thousand years ago that zoologist from outer space would have viewed us as just one more species of big mammal. Granted, we had a couple of curious behavioural habits, notably our control of fire and our dependence on tools, but those habits would have seemed no more curious to the extraterrestrial visitor than would the habits of beavers and bowerbirds. Somehow, within a few tens of thousands of years – a time that is almost infinitely long when measured against one person's memory but is only a tiny fraction of our species' separate history – we had begun to demonstrate the qualities that make us unique and fragile.

What were those few key ingredients that made us human? Since our unique properties appeared so recently and involved so few changes, those properties or at least their precursors must already be present in animals. What are those animal precursors of art and language, of genocide and drug abuse?

Our unique qualities have been responsible for our present biological success as a species. No other large animal is native to all the continents, or breeds in all habitats from deserts and the Arctic to tropical rainforests. No large wild animal rivals us in numbers. But among our unique qualities are two that now jeopardize our existence: our propensities to kill each other and to destroy our environment. Of course, both propensities occur in other species: lions and many other animals kill their own kind, while elephants and others damage their environment. However, these propensities are much more threatening in us than in other animals because of our technological power and exploding numbers.

There is nothing new about prophecies to the effect that the end of the world is near if we do not repent. What is new is that such a prophecy is now true, for two obvious reasons. First, nuclear weapons give us the means to wipe ourselves out quickly: no humans possessed this means before. Second, we already appropriate about forty per cent of the Earth's net productivity (that is, the net energy captured from sunlight). With the world's human population now doubling every forty-one years, we will soon have reached the biological limit to growth, at which point we will have to start fighting each other in deadly earnest for a slice of the world's fixed pie of resources. In addition, given the present rate at which we are exterminating species, most of the world's species will become extinct or endangered within the next century, but we depend on many species for our own life support.

Why rehearse these familiar depressing facts? Why try to trace the animal origins of our destructive qualities? If they really are part of our evolutionary heritage, that seems to imply that they are genetically fixed and hence unchangeable.

In fact, our situation is not hopeless. Perhaps the urge to murder strangers or sexual rivals is innate in us, but that still has not prevented human societies from attempting to thwart those instincts, and from succeeding in sparing most people the fate of being murdered. Even taking two world wars into account, proportionately far fewer people have suffered violent deaths in twentieth-century industrialized states than in stone-age tribal societies. Many modern populations enjoy longer lifespans than did humans of the past. Environmentalists do not always lose in battles with developers and destroyers. Even some genetic infirmities, such as phenylketonuria and juvenile-onset diabetes, can now be mitigated or cured. Therefore, my purpose in rehearsing our situation is to help us avoid repeating our mistakes – to use knowledge of our past and our propensities in order to change our behaviour. That is the hope behind the dedication of this book. My twin sons were born in 1987 and will reach my present age in the year 2040. What we are doing now is shaping their world.

It is not the goal of this book to propose specific solutions to our predicament, because the solutions we should adopt are already clear in broad outline. Some of those solutions include halting population growth, limiting or eliminating nuclear weapons, developing peaceful means for solving international disputes, reducing our impact on the environment, and preserving species and natural habitats. Many excellent books make detailed proposals on how to carry out these policies. Some of these policies are being implemented in some cases now; we 'just' need to implement them consistently. If we all became convinced today that they were essential, we would already know enough to start carrying them out tomorrow.

What is lacking is the necessary political will. Hence I seek to foster that will, by tracing in this book our history as a species. Our problems have deep roots tracing back to our animal ancestry. They have been growing for a long time with our increasing power and numbers, and are now steeply accelerating. We can convince ourselves of the inevitable outcome of our current short-sighted practices just by examining the many past societies that destroyed themselves by destroying their own resources, despite having less potent means of self-destruction than ours. Political historians justify the study of individual states and rulers by the opportunity to learn from the past. That justification applies even more so to the study of our history as a species, because the lessons of that study are simpler and clearer.

*

The story of our rise and fall divides into five natural parts. In the first part (Chapters One and Two) we shall follow our history from several million years ago until just before the appearance of agriculture ten thousand years ago. These two chapters deal with the evidence of bones, tools, and genes – the evidence that is preserved in the archaeological and biochemical record, and that gives us our most direct information about how we have changed. Fossilized bones and tools can often be dated, permitting us to deduce just when we changed. We shall examine the basis of the conclusion that we are still ninety-eight per cent chimps in our genes, and try to figure out what in the remaining two per cent was responsible for our great leap forward.

The second part (Chapters Three to Seven) deals with changes in the human life-cycle, which were as essential to the development of language and art as were the skeletal changes discussed in Part One. It is restating the obvious to mention that we feed our children after the age of weaning, instead of leaving them to find food on their own; that most adult men and women associate in couples; that most fathers as well as mothers care for their children; that many people live long enough to experience being grandparents; and that women undergo menopause. To us, these traits are the norm, but by the standards of our closest animal relatives they are bizarre. They constitute major changes from our ancestral condition, though they do not fossilize and so we do not know when they arose. For that reason they receive much briefer treatment in human paleontology texts than do our changes in brain size and pelvis, but they were crucial to our uniquely human cultural development, and merit equal attention.

With Parts One and Two having surveyed the biological underpinnings of our cultural flowering, Part Three (Chapters Eight to Twelve) considers the cultural traits that we believe distinguish us from animals. Those that come to mind first are the ones of which we are proudest: language, art, technology, and agriculture, the hallmarks of our rise. Yet our distinguishing cultural traits also include black marks on our record, such as abuse of toxic chemicals. While one can debate whether all these hallmarks rank as uniquely human, they at least constitute huge advances on animal precursors. But animal precursors there must have been, since these traits flowered only recently on an evolutionary time scale. What were those precursors? Was their flowering inevitable in the history of life on Earth, for example, so inevitable that we expect there to be many other planets out in space, inhabited by creatures as advanced as ourselves?

Besides chemical abuse, our self-destructive traits include two serious enough that they may lead to our fall. Part Four (Chapters Thirteen to Sixteen) considers the first of these: our propensity for xenophobic killing of other human groups. This trait has direct animal precursors – namely, the contests between competing individuals and groups that, in many species besides our own, may be resolved by murder. We have merely used our technological prowess to improve our killing power. In Part Four we shall consider the xenophobia and extreme isolation that marked the human condition before the rise of political states began to make us more homogenous culturally. We shall see how technology, culture, and geography affected the outcome of two of the most familiar historical sets of contests between human groups. We shall then survey the worldwide recorded history of xenophobic mass murder. This is painful material, but here above all is an example of how our refusal to face up to our history condemns us to repeat past mistakes on a more dangerous scale.

The other dark trait that now threatens our survival is our accelerating assault on our environment. This too has its direct animal precursors. Animal populations that for one reason or another escaped control by predators and parasites have in some cases also escaped their own internal controls on their numbers, multiplied until they damaged their resource base, and occasionally have eaten their way into extinction. Such a risk applies with special force to humans, because predation on us is now negligible, no habitat is beyond our influence, and our power to kill individual animals and destroy habitats is unprecedented.

Unfortunately, many people still cling to the Rousseau-esque fantasy that this tendency appeared in us only with the Industrial Revolution, before which we lived in harmony with Nature. If that were true, we would have nothing to learn from the past except how virtuous we once were, and how evil we have now become. Hence Part Five (Seventeen to Nineteen) seeks to dismantle this fantasy by facing up to our long history of environmental mismanagement. In Part Five as in Part Four, the emphasis is on recognizing that our present situation is not novel, except in degree. The experiment has already been run many times, and the outcome is there for us to learn from.

This book concludes with an epilogue that traces our rise from animal status. It also traces the acceleration in our means to bring about our fall. I would not have written this book if I thought that the risk was remote, but I also would not have written it if I considered our situation hopeless. Lest any readers get so discouraged by our track record and present predicament that they overlook this message, I point out the hopeful signs and the ways in which we can learn from the past. For those of you who would like suggestions for further reading, a section at the end will guide you to more books and articles on the material of each chapter.

*

A volume that ranges over such a broad canvas as this one has to be selective. Every reader is bound to find some absolutely crucial favourite subjects omitted and some other subjects pursued in inordinate detail. So that you will not feel you were misled, I shall lay out at the start my own particular interests, and where they come from.

My father is a physician, my mother a musician with a gift for languages. Whenever I was asked as a child about my career plans, my response was that I wanted to be a doctor like my father. By my last year in college, that goal had become gently transformed into the related goal of medical research, and so I trained in physiology, the area in which I now teach and do research at the University of California Medical School in Los Angeles.

However, I had also become interested at the age of seven in bird-watching, and I had been fortunate to go to a school that let me delve into languages and history. After I got my PhD., the prospect of devoting the rest of my life to the single professional interest of physiology began to look increasingly oppressive. At that point a happy constellation of events and people gave me the chance to spend a summer in the highlands of New Guinea. Ostensibly, the purpose of my trip was to measure nesting success of New Guinea birds, a project that collapsed dismally within a few weeks when I found myself unable to locate even a single bird's nest in the jungle. Yet the real purpose of the trip succeeded completely: to indulge my thirst for adventure and bird-watching in one of the wildest remaining parts of the world. What I saw then of New Guinea's fabulous birds, including its bowerbirds and birds of paradise, led me to develop a parallel second career in bird ecology, evolution, and biogeography. Since then, I have returned to New Guinea and the neighbouring Pacific islands a dozen times to pursue my bird research.

I found it hard to work in New Guinea amid the accelerating destruction of the birds and forests that I loved, without getting involved in conservation biology. So I began to combine my academic research with practical work as a consultant for governments, by applying what I knew about animal distributions to designing national park systems and surveying their proposed national parks. It was also hard to work in New Guinea, where languages replace each other every twenty miles, and where learning bird names in each local language proved to be the key to tapping New Guineans' encyclopedic knowledge of their birds, without returning to my earlier interest in languages. Most of all, it was hard to study the evolution and extinction of bird species without wanting to understand the evolution and possible extinction of *Homo sapiens*, by far the most interesting species of all. That interest, too, was especially difficult to ignore in New Guinea, with its enormous human diversity.

Those are the paths by which I came to be interested in the particular aspects of humans that are emphasized in this book. I do not feel as if I am thereby making excuses for inappropriately slanted coverage. Numerous excellent books by anthropologists and archaeologists already discuss human evolution in terms of tools and bones, which this book can therefore summarize more briefly. However, those other volumes devote much less space to my particular interests of the human life-cycle, human geography, human impact on the environment, and humans as animals. Those subjects are as central to human evolution as are the more traditional subjects involving tools and bones.

What may at first seem here to be a plethora of examples drawn from New Guinea is also, I believe, appropriate. Granted, New Guinea is just one island, located in a particular part of the world (the tropical Pacific), and hardly providing a random cross-section of modern humanity. But New Guinea harbours a much bigger slice of humanity than you would at first guess from its area. About a thousand of the world's approximately 5,000 languages are spoken only in New Guinea. Much of the cultural diversity that survives in the modern world is contained within New Guinea. All highland peoples in New Guinea's mountainous interior were stone-age farmers until very recently, while many lowland groups were nomadic hunter-gatherers and fishermen practising somewhat casual agriculture. Local xenophobia was extreme, cultural diversity correspondingly so, and travel outside one's tribal territory would have been suicidal. Many of the New Guineans who have worked with me are deadly and expert hunters who lived out their childhood in the days of stone tools and xenophobia. Thus, New Guinea is as good a model as we have left today of what much of the rest of the human world was like until recently.

PART ONE

JUST ANOTHER SPECIES
OF BIG MAMMAL

THE CLUES ABOUT WHEN, WHY, AND IN WHAT WAYS WE CEASED TO BE JUST another species of big mammal come from three types of evidence. Part One considers some of the traditional evidence from archaeology, which studies fossil bones and preserved tools, plus newer evidence from molecular biology. Other evidence from studies of living apes and people will be taken up in Parts Two and Three.

One basic question concerns just how extensive the genetic differences between ourselves and chimps are. That is, do we differ in ten, fifty, or ninety-nine per cent of our genes? Merely looking at humans and chimps or counting up visible traits would not be any help, because many genetic changes have no visible effects at all, while other changes have sweeping effects. For example, the visible differences between breeds of dogs such as great danes and pekinese are far greater than those between chimps and ourselves. Yet all dog breeds are interfertile, breed with each other (insofar as it is mechanically feasible) when given the opportunity, and belong to the same species. To a naive observer, the appearance of great danes and pekinese would suggest that they are genetically much further apart than chimps are from humans. Those visible differences among dog breeds in size, proportions, and hair colour depend on relatively few genes which have negligible consequences for reproductive biology.

How, then, can we estimate our genetic distance from chimps? Chapter One describes how this problem has been solved only within the past half a dozen years by molecular biologists. The answer is not just intellectually surprising but may also have some practical ethical implications for how we treat chimps. We shall see that gene differences between us and chimps, although large compared to those among living human populations or among breeds of dogs, are still small compared to differences among many other familiar pairs of related species. Evidently, changes in only a small percentage of chimpanzee genes had enormous consequences for our behaviour. It has also proved possible to work out a calibration between genetic distance and elapsed time, and thereby to get an approximate answer to the question of when we and chimps split apart from our common ancestor. That turns out to be somewhere around seven million years ago, give or take a few million years.

While the molecular biological story of the first chapter yields overall measures of genetic distance and elapsed time, it tells us nothing about how specifically we differ from chimps, and when those specific differences appeared. Hence Chapter Two will consider what more can be learned from bones and tools left by creatures variously intermediate between our ape-like ancestor and modern humans. The changes in bones constitute the traditional subject matter of physical anthropology. Especially important were our increase in brain size, skeletal changes associated with walking upright, and decreases in skull thickness, tooth size, and jaw muscles.

Our large brain was surely prerequisite for the development of human language and innovativeness. One might therefore expect the fossil record to show a close parallel between increased brain size and sophistication of tools. In fact, the parallel is not at all close. This proves to be the greatest surprise and puzzle of human evolution. Stone tools remained very crude for hundreds of thousands of years after we had undergone most of our expansion of brain size. As recently as 40,000 years ago, Neanderthals had brains even larger than those of modern humans, yet their tools show no signs of innovativeness and art. Neanderthals were still just another species of big mammal. Even for tens of thousands of years after some other human populations had achieved virtually modern skeletal anatomy, their tools too remained as boring as those of Neanderthals.

These paradoxes sharpen the conclusion drawn from Chapter One. Within the modest percentage of genes that differs between us and chimps, there must have been an even smaller percentage of genes which were not involved in the shapes of our bones, but which were responsible for the distinctively human traits of innovation, art, and complex tools. At least in Europe, those traits appear unexpectedly suddenly, at the time of the replacement of Neanderthals by Cro-Magnons. That is the time when we finally ceased to be just another species of big mammal. In Chapter Two I shall speculate about what those few changes were that triggered our steep rise to human status.

ONE

A Tale of Three Chimps

*By what percentage of our genes do we differ from (the other two)
chimpanzees? And what implications does that number have?
Darwin himself would have been surprised by the answers.*

*

The next time that you visit a zoo, make a point of walking past the ape
cages. Imagine that the apes had lost most of their hair, and imagine a cage
nearby holding some unfortunate people who had no clothes and
couldn't speak but were otherwise normal. Now try guessing how
similar those apes are to ourselves genetically. For instance, would you
guess that a chimpanzee shares ten, fifty, or ninety-nine per cent of its
genes with humans?

Then ask yourself why those apes are on exhibit in cages, and why other
apes are being used for medical experiments, while it is not permissible to do
either of those things to humans. Suppose it turned out that chimps shared
99.9% of their genes with us, and that the important differences between
humans and chimps were due to just a few genes. Would you still think it is
okay to put chimps in cages and to experiment on them? Consider those
unfortunate mentally-defective people who have much less capacity to solve
problems, to care for themselves, to communicate, to engage in social
relationships, and to feel pain, than do apes. What is the logic that forbids
medical experiments on those people, but not on apes?

You might answer that apes are 'animals', while humans are humans,
and that is enough. An ethical code for treating humans should not be
extended to an 'animal', no matter what percentage of its genes it shares
with us, and no matter what its capacity for social relationships or for
feeling pain. That is an arbitrary but at least self-consistent answer that
cannot be lightly dismissed. In that case, learning more about our
ancestral relationships will not have any ethical consequences, but it will
still satisfy our intellectual curiosity to understand where we come from.
Every human society has felt a deep need to make sense of its origins, and
has answered that need with its own story of the Creation. The Tale of
Three Chimps is the creation story of our time.

*

For centuries it has been clear approximately where we fit into the animal kingdom. We are obviously mammals, the group of animals characterized by having hair, nursing their young, and other features. Among mammals we are obviously primates, the group of mammals including monkeys and apes. We share with other primates numerous traits lacking in most other mammals, such as flat fingernails and toenails rather than claws, hands for gripping, a thumb that can be opposed to the other four fingers, and a penis that hangs free rather than being attached to the abdomen. Already by the Second Century AD, the Greek physician Galen deduced our approximate place in Nature correctly when he dissected various animals and found that a monkey was 'most similar to man in viscera, muscles, arteries, veins, nerves and in the form of bones'.

It is also easy to place us within the primates, among which we are obviously more similar to apes than to monkeys. To name only one of the most visible signs, monkeys sport tails, which we lack along with apes. It is also clear that gibbons, with their small size and very long arms, are the most distinctive apes, and that orangutans, chimpanzees, gorillas, and humans are all more closely related to each other than any of them is to gibbons. But to go further with our relationships proves unexpectedly difficult. It has provoked an intense scientific debate, which revolves around three questions including the one that I posed in the first paragraph of this chapter:

What is the detailed family tree of relationships among humans, the living apes, and extinct ancestral apes? For example, which of the living apes is our closest relative?

When did we and that closest living relative, whichever ape it is, last share a common ancestor?

What fraction of our genes do we share with that closest living relative?

At first, it would seem natural to assume that comparative anatomy had already solved the first of those three questions. We look especially like chimpanzees and gorillas, but differ from them in obvious features such as our larger brains, upright posture, and much sparser body hair, as well as in many more subtle points. However, on closer examination these anatomical facts are not decisive. Depending on what anatomical characters one considers most important and how one interprets them, biologists differ on whether we are most closely related to the orangutan (the minority view), with chimps and gorillas having branched off our family tree before we split off from orangutans, or whether we are instead closest to chimps and gorillas (the majority view), with the ancestors of orangutans having gone their separate way earlier.

Within the majority, most biologists have thought that gorillas and chimps are more like each other than either is like us, implying that we branched off before the gorillas and chimps diverged from each other. This conclusion reflects the common-sense view that chimps and gorillas can be lumped in a category termed 'apes', while we are something different. However, it is also conceivable that we look distinct only because chimps and gorillas have not changed much since we shared a common ancestor with them, while we were changing greatly in a few important and highly visible features like upright posture and brain size. In that case, humans might be most similar to gorillas, or humans might be most similar to chimps, or humans and gorillas and chimps might be roughly equidistant from each other, in overall genetic make-up.

Hence, anatomists have continued to argue about the first question, the details of our family tree. Whichever tree one prefers, anatomical studies by themselves tell us nothing about the second and third questions, our time of divergence and genetic distance from apes. Perhaps fossil evidence might in principle solve the questions of the correct ancestral tree and of dating, though not the question of genetic distance. If we had abundant fossils, we might hope to find a series of dated proto-human fossils and another series of dated proto-chimp fossils converging on a common ancestor around ten million years ago, converging in turn on a series of proto-gorilla fossils twelve million years ago. Unfortunately, that hope for insight from the fossil record has also been frustrated, because almost no ape fossils of any sort have been found for the crucially relevant period between five and fourteen million years ago in Africa.

*

The solution to these questions about our origins came from an unexpected direction: molecular biology as applied to bird taxonomy. About thirty years ago, molecular biologists began to realize that the chemicals of which plants and animals are composed might provide 'clocks' by which to measure genetic distances and to date times of evolutionary divergence. The idea is as follows. Suppose there is some class of molecules that occurs in all species, and whose particular structure in each species is genetically determined. Suppose further that that structure changes slowly over the course of millions of years because of genetic mutations, and that the rate of change is the same in all species. Two species derived from a common ancestor would start off with identical forms of the molecule, which they inherited from that ancestor, but mutations would then occur independently and produce structural changes between the molecules of the two species. The two species' versions of the molecule would gradually diverge in structure. If we knew how many structural changes occur on the average every million years, we could then use the difference today in the molecule's structure between any two related animal species as a clock, to calculate how much time had passed since the species shared a common ancestor.

For instance, suppose one knew from fossil evidence that lions and tigers diverged five million years ago. Suppose the molecule in lions were ninety-nine per cent identical in structure to the corresponding molecule in tigers and differed only by one per cent. If one then took a pair of species of unknown fossil history and found that the molecule differed by three per cent between those two species, the molecular clock would say that they had diverged three times five million, or fifteen million, years ago.

Neat as this scheme sounds on paper, testing whether it succeeds in practice has cost biologists much effort. Four things had to be done before molecular clocks could be applied: find the best molecule; find a quick way of measuring changes in its structure; prove that the clock runs steady (that is, that the molecule's structure really does evolve at the same rate among all species that one is studying); and measure what that rate is.

Molecular biologists worked out the first two of these problems by around 1970. The best molecule proved to be deoxyribonucleic acid (abbreviated to DNA), the famous substance whose structure James Watson and Francis Crick showed to consist of a double helix, thereby revolutionizing the study of genetics. DNA is made up of two complementary and extremely long chains, each made up of four types of small molecules whose sequence within the chain carries all the genetic information transmitted from parents to offspring. A quick method of measuring changes in DNA structure is to mix the DNA from two species, then to measure by how many degrees of temperature the melting point of the mixed (hybrid) DNA is reduced below the melting point of pure DNA from a single species. Hence the method is generally referred to as DNA hybridization. As it turns out, a melting point lowered by one degree centigrade (abbreviated: delta $T = 1°C$) means that the DNA's of the two species differ by roughly one per cent.

*

In the 1970s most molecular biologists and most taxonomists had little interest in each other's work. Among the few taxonomists who appreciated the potential power of the new DNA hybridization technique was Charles Sibley, an ornithologist then serving as Professor of Ornithology and Director at Yale's Peabody Museum of Natural History. Bird taxonomy is a difficult field because of the severe anatomical constraints imposed by flight. There are only so many ways to design a bird capable, say, of catching insects in mid-air, with the result that birds of similar habits tend to have very similar anatomies, whatever their ancestry. For example, American vultures look and behave much like Old World vultures, but biologists have come to realize that the former are related to storks, the latter to hawks, and that their resemblances result from their common lifestyle. Frustrated by the shortcomings of traditional methods for deciphering bird relationships, Sibley and Jon Ahlquist turned in 1973 to the DNA clock, in the most massive application to date of the methods of molecular biology to taxonomy. Not until 1980 were Sibley and Ahlquist ready to begin publishing their results, which eventually came to encompass applying the DNA clock to about 1,700 bird species – nearly one-fifth of all living birds.

While Sibley's and Ahlquist's achievement was a monumental one, it initially caused much controversy because so few other scientists possessed the blend of expertise required to understand it. Here are typical reactions I heard from my scientist friends:

'I'm sick of hearing about that stuff. I no longer pay attention to anything those guys write,' (an anatomist).

'Their methods are okay, but why would anyone want to do something so boring as all that bird taxonomy?' (a molecular biologist).

'Interesting, but their conclusions need a lot of testing by other methods before we can believe them,' (an evolutionary biologist).

'Their results are The Revealed Truth, and you better believe it,' (a geneticist).

My own assessment is that the last view will prove to be the most nearly correct one. The principles on which the DNA clock rests are unassailable; the methods used by Sibley and Ahlquist are state-of-the-art; and the internal consistency of their genetic-distance measurements from over 18,000 hybrid pairs of bird DNA testifies to the validity of their results.

Just as Darwin had the good sense to marshal his evidence for variation in barnacles before discussing the explosive subject of human variation, Sibley and Ahlquist similarly stuck to birds for most of the first decade of their work with the DNA clock. Not until 1984 did they publish their first conclusions from applying the same DNA methods to human origins, and they refined their conclusions in later papers. Their study was based on DNA from humans and from all of our closest relatives: the common chimpanzee, pygmy chimpanzee, gorilla, orangutan, two species of gibbons, and seven species of Old World monkeys. The figure on this page summarizes the results.

As any anatomist would have predicted, the biggest genetic difference,

FAMILY TREE OF THE HIGHER PRIMATES

Trace back each pair of modern higher primates to the black dot connecting them. The numbers to the left then give the percentage difference between the DNAs of those modern primates, while the numbers to the right give the estimated number of millions of years since they last shared a common ancestor. For example, the common and pygmy chimps differ in about 0.7% of their DNA and diverged about three million years ago; we differ in 1.6% of our DNA from both chimps and diverged from their common ancestor about seven million years ago; gorillas differ in about 2.3% of their DNA from us or from chimps and diverged from the common ancestor leading to us and the two chimps about ten million years ago.

expressed in a big DNA melting point lowering, is between monkey DNA and the DNA of humans or of any ape. This simply puts a number on what everybody has agreed ever since apes first became known to science: that humans and apes are more closely related to each other than either are to monkeys. The actual statistic is that monkeys share ninety-three per cent of their DNA structure with humans and apes, and differ in seven per cent.

Equally unsurprising is the next biggest difference, one of five per cent between gibbon DNA and the DNA of other apes or humans. This too confirms the accepted view that gibbons are the most distinct apes, and that our affinities are instead with gorillas, chimpanzees, and orangutans. Among those latter three groups of apes, most recent anatomists have considered the orangutan as somewhat separate, and that conclusion too fits the DNA evidence: a difference of 3.6% between orangutan DNA and that of humans, gorillas, or chimpanzees. Geography confirms that the latter three species parted from gibbons and orangutans quite some time ago: living and fossil gibbons and orangutans are confined to Southeast Asia, while living gorillas and chimpanzees plus early fossil humans are confined to Africa.

At the opposite extreme but equally unsurprising, the most similar DNAs are those of common chimpanzees and pygmy chimpanzees, which are 99.3% identical and differ by only 0.7%. So similar are these two chimp species in appearance that it was not until 1929 that anatomists even bothered to give them separate names. Chimps living on the equator in central Zaire rate the name 'pygmy chimps' because they are on average slightly smaller (and have more slender builds and longer legs) than the widespread 'common chimps' ranging across Africa just north of the equator. However, with the increased knowledge of chimp behaviour acquired in recent years, it has become clear that the modest anatomical differences between pygmy and common chimps mask considerable differences in reproductive biology. Unlike common chimps but like ourselves, pygmy chimps assume a wide variety of positions for copulation, including face-to-face; copulation can be initiated by either sex, not just by the male; females are sexually receptive for much of the month, not just for a briefer period in mid-month; and there are strong bonds among females or between males and females, not just among males. Evidently, those few genes (0.7%) that differ between pygmy and common chimps have big consequences for sexual physiology and roles. That same theme – a small percentage of gene differences having great consequences – will recur later in this and the next chapter in regard to the gene differences between humans and chimps.

In all the cases that I have discussed so far, anatomical evidence of relationships was already convincing, and the DNA-based conclusions confirmed what the anatomists had already concluded. But DNA was also able to resolve the problem at which anatomy had failed – the relationships between humans, gorillas, and chimpanzees. As the figure on page 17 shows, humans differ from both common chimps and pygmy chimps in about 1.6% of their (our) DNA, and share 98.4%. Gorillas differ somewhat more, by about 2.3%, from us and from both of the chimps.

Let us pause to let some of the implications of these momentous numbers sink in.

*

The gorilla must have branched off from our family tree slightly before we separated from the common and pygmy chimpanzees. The chimpanzees, not the gorilla, are our closest relatives. Put another way, the chimpanzees' closest relative is not the gorilla but the human. Traditional taxonomy has reinforced our anthropocentric tendencies by claiming to see a fundamental dichotomy between mighty man, standing alone on high, and the lowly apes all together in the abyss of bestiality. Now future taxonomists may see things from the chimpanzees' perspective: a weak dichotomy between slightly higher apes (the *three* chimpanzees, including the 'human chimpanzee') and slightly lower apes (gorilla, orangutan, gibbons). The traditional distinction between 'apes' (defined as chimps, gorillas, etc.) and humans misrepresents the facts.

The genetic distance (1.6%) separating us from pygmy or common chimps is barely double that separating pygmy from common chimps (0.7%). It is less than that between two species of gibbons (2.2%), or between such closely related North American bird species as red-eyed vireos and white-eyed vireos (2.9%), or between such closely related and hard-to-distinguish European bird species as willow warblers and chiffchaffs (2.6%). The remaining 98.4% of our genes are just normal chimp genes. For example, our principal haemoglobin, the oxygen-carrying protein that gives blood its red colour, is identical in all 287 units with chimp haemoglobin. In this respect as in most others, we are just a third species of chimpanzee, and what is good enough for common and pygmy chimps is good enough for us. Our important visible distinctions from the other chimps – our upright posture, large brains, ability to speak, sparse body hair, and peculiar sexual lives (of which I will say more in Chapter Three) – must be concentrated in a mere 1.6% of our genes.

If genetic distances between species accumulated at a uniform rate with time, they would function as a smoothly ticking clock. All that would be required to convert genetic distance into absolute time since the last common ancestor would be a calibration, furnished by a pair of species for which we know *both* the genetic distance *and* the time of divergence as dated independently by fossils. In fact, two independent calibrations are available for higher primates. On the one hand, monkeys diverged from apes between twenty-five and thirty million years ago according to fossil evidence, and now differ in about 7.3% of their DNA. On the other hand, orangutans diverged from chimps and gorillas between twelve and sixteen million years ago and now differ in about 3.6% of their DNA. Comparing these two examples, a doubling of evolutionary time, as one goes from twelve or sixteen to twenty-five or thirty million years, leads to a doubling of genetic distance (3.6 to 7.3% of DNA). Thus, the DNA clock has ticked relatively steadily among higher primates.

With those calibrations, Sibley and Ahlquist estimated the following time scale for our evolution. Since our own genetic distance from chimps (1.6%) is about half the distance of orangutans from chimps (3.6%), we must have been going our separate way for about half of the twelve to sixteen million years that orangutans had to accumulate their genetic distinction from chimps. That is, the human and 'other chimp' evolutionary lines diverged around six to eight million years ago. By the same reasoning, gorillas parted from the common ancestor of us three chimpanzees around nine million years ago, and the pygmy and common chimps diverged around three million years ago. In contrast, when I took physical anthropology as a college freshman in 1954, the assigned textbooks said that humans diverged from apes fifteen to thirty million years ago. Thus, the DNA clock strongly supports a controversial conclusion also drawn from several other molecular clocks based on amino acid sequences of proteins, mitochondrial DNA, and globin pseudogene DNA. Each clock indicates that humans have had only a short history as a species distinct from other apes, much shorter than paleontologists used to assume.

*

What do these results imply about our position in the animal kingdom? Biologists classify living things in hierarchical categories, each less distinct than the next: subspecies, species, genus, family, superfamily, order, class, and phylum. The *Encyclopaedia Britannica* and all the biology texts on my shelf say that humans and apes belong to the same order, called Primates, and the same superfamily, called Hominoidea, but to separate families, called Hominidae and Pongidae. Whether Sibley's and Ahlquist's work changes this classification depends on one's philosophy of taxonomy. Traditional taxonomists group species into higher categories by making somewhat subjective evaluations of how important the differences between species are. Such taxonomists place humans in a separate family because of distinctive functional traits like large brain and bipedal posture, and this classification would remain unaffected by measures of genetic distance.

However, another school of taxonomy, called cladistics, argues that classification should be objective and uniform, based on genetic distance or times of divergence. All taxonomists agree now that red-eyed and white-eyed vireos belong together in the genus *Vireo*, willow warblers and chiffchaffs in the genus *Phylloscopus*, the various species of gibbons in the genus *Hylobates*. Yet the members of each of these pairs of species are genetically more distant from each other than are humans from the other two chimpanzees, and diverged longer ago. On this basis, then, humans do not constitute a distinct family, nor even a distinct genus, but belong in the same genus as common and pygmy chimps. Since our genus name *Homo* was proposed first, it takes priority, by the rules of zoological nomenclature, over the genus name *Pan* coined for the 'other' chimps. Thus, there are not one but three species of genus *Homo* on Earth today: the common chimpanzee, *Homo troglodytes*; the pygmy chimpanzee, *Homo paniscus*; and the third chimpanzee or human chimpanzee, *Homo sapiens*. Since the gorilla is only slightly more distinct, it has almost equal right to be considered a fourth species of *Homo*.

Even taxonomists espousing cladistics are anthropocentric, and the lumping of humans and chimps into the same genus will undoubtedly be a bitter pill for them to swallow. There is no doubt, however, that whenever chimpanzees learn cladistics, or whenever taxonomists from outer space visit Earth to inventory its inhabitants, they will unhesitatingly adopt the new classification.

*

Which particular genes are the ones that differ between humans and chimps? Before we can consider this question, we need first to understand what it is that DNA, our genetic material, does.

Much or most of our DNA has no function and may just constitute 'molecular junk': that is, DNA molecules that have become duplicated or have lost former functions, and that natural selection has not eliminated from us because they do us no harm. Of our DNA that does have known functions, the main ones have to do with the long chains of amino acids called proteins. Certain proteins make up much of our body's structure (such as the proteins keratin, of hair, or collagen, of connective tissue), while other proteins, termed enzymes, synthesize and break down most of our body's remaining molecules. The sequences of the component small molecules (nucleotide bases) in DNA specify the sequence of amino acids in our proteins. Other parts of our functional DNA regulate protein synthesis.

Those of our observable features that are easiest to understand genetically are ones arising from single proteins and single genes. For instance, our blood's oxygen-carrying protein haemoglobin, already mentioned, consists of two amino acid chains, each specified by a single chunk of DNA (a single 'gene'). These two genes have no observable effects except through specifying the structure of haemoglobin, which is confined to our red blood cells. Conversely, haemoglobin's structure is totally specified by those genes. What you eat or how much you exercise may affect how much haemoglobin you make, but not the details of its structure.

That is the simplest situation, but there are also genes influencing many observable traits. For example, the fatal genetic disorder known as Tay–Sachs disease involves many behavioural as well as anatomical anomalies: excessive drooling, rigid posture, yellowish skin, abnormal head growth, and other changes. We know in this case that all these observable effects result somehow from changes in a single enzyme specified by the Tay–Sachs gene, but we do not know exactly how. Since that enzyme occurs in many tissues of our bodies and breaks down a widespread cellular constituent, changes in that one enzyme have wide-ranging and ultimately fatal consequences. Conversely, some traits, such as your height as an adult, are influenced simultaneously by many genes and also by environmental factors (for example, your nutrition as a child).

While scientists understand well the function of numerous genes that specify known individual proteins, we know much less about the function of genes involved in more complex determinations of traits, such as most behavioural features. It would be absurd to think that human hallmarks such as art, language, or aggression depend on a single gene. Behavioural differences among individual humans are obviously subject to enormous environmental influences, and what role genes play in such individual differences is a controversial question. However, for those consistent behavioural differences between chimps and humans, genetic differences are likely to be involved in those species' differences, even though we cannot yet specify the genes responsible. For instance, the ability of humans but not chimps to speak surely depends on differences in genes specifying the anatomy of the voice box and the wiring of the brain. A young chimpanzee brought up in a psychologist's home along with the psychologist's human baby of the same age still continued to look like a chimp and did not learn to talk or walk erect. But whether an individual human grows up to be fluent in English or Korean is independent of genes and dependent solely on its childhood linguistic environment, as proved by the linguistic attainments of Korean infants adopted by English-speaking parents.

With this as background, what can we say about the 1.6% of our DNA that differs from chimp DNA? We know that the genes for our principal haemoglobin do not differ, and that certain other genes do exhibit minor differences. In the nine protein chains studied to date in both humans and common chimps, only five out of a total of 1,271 amino acids differ: one amino acid in a muscle protein called myoglobin, one in a minor haemoglobin chain called the delta chain, and three in an enzyme called carbonic anhydrase. But we do not yet know which chunks of our DNA are responsible for the functionally significant differences between humans and chimps to be discussed in Chapters Two to Seven: the differences in brain size, anatomy of the pelvis and voice box and genitalia, amount of body hair, female menstrual cycle, menopause, and other traits. Those important changes certainly do not arise from the five amino acid differences detected to date. At present, all we can say with confidence is this: much of our DNA is junk; at least some of the 1.6% that differs between us and chimps is already known to be junk; and the functionally significant differences must be confined to some as-yet-unidentified small fraction of 1.6%.

While we do not know which particular genes are the crucial ones, there are numerous precedents for one or a few genes having a big impact. I just mentioned the many large and visible differences between Tay–Sachs patients and normal people, all somehow arising from a single change in one enzyme. That is an example of differences among individuals of the same species. As for differences between related species, a good example is provided by the cichlid fishes of Africa's Lake Victoria. Cichlids are popular aquarium species, of which about two hundred are confined to that one lake, where they evolved from a single ancestor within perhaps the last 200,000 years. Those two hundred species differ among themselves in their food habits as much as do tigers and cows. Some graze on algae, others catch other fish, and still others variously crush snails, feed on plankton, catch insects, nibble the scales off other fish, or specialize in grabbing fish embryos from brooding mother fish. Yet all those Lake Victoria cichlids differ from each other on the average by only about 0.4% of their DNA studied. Thus it took even fewer genetic mutations to change a snail-crusher into a specialized baby-killer than it took to produce us from an ape.

*

Do the new results about our genetic distance from chimps have any broader implications, besides technical questions of taxonomic names? Probably the most important implications concern how we think about the place of humans and apes in the universe. Names are not just technical details but express and create attitudes. (To convince yourself, try greeting your spouse this evening either as 'my darling' or as 'you swine', using the same expression and tone of voice.) The new results do not specify how we *should* think about humans and apes, but, just as did Darwin's *On the Origin of Species*, they will probably influence how we *do* think, and it will probably take us many years to readjust our attitudes. I shall mention just one example of a controversial area that might be affected: our use of apes.

At present we make a fundamental distinction between animals (including apes) and humans, and this distinction guides our ethical code and actions. For instance, as I noted at the start of this chapter, it is considered acceptable to exhibit caged apes in zoos, but it is not acceptable to do the same with humans. I wonder how the public will feel when the identifying label on the chimp cage in the zoo reads 'Homo troglodytes'. Yet if it were not for the sympathetic interest in apes that many people gain at zoos, there might be much less public financial support for conservationists' efforts to protect apes in the wild.

I also noted earlier that it is considered acceptable to subject apes, but not humans, without their consent to lethal experiments for purposes of medical research. The motive for doing so is precisely because apes are so similar to us genetically. They can be infected with many of the same diseases as we can, and their bodies respond similarly to the disease organisms. Thus, experiments on apes offer a far better way to devise improved medical treatments for humans than would experiments on any other animals.

This ethical choice poses an even more difficult problem than does caging apes in zoos. After all, we regularly cage millions of human criminals under worse conditions than zoo apes, but there is no socially accepted human analogue of medical research on animals, even though lethal experiments on humans would provide medical scientists with far more valuable information than do lethal experiments on chimps. Yet the human experiments performed by Nazi concentration camp physicians are widely viewed as one of the most abominable of all the Nazis' abominations. Why is it all right to perform such experiments on chimps?

Somewhere along the scale from bacteria to humans, we have to decide where killing becomes murder, and where eating becomes cannibalism. Most people draw those lines between humans and all other species. However, quite a few people are vegetarians, unwilling to eat any animal (yet willing to eat plants). An increasingly vocal minority, belonging to the animal rights movement, objects to medical experiments on animals – or at least on certain animals. That movement is especially indignant at research on cats and dogs and primates, less concerned about mice, and generally silent about insects and bacteria.

If our ethical code makes a purely arbitrary distinction between humans and all other species, then we have a code based on naked selfishness devoid of any higher principle. If our code instead makes distinctions based on our superior intelligence, social relationships, and capacity for feeling pain, then it becomes difficult to defend an all-or-nothing code that draws a line between all humans and all animals.

Instead, different ethical constraints should apply to research on different species. Perhaps it is just our naked selfishness, re-emerging in a new disguise, that would advocate granting special rights to those animal species genetically closest to us. But an objective case, based on the considerations I have just mentioned (intelligence, social relationships, etc.), can be made that chimps and gorillas qualify for preferred ethical consideration over insects and bacteria. If there is any animal species currently used in medical research for which a total ban on medical experimentation can be justified, that species is surely the chimpanzee.

The ethical dilemma posed by animal experiments is compounded for chimps by the fact that they are endangered as a species. In this case, medical research not only kills individuals but threatens to kill the species itself. That is not to say that demands for research have been the sole threat to wild chimp populations – habitat destruction and capture for zoos have also been major threats – but it is enough that research demands have been a significant threat. The ethical dilemma is further compounded by other considerations: that on the average several wild chimps are killed in the process of capturing one (often a young animal with its mother) and delivering it to a medical research laboratory; that medical scientists have played little role in the struggle to protect wild chimp populations, despite their obvious self-interest in doing so; and that chimps used for research are often caged under cruel conditions. The first chimp that I saw being used for medical research had been injected with a slow-acting lethal virus and was being kept alone, for the several years until it died, in a small, empty, indoor cage at the US National Institutes of Health.

Breeding chimps in captivity for research use avoids objections based on depleting wild chimp populations, but that still does not get around the basic dilemma, any more than enslaving children of US-born blacks after abolition of the African slave trade made black slavery in the nineteenth-century US acceptable. Why is it all right to experiment on *Homo troglodytes*, but not on *Homo sapiens*? Conversely, how should we explain to parents, whose children are at risk of dying from diseases now being studied in captive chimps, that their children are less important than chimps? Ultimately, we the public, not just scientists, will have to make these terrible choices. All that is certain is that our view of man and apes will determine our decision.

Finally, changes in our attitudes about apes may be crucial in determining whether apes will survive at all in the wild. At present, their populations are threatened especially by destruction of their rainforest habitats in Africa and Asia, and by legal and illegal capture and killing. If present trends continue, the mountain gorilla, orangutan, pileated gibbon, Kloss's gibbon, and possibly some other apes as well will exist only in zoos by the time that this year's crop of human babies enters college. It is not enough for us to preach to the governments of Uganda, Zaire, and Indonesia about their moral obligation to protect their wild apes. These are impoverished countries, and national parks are expensive to create and maintain. If we as the third chimpanzee decide that the other two chimpanzees are worth saving, those of us in the richer countries will have to bear most of the expense. From the point of view of the apes themselves, the most important effect of what we have recently learned about the Tale of Three Chimps will be on how we feel about footing that bill.

TWO

THE GREAT LEAP FORWARD

*What happened at that magic moment in evolution around 40,000
years ago, when we suddenly became human?*

*

As we saw in Chapter One, our lineage diverged from that of apes
millions of years ago. For most of the time since then, we have remained
little more than glorified chimpanzees in the ways we have made our
living. As recently as 40,000 years ago, Western Europe was still
occupied by Neanderthals, primitive beings for whom art and progress
scarcely existed. Then there was an abrupt change, as anatomically
modern people appeared in Europe, bringing with them art, musical
instruments, lamps, trade, and progress. Within a short time, the
Neanderthals were gone.

That Great Leap Forward in Europe was probably the result of a similar
leap that had occurred over the course of the preceding few tens of
thousands of years in the Near East and Africa. Even a few dozen
millenia, though, is a trivial fraction (less than one per cent) of our
millions of years of history separate from that of the apes. Insofar as there
was any single point in time when we could be said to have become
human, it was at the time of that leap. Only a few more dozen millenia
were needed for us to domesticate animals, develop agriculture and
metallurgy, and invent writing. It was then but a short further step to
those monuments of civilization that distinguish humans from animals
across what used to seem an unbridgeable gulf – monuments such as the
'Mona Lisa' and the *Eroica* Symphony, the Eiffel Tower and Sputnik,
Dachau's ovens and the bombing of Dresden.

This chapter will confront the questions posed by our abrupt rise to
humanity. What made it possible, and why was it so sudden? What held
back the Neanderthals, and what was their fate? Did Neanderthals and
modern peoples ever meet, and if so, how did they behave towards each
other?

Understanding the Great Leap Forward is not easy, and writing about
it is not easy either. The immediate evidence comes from technical details

of preserved bones and stone tools. Archaeologists' reports are full of terms obscure to the rest of us, such as 'transverse occipital torus', 'receding zygomatic arches', and 'Châtelperronian backed knives'. What we really want to understand – the way of life and the humanity of our various ancestors – is not directly preserved but only inferred from those technical details of bones and tools. Much of the evidence is missing, and archaeologists often disagree over the meaning of such evidence as has survived. Since the books and articles listed on pages 334–5 will slake the interest of readers curious to learn more about receding zygomatic arches, I shall emphasize instead the inferences from bones and tools.

*

Our ancestors were confined to Africa for millions of years, where, as we have already discussed, they diverged from the ancestors of chimps and gorillas between about six and ten million years ago. For comparison, life originated on Earth several billion years ago, and the dinosaurs became extinct around sixty-five million years ago. (Science-fiction films that depict cavemen fleeing from dinosaurs are just that, science fiction.) Initially, our ancestors would have been classified as merely another species of ape, but a sequence of three changes launched us in the direction of modern humans.

The first of these changes had occurred by around four million years ago, when the structure of fossilized limb bones shows that our ancestors were habitually walking upright on the two hindlimbs. In contrast, gorillas and chimps walk upright only occasionally, and usually proceed on all fours. The upright posture freed our ancestors' forelimbs to do other things, among which tool-making proved the most important.

The second change occurred around three million years ago, when our lineage split into at least two distinct species. Recall that members of two animal species living in the same area must fill different ecological roles and do not normally interbreed with each other. For example, coyotes and wolves are obviously closely related and (until wolves were exterminated in most of the US) lived in many of the same areas of North America. However, wolves are larger, mainly hunt big mammals like deer and moose, and often live in large packs, whereas coyotes are smaller, mainly hunt small mammals like rabbits and mice, and usually live in pairs or small groups. Similarly, Europe's wildcat and lynx are closely related and overlap widely in range but differ ecologically and do not interbreed.

Every human population living today has interbred with every other human population with which it has had extensive contact. Ecological differences among existing humans are entirely a product of childhood education: it is not the case that some of us are born with sharp teeth and equipped to hunt deer, while others are born with grinding teeth, gather berries, and do not marry the deerhunters. Therefore all modern humans belong to the same species.

On perhaps two occasions in the past, however, the human lineage split into separate species, as distinct as wolves and coyotes. The most recent such occasion, which I shall describe later, may have been at the time of the Great Leap Forward. The earlier occasion was around three million years ago, when our lineage split into two: a man-ape with a robust skull and very big cheek teeth, assumed to eat coarse plant food, and often referred to as *Australopithecus robustus* (meaning 'the robust southern ape'); and a man-ape with a more lightly built skull and smaller teeth, assumed to have an omnivorous diet, and known as *Australopithecus africanus* ('the southern ape of Africa'). The latter man-ape evolved into a larger-brained form termed *Homo habilis* ('man the handyman'). However, fossil bones often attributed to male and female *Homo habilis* differ so much in skull size and tooth size that they may actually imply another fork in our lineage yielding two distinct *habilis*-like species: *Homo habilis* himself, and a mysterious 'Third Man'. Thus, two million years ago there were at least two, and possibly three, proto-human species.

The third and last of the big changes that began to make our ancestors more human and less ape-like was the regular use of stone tools. This is a human hallmark with clear animal precedents: woodpecker finches, Egyptian vultures, and sea otters are among the other animal species that evolved independently to employ tools in capturing or processing food, though none of these species is as heavily dependent on implements as we are now. Common chimpanzees also use tools, occasionally of stone, but not in numbers sufficient to litter the landscape. But by around two-and-a-half million years ago, very crude stone tools appear in numbers in areas of East Africa occupied by the proto-humans. Since there were two or three proto-human species, who made the tools? Probably the light-skulled species, since both it and the tools persisted and evolved.

With only one human species surviving today but two or three a few million years ago, it is clear that one or two species must have become extinct. Who was our ancestor; which species ended up instead as a discard in the rubbish-heap of evolution; and when did this shakedown occur? The winner was the light-skulled *Homo habilis*, who went on to increase in brain size and body size. By around 1,700,000 years ago the differences were sufficient that anthropologists give our lineage a new name, *Homo erectus*, meaning 'the man that walks upright'. (*Homo erectus* fossils were discovered before all the earlier fossils I have been discussing, so anthropologists did not realize that *Homo erectus* was not the first

THE HUMAN FAMILY TREE

Several branches of our family tree have become extinct, including those belonging to the robust australopithecines, Neanderthals, and possibly a poorly understood 'Third Man' and an Asian population contemporary with Neanderthals. Some descendants of *Homo habilis* survived to evolve into modern humans. To recognize by different names the changes in fossils representing this line, they are somewhat arbitrarily divided into *Homo habilis*, then *Homo erectus* appearing about 1.7 million years ago, and *Homo sapiens* appearing about 500,000 years ago. *A.* stands for the genus name of *Australopithecus*, *H.* for *Homo*.

proto-human to walk upright.) The robust man-ape disappeared around 1,200,000 years ago, and the 'Third Man' (if he ever existed) must have disappeared by then also. As for why *Homo erectus* survived and the robust man-ape didn't, we can only speculate. A plausible guess is that the robust man-ape could no longer compete, since *Homo erectus* ate both meat and plant food, and since tools and a larger brain made *Homo erectus* more efficient at getting even the plant food on which his robust sibling depended. It is also possible that *Homo erectus* gave his sibling a direct push into oblivion, by killing him for meat.

All the developments that I have been discussing so far were played out within the continent of Africa, to which our closest living relatives (the chimps and gorilla) are still confined. The shakedown had left *Homo erectus* as the sole proto-human on the African stage. Around one million years ago *Homo erectus* expanded his horizons. His stone tools and bones show that he reached the Near East, then the Far East (where he is represented by the famous fossils known as Peking Man and Java Man) and Europe. He continued to evolve in our direction by an increase in brain size and in skull roundness. By around 500,000 years ago, some of our ancestors looked sufficiently like us, and different enough from earlier *Homo erectus*, to be classified as our own species (*Homo sapiens*, meaning 'the wise man'), though they still had thicker skulls and brow ridges than we do today.

Readers unfamiliar with details of our evolution might be forgiven for assuming that the appearance of *Homo sapiens* constituted the Great Leap Forward. Was our meteoric ascent to *sapiens* status half-a-million years ago the brilliant climax of Earth's history, when art and sophisticated technology finally burst upon our previously dull planet? Not at all: the appearance of *Homo sapiens* was a non-event. Cave paintings, houses, and bows and arrows still lay hundreds of thousands of years off in the future. Stone tools continued to be the crude ones that *Homo erectus* had been making for nearly a million years. The extra brain size of those early *Homo sapiens* had no dramatic effect on our way of life. That whole long tenure of *Homo erectus* and early *Homo sapiens* outside Africa was a period of infinitesimally slow cultural change. In fact, the sole candidate for a major advance was possibly the control of fire, of which caves occupied by Peking Man provide one of the earliest indications in the form of ash, charcoal, and burnt bones. Even that advance – if those cave fires really were man-lit rather than natural – would belong to *Homo erectus*, not *Homo sapiens*.

Thus, the emergence of *Homo sapiens* illustrates the paradox discussed in Chapter One: that our rise to humanity was not directly proportional to the changes in our genes. Early *Homo sapiens* had progressed much further in anatomy than in cultural attainments along the road up from chimpanzeehood. Some crucial ingredients still had to be added before the Third Chimpanzee could conceive of painting the Sistine Chapel.

*

How did our ancestors make their living during the one-and-a-half million years that spanned the emergence of Homo erectus and Homo sapiens?

The only surviving tools from this period are stone tools that can charitably be described as very crude, in comparison with the beautiful, polished stone tools made until recently by Polynesians, American Indians, and other modern stone-age peoples. Early stone tools vary in size and shape, and archaeologists have used those differences to give the tools different names, such as 'hand-axe', 'chopper', and 'cleaver'. These names conceal the fact that none of those early tools had a sufficiently consistent or distinctive shape to suggest any specific function, as do the obvious needles and spear-points left by the much later Cro-Magnons. Wear-marks on the tools show that they were variously used to cut meat, bone, hides, wood, and non-woody parts of plants, but any size or shape of tool seems to have been used to cut any of those things, and the tool names applied by archaeologists may be little more than arbitrary divisions of a continuum of stone forms.

Negative evidence is also significant here. Many advances in tools that appear after the Great Leap Forward were unknown to Homo erectus and early Homo sapiens. There were no bone tools, no ropes to make nets, and no fishhooks. All the early stone tools may have been held directly in the hand; they show no signs of being mounted on other materials for increased leverage, as we mount steel axe-blades on wooden handles.

What food did our early ancestors get with those crude tools, and how did they get it? At this point, anthropology textbooks usually insert a long chapter entitled something like 'Man the Hunter'. The point here is that baboons, chimps, and some other primates occasionally prey on small vertebrates, but recently surviving stone-age people (like Bushmen) did a lot of big-game hunting. So did Cro-Magnons, according to abundant archaeological evidence. There is no doubt that our early ancestors also ate some meat, as shown by marks of their stone tools on animal bones and by wear-marks on their stone tools caused by cutting meat. The real question is: how *much* big-game hunting did our early ancestors do? Did big-game hunting skills improve gradually over the past one-and-a-half million years, or was it only since the Great Leap Forward that they made a large contribution to our diet?

Anthropologists routinely reply that we have been successful big-game hunters for a long time. The supposed evidence comes mainly from three archaeological sites occupied around 500,000 years ago: a cave at Zhoukoudian near Beijing, containing bones and tools of Homo erectus ('Peking Man') and bones of many animals; and two non-cave (open-air) sites at Torralba and Ambrona in Spain, with stone tools and bones of elephants and other large animals. It is usually assumed that the people who left the tools killed the animals, brought their carcasses to the site, and ate them there, but all three sites also have bones and faecal remains of hyenas, which could equally well have been the hunters. The bones of the Spanish sites in particular look like they came from a collection of scavenged, water-washed, trampled carcasses such as one can find around African water-holes today, rather than from a human hunters' camp.

Thus, while early humans ate some meat, we do not know how much meat they ate, nor whether they got the meat by hunting or scavenging. It is not until much later, around 100,000 years ago, that we have good evidence about human hunting skills, and it is clear that humans then were still very *ineffective* big-game hunters. Human hunters of 500,000 years ago and earlier must have been even more ineffective.

The mystique of Man the Hunter is now so rooted in us that it is hard to abandon our belief in its long-standing importance. Today, shooting a big animal is regarded as an ultimate expression of macho masculinity. Trapped in this mystique, male anthropologists like to stress the key role of big-game hunting in human evolution. Supposedly, big-game hunting was what induced proto-human males to cooperate with each other, develop language and big brains, join into bands, and share food. Even women were supposedly moulded by men's big-game hunting: women suppressed the external signs of monthly ovulation that are so conspicuous in chimps, so as not to drive men into a frenzy of sexual competition and thereby spoil men's cooperation at hunting.

As an example of the purple prose spawned by this men's locker-room mentality, consider the following account of human evolution by Robert Ardrey in his book *African Genesis*:

> In some scrawny troop of beleagured not-yet-men on some scrawny forgotten plain a radian particle from an unknown source fractured a never-to-be-forgotten gene, and a primate carnivore was born. For better or worse, for tragedy or for triumph, for ultimate glory or ultimate damnation, intelligence made alliance with the way of the killer, and Cain with his sticks and his stones and his quickly running feet emerged on the high savannah.

What pure fantasy!

Western male writers and anthropologists are not the only men with an exaggerated view of hunting. In New Guinea I have lived with real hunters, men who recently emerged from the Stone Age. Conversations at campfires go on for hours over each species of game animal, its habits, and how best to hunt it. To listen to my New Guinea friends, you would think that they eat fresh kangaroo for dinner every night and do little each day except hunt. In fact, when pressed for details, most New Guinea hunters admit that they have bagged only a few kangaroos in their whole life.

I still recall my first morning in the New Guinea highlands, when I set out with a group of a dozen men, armed with bows and arrows. As we passed a fallen tree, there was suddenly much excited shouting, men surrounded the tree, some spanned their bows, and others pressed forward into the brushpile. Convinced that an enraged boar or kangaroo was about to come out fighting, I looked for a tree that I could climb to a perch of safety. Then I heard triumphant shrieks, and out of the brushpile came two mighty hunters holding aloft their prey: two baby wrens, not quite able to fly, weighing about a third of an ounce each, and promptly plucked, roasted, and eaten. The rest of that day's catch consisted of a few frogs and many mushrooms.

Studies of most modern hunter-gatherers with far more effective weapons than early *Homo sapiens* show that most of a family's calories come from plant food gathered by women. Men catch rabbits and other small game never mentioned in the heroic campfire stories. Occasionally the men do bag a large animal, which does indeed contribute significantly to protein intake. But it is only in the Arctic, where little plant food is available, that big-game hunting becomes the dominant food source, and humans did not reach the Arctic until within the last few dozen millenia.

Thus I would guess that big-game hunting contributed only modestly to our food intake until *after* we had evolved fully modern anatomy and behaviour. I doubt the usual view that hunting was the driving force behind our uniquely human brain and societies. For most of our history we were not mighty hunters but skilled chimps, using stone tools to acquire and prepare plant food and small animals. Occasionally, men did bag a large animal, and then retold the story of that rare event incessantly.

*

In the period just before the Great Leap Forward, at least three distinct human populations occupied different parts of the Old World. These were the last truly primitive humans, supplanted by fully modern people at the time of the Great Leap. Let's consider those among the last primitives whose anatomy is best known and who have become a metaphor for brutish subhumans: the Neanderthals.

Where and when did they live? Their geographic range extended from Western Europe, through southern European Russia and the Near East, to Uzbekhistan in Central Asia near the border of Afghanistan. (The name 'Neanderthal' comes from Germany's Neander Valley (valley = Thal in German), where one of the first skeletons was discovered.) The time of their origin is a matter of definition, since some old skulls have characteristics anticipating later full-blown Neanderthals. The earliest 'full-blown' examples date to around 130,000 years ago, and most specimens postdate 74,000 years ago. While their start is thus arbitrary, their end is abrupt: the last Neanderthals died around 40,000 years ago.

During the time that Neanderthals flourished, Europe and Asia were in the grip of the last Ice Age. Neanderthals must have been a cold-adapted people – but only within limits. They got no further north than southern Britain, northern Germany, Kiev, and the Caspian Sea. The first penetration of Siberia and the Arctic was left to later, fully modern humans.

Neanderthals' head anatomy was so distinctive that, even if a Neanderthal dressed in a business suit or a designer dress were to walk down the streets of New York or London today, everybody else (all the *homines sapientes*) on the street would be staring in shock. Imagine converting a modern face to soft clay, gripping the middle of the face from the bridge of the nose to the jaws in a vice, pulling the whole mid-face forward, and letting it harden again. You will then have some idea of a Neanderthal's appearance. Their eyebrows rested on prominently-bulging bony ridges, and their nose and jaws and teeth protruded far forward. Their eyes lay in deep sockets, sunk behind the protruding nose and brow ridges. Their foreheads were low and sloping, unlike our high vertical modern foreheads, and their lower jaws sloped back without a chin. Despite these startlingly primitive features, Neanderthals' brain size was nearly ten per cent *greater* than ours!

A dentist who examined a Neanderthal's teeth would have been in for a further shock. In adult Neanderthals, the incisors (front teeth) were worn down on the outer-facing surface, in a way found in no modern people. Evidently, this peculiar wear-pattern somehow resulted from a use of their teeth as tools, but what exactly was that function? As one possibility, they may have routinely used their teeth as a vice to grip objects, like my baby sons, who gripped their milk bottle in their teeth and ran around with their hands free. Alternatively, Neanderthals may have bitten hides with their teeth to make leather, or bitten wood to make wooden tools.

While a Neanderthal in a business suit or dress would attract attention today, one in shorts or a bikini would have drawn gasps. Neanderthals were more heavily muscled, especially in their shoulders and neck, than all but the most avid modern bodybuilders. Their limb-bones, which took the force of those big muscles when they were contracting, had to be considerably thicker than ours to withstand the stress. Their arms and legs would have looked stubby to us, because the lower leg and forearm were relatively shorter than ours. Even their hands were much more powerful than ours; a Neanderthal's handshake would have been literally bone-crushing. While their average height was only around 5 feet 4 inches, their weight would have been at least 20 pounds more than that of a modern person of that height, and this excess was mostly in the form of lean muscle.

One other possible anatomical difference is intriguing, though its reality as well as its interpretation are quite uncertain. A Neanderthal woman's birth canal may have been wider than a modern woman's, permitting her baby to grow inside her to a bigger size before birth. If so, a Neanderthal pregnancy might have lasted a year, instead of our nine months.

Besides their bones, our other main source of information about Neanderthals is their stone tools. Like the earlier human tools, Neanderthal tools may have been simple hand-held stones not mounted on separate parts such as handles. The tools do not fall into distinct types with unique functions. There were no standardized bone tools, no bows and arrows. Some of the stone tools were undoubtedly used to make wooden tools, which rarely survive. One notable exception is a wooden thrusting spear 8 feet long, found in the ribs of a long-extinct species of elephant at an archaeological site in Germany. Despite that (lucky?) success, Neanderthals were probably not very good at big-game hunting, because Neanderthal numbers (to judge from the number of their sites) were much lower than those of later Cro-Magnons, and because (as I will explain later) even anatomically more modern people living in Africa at the same time as the Neanderthals were undistinguished as hunters.

If you say 'Neanderthal' to friends and ask for their first association, you will probably get back the answer 'caveman'. While most excavated Neanderthal remains do come from caves, that is surely an artifact of preservation, since open-air sites would be eroded much more quickly. Among my hundreds of campsites in New Guinea, one was in a cave, and that is the only site where future archaeologists are likely to find my pile of discarded tin cans intact. So archaeologists will also be deceived into considering me a caveman. Neanderthals must have constructed some type of shelter against the cold climate in which they lived, but those shelters must have been crude. All that remains are a few piles of stones and a posthole, compared to the elaborate remains of houses built by the later Cro-Magnons.

The list of other quintessentially modern human things that Neanderthals lacked is a long one. They left no unequivocal art objects. They must have worn some clothing in their cold environment, but it had to be crude, as they lacked needles and other evidence of sewing. They evidently lacked boats, as no Neanderthal remains are known from Mediterranean islands nor even from North Africa, just eight miles across the Straits of Gibraltar from Neanderthal-populated Spain. There was no long-distance overland trade: Neanderthal tools are made of stones available within a few miles of the site.

Today we take cultural differences among people inhabiting different areas for granted. Every human population alive today has its characteristic house-style, implements, and art. If you were shown chopsticks, a Guinness beer bottle, and a blowgun and asked to associate one object each with China, Ireland, and Borneo, you would have no trouble giving the right answers. No such cultural variation is apparent for Neanderthals, whose tools look much the same whether they come from France or Russia.

We also take cultural progress with time for granted. The wares from a Roman villa, medieval castle, and 1990 New York apartment differ obviously. In the year 2000 my sons will look with astonishment at the slide rule I used for calculations throughout the 1950s: 'Daddy, are you *really* that old?' But Neanderthal tools from 100,000 and 40,000 years ago look essentially the same. In short, Neanderthal tools had no variation in either time or space to suggest that most human of characteristics, *innovation*. As one archaeologist put it, Neanderthals had 'beautiful tools stupidly made'. Despite Neanderthals' big brains, something was still missing.

Grandparenting, and what we consider old age, must also have been rare among Neanderthals. Their skeletons make clear that adults might live to their thirties or early forties, but not beyond forty-five. If we lacked writing *and* if none of us lived past forty-five, just think how the ability of our society to accumulate and transmit information would suffer.

I have had to mention all these subhuman qualities of Neanderthals, but there are three respects in which we can relate to their humanity. First, virtually all well-preserved Neanderthal caves have small areas of ash and charcoal indicating a simple fireplace. Hence, although Peking Man may have already used fire hundreds of thousands of years earlier, Neanderthals were the first people to leave undisputed evidence of the regular use of fire. Neanderthals may also have been the first people who regularly buried their dead, but that is disputed, and whether it would imply religion is a matter of pure speculation. Finally, they regularly took care of their sick and aged. Most skeletons of older Neanderthals show signs of severe impairment, such as withered arms, healed but incapacitating broken bones, tooth loss, and severe osteoarthritis. Only care by young Neanderthals could have enabled such older Neanderthals to stay alive to the point of such incapacitation. After my long litany of what Neanderthals lacked, we have finally found something that lets us feel a spark of kindred spirit in these strange creatures of the last Ice Age – nearly human in form, and yet not really human in spirit.

Did Neanderthals belong to the same species as we do? That depends on whether we could and would have mated and reared a child with a Neanderthal man or woman, given the opportunity. Science-fiction novels love to imagine the scenario. You may remember the blurb on many a back cover:

> A team of explorers stumbles on a steep-walled valley in the centre of deepest Africa, a valley that time forgot. Here they find a tribe of incredibly primitive people, living in ways that our stone-age ancestors discarded thousands of years ago. Do they belong to the same species as we do? There's only one way to find out, but who among the intrepid explorers can bring himself [male explorers, of course] to make the test?

At this point one of the bone-chewing cavewomen suddenly is described as beautiful and sexy in a primitively erotic way, so that modern novel readers will find the brave explorer's dilemma believable: does he or doesn't he have sex with her?

Believe it or not, something like that experiment actually took place. As we shall now see, it happened repeatedly around 40,000 years ago, at the time of the Great Leap Forward.

*

I mentioned that the Neanderthals of Europe and Western Asia were just one of at least three human populations occupying different parts of the Old World around 100,000 years ago. A few fossils from Eastern Asia suffice to show that people there differed from Neanderthals as well as from us moderns, but too few bones have been found to describe these Asians in more detail. The best characterized contemporaries of the Neanderthals are those from Africa, some of whom were virtually modern in their skull anatomy. Does this mean that, 100,000 years ago in Africa, we have at last arrived at the watershed of human cultural development?

Surprisingly, the answer is still 'no'. The stone tools of these modern-looking Africans were very similar to those of the decidedly unmodern-looking Neanderthals, hence we refer to them as 'Middle Stone Age Africans'. They still lacked standardized bone tools, bows and arrows, nets, fishhooks, art, and cultural variation in tools from place to place.

Despite their almost modern bodies, these Africans were still missing that vital something necessary to endow them with full humanity. Once again, we face the paradox that almost modern bones, and presumably almost modern genes, are not enough by themselves to produce modern behaviour.

Some South African caves occupied around 100,000 years ago provide us – for the first time in human evolution – with detailed information about what people actually were eating. Our confidence stems from the fact that the African caves are full of stone tools, animal bones with cut-marks from stone tools, and human bones, but few or no bones of carnivores like hyenas. Thus, it is clear that people, not hyenas, brought the bones to the caves. Among the bones are many of seals and penguins, as well as shellfish such as limpets. Hence Middle Stone Age Africans are the first people for whom there is even a hint that they exploited the seashore. However, the caves contain very few remains of fish or flying seabirds, undoubtedly because people still lacked the fishhooks and nets needed to catch fish and birds.

The mammal bones from the caves include those of quite a few medium-sized species, among which those of eland, an antelope, predominate by far. Eland bones in the caves represent eland of all ages, as if people had somehow managed to capture a whole herd and kill every individual. At first, the relative abundance of eland among hunters' prey is surprising, since the caves' environment 100,000 years ago was much as it is today and since eland is now one of the least common large animals in the area. The secret to the hunters' success with eland probably lay in the fact that eland are rather tame, not dangerous, and easy to drive in herds. This suggests that hunters occasionally managed to drive a whole herd over a cliff, explaining why the distribution of eland age groups among the cave kills is like that in a living herd. In contrast, remains of more dangerous prey such as Cape buffalo, pigs, elephants, and rhinos yield a very different picture. Buffalo bones in the caves are mainly of very young or very old individuals, while pigs, elephants, and rhinos are virtually unrepresented.

Middle Stone Age Africans can be considered big-game hunters, but only barely. They either avoided dangerous species entirely or confined themselves to old, weak animals or babies. Those choices reflect sound prudence on the hunters' part, since their weapons were still spears, for thrusting, rather than bows and arrows. Along with drinking a strychnine cocktail, poking an adult rhinoceros or Cape buffalo with a spear ranks as one of the most effective means of suicide that I know. Nor could the hunters have succeeded often at driving eland herds over a cliff, since elands were not exterminated but continued to coexist with hunters. As with earlier peoples and modern stone-age hunters, I suspect that plants and small game made up most of the diets of these not-so-great Middle Stone Age hunters. They were definitely more effective than chimpanzees, but not up to the skill of modern Bushmen and Pygmies.

Thus, the scene that the human world presented from around 100,000 to somewhat before 50,000 years ago was this. Northern Europe, Siberia, Australia, the oceanic islands, and the whole New World were still empty of people. In Europe and Western Asia lived the Neanderthals; in Africa, people increasingly like us moderns in their anatomy; and in Eastern Asia, people unlike either the Neanderthals or Africans but known from only a few bones. All three of these populations were, at least initially, still primitive in their tools, behaviour, and limited innovativeness. The stage was set for the Great Leap Forward. Which among these three contemporary populations would take that leap?

*

The evidence for an abrupt rise is clearest in France and Spain, in the Late Ice Age around 40,000 years ago. Where there had previously been Neanderthals, anatomically fully modern people (often known as Cro-Magnons, from the French site where their bones were first identified) now appear. Had one of those gentlemen or ladies strolled down the Champs Élysées in modern attire, he or she would not have stood out from the Parisian crowds in any way. As dramatic to archaeologists as the Cro-Magnons' skeletons are their tools, which are far more diverse in form and obvious in function than any in the earlier archaeological record. The tools suggest that modern anatomy had at last been joined by modern innovative behaviour.

Many of the tools continued to be of stone, but they were now made from thin blades struck off a larger stone, thereby yielding ten times more cutting edge from a given quantity of raw stone than previously obtainable. Standardized bone and antler tools appeared for the first time. So did unequivocal compound tools of several parts tied or glued together, such as spear points set in shafts or axe-heads fitted on to wooden handles. Tools fall into many distinct categories whose function is often obvious, such as needles, awls, mortars and pestles, fishhooks, net-sinkers, and rope. The rope (used in nets or snares) accounts for the frequent bones of foxes, weasels, and rabbits at Cro-Magnon sites, while the rope, fishhooks, and net-sinkers explain the bones of fish and flying birds at contemporary South African sites.

Sophisticated weapons for safely killing dangerous large animals at a distance now appear – weapons such as barbed harpoons, darts, spear-throwers, and bows and arrows. South African caves occupied by people now yield bones of such vicious prey as adult Cape buffalo and pigs, while European caves were full of bones of bison, elk, reindeer, horse, and ibex. Even today, hunters armed with high-powered telescopic rifles find it hard to bag some of these species, which must have required highly skilled communal hunting methods based on detailed knowledge of each species' behaviour.

Several types of evidence testify to the effectiveness of Late Ice Age people as big-game hunters. Their sites are much more numerous than those of earlier Neanderthals or Middle Stone Age Africans, implying more success at obtaining food. Numerous species of big animals that had survived many previous ice ages became extinct towards the end of the last Ice Age, suggesting that they were exterminated by human hunters' new skills. These likely victims include the mammoths of North America (Chapter Eighteen), Europe's woolly rhino and giant deer, southern Africa's giant buffalo and giant Cape horse, and Australia's giant kangaroos (Chapter Nineteen). Thus, the most brilliant moment of our rise already contained the seeds of what may yet prove a cause of our fall.

Improved technology now allowed humans to occupy new environments, as well as to multiply in previously occupied areas of Eurasia and Africa. Australia was first reached by humans around 50,000 years ago, implying watercraft capable of crossing stretches of water as much as sixty miles wide between eastern Indonesia and Australia. The occupation of northern Russia and Siberia by at least 20,000 years ago depended on many advances: tailored clothing, whose existence is reflected in eyed needles, cave paintings of parkas, and grave ornaments marking outlines of shirts and trousers; warm furs, indicated by fox and wolf skeletons minus the paws (removed in skinning and found in a separate pile); elaborate houses (marked by postholes, pavements, and walls of mammoth bones), with elaborate fireplaces; and stone lamps to hold animal fat and light the long Arctic nights. The occupation of Siberia and Alaska in turn led to the occupation of North America and South America around 11,000 years ago (Chapter Eighteen).

Whereas Neanderthals obtained their raw materials within a few miles of home, Cro-Magnons and their contemporaries throughout Europe practised long-distance trade, not only for raw materials for tools but also for 'useless' ornaments. Tools of high-quality stone such as obsidian, jasper, and flint are found hundreds of miles from where those stones were quarried. Baltic amber reached southeastern Europe, while Mediterranean shells were carried to inland parts of France, Spain, and the Ukraine. I saw very similar patterns in modern stone-age New Guinea, where cowry shells prized as decorations were traded up to the highlands from the coast, bird-of-paradise plumes were traded back down to the coast, and obsidian for stone axes was traded out from a few highly valued quarries.

WORLD CONQUEST

This map illustrates stages in the spread of our ancestors from their
African origins to conquer the world. Numbers stand for estimated
number of years before the present. Further discoveries of older
archaeological sites may well show that some regions, such as Siberia or
the Solomon Islands, were colonized earlier than the estimated dates
shown here.

The evident aesthetic sense reflected in the Late Ice Age trade in
ornaments relates to the achievements for which we most admire the
Cro-Magnons – their art. Best known, of course, are the rock paintings
from caves like Lascaux, with stunning polychrome depictions of now-
extinct animals, but equally impressive are the bas-reliefs, necklaces and
pendants, fired-clay ceramic sculptures, Venus figurines of women with
enormous breasts and buttocks, and musical instruments ranging from
flutes to rattles.

Unlike Neanderthals, few of whom lived past the age of forty, some
Cro-Magnon skeletons indicate survival to sixty years of age. Many
Cro-Magnons, but few Neanderthals, lived to enjoy their grandchildren.
Those of us accustomed to getting our information from the printed page
or television will find it hard to appreciate how important even just one or
two elderly people are in a pre-literate society. In New Guinea villages it
often happens that younger men lead me to the oldest person in the village
when I stump them with a question about some uncommon bird or fruit.
For example, when I visited Rennell Island in the Solomons in 1976,
many islanders told me what wild fruits were good to eat, but only one
old man could tell me what other wild fruits could be eaten in an
emergency to avoid starvation. He remembered that information from a

cyclon' chat had hit Rennell in his childhood (around 1905), destroying gardens and reducing his people to a state of desperation. One such person in a pre-literate society can thus spell the difference between death and survival for the whole society. Hence the fact that some Cro-Magnons survived twenty years longer than any Neanderthal probably played a big role in Cro-Magnon success. As we shall see in Chapter Seven, living to an older age required not only improved survival skills but also some biological changes, possibly including the evolution of human female menopause.

I have described the Great Leap Forward as if all those advances in tools and art appeared simultaneously 40,000 years ago. In fact, different innovations appeared at different times. Spear-throwers appeared before harpoons or bows and arrows, while beads and pendants appeared before cave paintings. I have also described the changes as if they were the same everywhere, but they were not. 'Among Late Ice Age Africans, Ukrainians, and French, only the Africans made beads out of ostrich eggs, only the Ukrainians built houses out of mammoth bones, and only the French painted woolly rhinos on cave walls.

These variations of culture in time and space are totally unlike the unchanging monolithic Neanderthal culture. They constitute the most important innovation that came with our rise to humanity: namely, the capacity for innovation itself. To us today, who cannot picture a world in which Nigerians and Latvians in 1991 have virtually the same possessions as each other and as the Romans in 50 BC, innovation is utterly natural. To Neanderthals, it was evidently unthinkable.

Despite our instant sympathy with Cro-Magnon art, their stone tools and hunter-gatherer lifestyle make it hard for us to view them as other than primitive. Stone tools evoke cartoons of club-waving cavemen uttering grunts as they drag a woman off to their cave. We can form a more accurate impression of Cro-Magnons if we imagine what future archaeologists will conclude after excavating a New Guinea village site from as recently as the 1950s. The archaeologists will find a few simple types of stone axes. Virtually all other material possessions were made of wood and will have perished. Nothing will remain of the multi-storey houses, beautifully woven baskets, drums and flutes, outrigger canoes, and world-quality painted sculpture. There will be no trace of the village's complex language, songs, social relationships, and knowledge of the natural world.

New Guinea material culture was until recently 'primitive' (that is, stone-age) for historical reasons, but New Guineans are fully modern humans. New Guineans whose fathers lived in the Stone Age now pilot aeroplanes, operate computers, and govern a modern state. If we could carry ourselves back 40,000 years in a time machine, I expect that we would find Cro-Magnons to be equally modern people, capable of learning to fly a jet plane. They made stone and bone tools only because no other tools had yet been invented; that is all that they had the opportunity to learn.

*

It used to be argued that Neanderthals evolved into Cro-Magnons within Europe. That possibility now seems increasingly unlikely. The last Neanderthal skeletons from around 40,000 years ago were still 'full-blown' Neanderthals, while the first Cro-Magnons appearing in Europe at the same time were already anatomically fully modern. Since anatomically modern people were already present in Africa and the Near East tens of thousands of years earlier, it seems much more likely that anatomically modern people invaded Europe from that direction than that they evolved within Europe.

What happened when invading Cro-Magnons met the resident Neanderthals? We can be certain only of the end result: within a short time, no more Neanderthals. The conclusion seems to me inescapable that Cro-Magnon arrival somehow caused Neanderthal extinction. Yet many archaeologists recoil at this conclusion and invoke environmental changes instead. For example, the *Encyclopaedia Britannica*'s fifteenth edition concludes its entry for Neanderthals with the sentence, 'The disappearance of the Neanderthals, although it cannot yet be fixed in time, was probably the result of being creatures of an interglacial period unable to avoid the ravages of another Ice Age.' In fact, Neanderthals thrived during the last Ice Age, and suddenly disappeared over 30,000 years after its start and an equal time before its end.

My guess is that events in Europe at the time of the Great Leap Forward were similar to events that have occurred repeatedly in the modern world, whenever a numerous people with more advanced technology invades the lands of a much less numerous people with less advanced technology. For instance, when European colonists invaded North America, most North American Indians proceeded to die of introduced epidemics; most of the survivors were killed outright or driven off their land (Chapter Sixteen); some of the survivors adopted European technology (horses and guns) and resisted for some time; and many of the remaining survivors were pushed on to lands that Europeans did not want, or else intermarried with Europeans (Chapter Fifteen). The displacement of Aboriginal Australians by European colonists, and of southern African San populations (Bushmen) by invading iron-age Bantu-speakers, followed a similar course.

By analogy, I guess that Cro-Magnon diseases, murders, and dis-

placements did in the Neanderthals. If so, then the Cro-Magnon/
Neanderthal transition was a harbinger of what was to come, when the
victors' descendants began squabbling among themselves. It may at first
seem paradoxical that Cro-Magnons prevailed over the far more
muscular Neanderthals, but weaponry rather than strength would have
been decisive. Similarly, it's not gorillas that are now threatening to
exterminate humans in Central Africa, but vice versa. People with huge
muscles require lots of food, and they therefore gain no advantage if
slimmer, smarter people can use tools to do the same work.

Like the Great Plains Indians of North America, some Neanderthals
may have learned some Cro-Magnon ways and resisted for a while. This
is the only sense I can make of a puzzling culture called the Châtelperronian, which coexisted in Western Europe along with a typical Cro-
Magnon culture (the so-called Aurignacian culture) for a short time after
Cro-Magnons arrived. Châtelperronian stone tools are a mixture of
typical Neanderthal and Cro-Magnon tools, but the bone tools and art
typical of Cro-Magnons are usually lacking. The identity of the people
who produced Châtelperronian culture was debated by archaeologists,
until a skeleton unearthed with Châtelperronian artifacts at Saint–Césaire
in France proved to be Neanderthal. Perhaps, then, some Neanderthals
managed to master some Cro-Magnon tools and hold out longer than
their fellows.

What remains unclear is the outcome of the interbreeding experiment
posed in science-fiction novels. Did some invading Cro-Magnon men
mate with some Neanderthal women? No skeletons that could reasonably be considered Neanderthal/Cro-Magnon hybrids are known. If
Neanderthal behaviour was as relatively rudimentary, and Neanderthal
anatomy as distinctive, as I suspect, few Cro-Magnons may have wanted
to mate with Neanderthals. Similarly, although humans and chimps
continue to coexist today, I am not aware of any matings. While Cro-
Magnons and Neanderthals were not nearly as different, the differences
may still have been a mutual turn-off. And if Neanderthal women were
geared for a twelve-month pregnancy, a hybrid foetus might not have
survived. My inclination is to take the negative evidence at face value, to
accept that hybridization occurred rarely if ever, and to doubt that living
people of European descent carry any Neanderthal genes.

So much for the Great Leap Forward in Western Europe. The
replacement of Neanderthals by modern people occurred somewhat
earlier in Eastern Europe, and still earlier in the Near East, where
possession of the same area apparently shifted back and forth between
Neanderthals and modern people from 90,000 to 60,000 years ago. The
slowness of the transition in the Near East, compared to its speed in
Western Europe, suggests that the anatomically modern people living

around the Near East before 60,000 years ago had not yet developed the modern behaviour that ultimately let them drive out the Neanderthals.

Thus, we have a tentative picture of anatomically modern people arising in Africa over 100,000 years ago, but initially making the same tools as Neanderthals and having no advantage over them. By perhaps 60,000 years ago, some magic twist of behaviour had been added to the modern anatomy. That twist (of which more in a moment) produced innovative, fully modern people who proceeded to spread westward into Europe, quickly supplanting Europe's Neanderthals. Presumably, those modern people also spread east into Asia and Indonesia, supplanting the earlier people there of whom we know little. Some anthropologists think that skull remains of those earlier Asians and Indonesians show traits recognizable in modern Asians and Aboriginal Australians. If so, the invading moderns may not have exterminated the original Asians without issue, as they did the Neanderthals, but instead interbred with them.

Two million years ago, several proto-human lineages had coexisted side by side until a shake-up left only one. It now appears that a similar shake-up occurred within the last 60,000 years, and that all of us alive in the world today are descended from the winner of that upheaval. What was the last missing ingredient whose acquisition helped our ancestor to win?

*

The identity of that ingredient poses an archaeological puzzle without an accepted answer. To help focus our speculations, let me recapitulate the pieces of the puzzle.

Some groups of humans who lived in Africa and the Near East over 60,000 years ago were quite modern in their anatomy, as far as can be judged from their skeletons, but they were not modern in their behaviour. They continued to make Neanderthal-like tools and to lack innovation. The ingredient that produced the Great Leap Forward does not show up in fossil skeletons.

There is another way to restate that puzzle. We share ninety-eight per cent of our genes with chimpanzees (Chapter One). The Africans making Neanderthal-like tools just before our sudden rise to humanity had covered almost all of the remaining genetic distance between us and chimps, to judge from their skeletons. Perhaps they shared 99.9% of their genes with us. Their brains were as large as ours, and Neanderthals' brains were even slightly larger. The missing ingredient may have been a change in only 0.1% of our genes. What tiny change in genes could have had such enormous consequences?

Like some other scientists who have speculated about this question, I can think of only one plausible answer: the anatomical basis for spoken complex language. Chimpanzees, gorillas, and even monkeys are capable of symbolic communication not dependent on spoken words. Both chimpanzees and gorillas have been taught to communicate by means of sign language, and chimpanzees have learned to communicate via the keys of a large computer-controlled console. Individual apes have thus mastered 'vocabularies' of hundreds of symbols. While scientists argue over the extent to which such communication resembles human language, there is little doubt that it constitutes a form of symbolic communication. That is, a particular sign or computer key symbolizes a particular something else.

Primates can use not only signs and computer keys, but also sounds, as symbols. For instance, wild vervet monkeys have a natural form of symbolic communication based on grunts, with slightly different grunts to mean 'leopard', 'eagle', and 'snake'. A month-old chimpanzee named Viki, adopted by a psychologist and his wife and reared virtually as their daughter, learned to 'say' approximations of four words: 'papa', 'mama', 'cup', and 'up'. (The chimp breathed rather than spoke those words.) Given this capability for symbolic communication using sounds, why have apes not gone on to develop much more complex natural languages of their own?

The answer seems to involve the structure of the larynx, tongue, and associated muscles that give us fine control over spoken sounds. Like a Swiss watch, all of whose many parts have to be well-designed for the watch to keep time at all, our vocal tract depends on the precise functioning of many structures and muscles. Chimps are thought to be physically incapable of producing several of the commonest human vowels. If we too were limited to just a few vowels and consonants, our own vocabulary would be greatly reduced. For example, take this paragraph, convert all vowels other than 'a' or 'i' to either of those two, convert all consonants other than 'd' or 'm' or 's' to one of those three, and then see how much of the paragraph you can still understand.

Therefore, the missing ingredient may have been some modifications of the proto-human vocal tract to give us finer control and permit formation of a much greater variety of sounds. Such fine modifications of muscles need not be detectable in fossil skulls.

It is easy to appreciate how a tiny change in anatomy resulting in capacity for speech would produce a huge change in behaviour. With language, it takes only a few seconds to communicate the message, 'Turn sharp right at the fourth tree and drive the male antelope towards the reddish boulder, where I'll hide to spear it.' Without language, that message could be communicated only with difficulty, if at all. Without language, two proto-humans could not brainstorm together about how to devise a better tool, or about what a cave painting might mean. Without language, even one proto-human would have had difficulty thinking out for himself or herself how to devise a better tool.

I do not suggest that the Great Leap Forward began as soon as the mutations for altered tongue and larynx anatomy arose. Given the right anatomy, it must have taken humans thousands of years to perfect the structure of language as we know it – to arrive at the concepts of word order and case endings and tenses, and to develop vocabulary. In Chapter Eight I shall consider some possible stages by which our language might have become perfected. But if the missing ingredient did consist of changes in our vocal tract that permitted fine control of sounds, then the capacity for innovation would follow eventually. It was the spoken word that made us free.

This interpretation seems to me to account for the lack of evidence for Neanderthal/Cro-Magnon hybrids. Speech is of overwhelming import-ance in the relations between men and women and their children. That is not to deny that mute or deaf people learn to function well in our culture, but they do so by learning to find alternatives for a spoken language that already exists. If Neanderthal language was much simpler than ours or non-existent, it is not surprising that Cro-Magnons did not choose to marry Neanderthals.

I have argued that we were fully modern in anatomy and behaviour and language by 40,000 years ago, and that a Cro-Magnon could have been taught to fly a jet aeroplane. If so, why did it take so long after the Great Leap Forward for us to invent writing and build the Parthenon? The answer may be similar to the explanation why the Romans, great engineers that they were, didn't build atomic bombs. To reach the point of building an A-bomb required two thousand years of technological advances beyond Roman levels, such as the invention of gunpowder and calculus, the development of atomic theory, and the isolation of uranium. Similarly, writing and the Parthenon depended on tens of thousands of years of cumulative developments after the arrival of Cro-Magnons – developments that included the bow and arrow, pottery, domestication of plants and animals, and many others.

Until the Great Leap Forward, human culture had developed at a snail's pace for millions of years. That pace was dictated by the slow rate of genetic change. After the Leap, cultural development no longer depended on genetic change. Despite negligible changes in our anatomy, there has been far more cultural evolution in the past 40,000 years than in the millions of years before. Had a visitor from outer space come to the Earth in Neanderthal times, humans would not have stood out as unique among the world's species. At most, the visitor might have mentioned humans along with beavers, bowerbirds, and army ants as examples of species with curious behaviour. Would the visitor have foreseen the change that would soon make us the first species, in the history of life on Earth, capable of destroying all life?

PART TWO

AN ANIMAL WITH A
STRANGE LIFE-CYCLE

CHAPTER TWO TRACED OUR EVOLUTIONARY HISTORY THROUGH THE appearance of humans with fully modern anatomy and behavioural capabilities, but that chapter does not prepare us to go straight on to consider in more detail the development of human cultural hallmarks, such as language and art. That is because Chapter Two took up only the evidence of bones and tools. Yes, our evolution of large brains and upright posture was prerequisite to language and art, but that was not enough by itself. Human bones alone do not guarantee humanity. Instead, our rise to humanity also required drastic changes in our life-cycle, which will be the subject of Part Two of this book.

For any species one can describe what biologists term its 'life-cycle'. That means traits such as the number of offspring produced per litter or birth; the interval between births; the parental care (if any) that offspring receive from the mother or father; social relations between adult individuals; how a male and female select each other to mate with; frequency of sexual relations; and longevity. We take the forms of these traits as they exist in humans for granted, as the norm, but our life-cycle is actually bizarre by animal standards. All the traits that I have just mentioned vary greatly between species, and we are extreme in most respects. To mention only some obvious examples, most animals produce litters much larger than one baby at a time, most animal fathers provide no parental care, and few other animal species live even a small fraction of three-score years and ten.

Of these exceptional features of ours, some are shared by apes, suggesting that we merely retained traits already acquired by our ape-like ancestors. For instance, apes too usually give birth to one baby at a time, have births spaced several years apart, and live for several decades. None of these things is true of the other animals most familiar (but less closely related) to us, such as cats, dogs, songbirds, and goldfish.

In others of these respects, we are greatly different even from apes. Here are some obvious differences whose functions are well understood. Human babies continue to have all food brought to them by their parents even after weaning, whereas weaned apes gather their own food. Most human fathers as well as mothers, but only chimpanzee mothers, are closely involved in caring for their young. Like seagulls but unlike apes or most other mammals, we live in dense breeding colonies of nominally monogamous couples, some of whom also pursue extramarital sex. All these traits are as essential as large brain-cases for the survival and education of human offspring. That is because our elaborate, tool-dependent methods of obtaining food make weaned human infants incompetent to feed themselves. They first require a long period of food-provisioning, training, and protection – an investment much more taxing than that facing the ape mother. Hence human fathers who want their offspring to survive to maturity have generally assisted their mate with more than just sperm, the sole parental input of an orangutan father.

Our life-cycle also differs from that of wild apes in more subtle respects whose functioning is nevertheless still discernible. Many of us live longer than most wild apes: even hunter–gatherer tribes include some elderly individuals who are enormously important as repositories of experience. Men's testes are much larger than those of gorillas but smaller than those of chimps, for reasons that will become apparent in Chapter Three. We regard human female menopause as inevitable, and Chapter Seven will show why it makes good sense for humans, but it is almost unprecedented among other animals. The closest mammalian parallel is among some tiny mouse-like marsupials in Australia, and it is their males, not their females, that undergo menopause. Our longevity, testis size, and menopause were thus also prerequisites to our humanity.

Still other features of our life-cycle differ far more drastically from those of apes than do our testes, yet the functions of those remaining novel features of ours remain hotly debated. We are unusual in having sex mainly in private and for fun, rather than mainly in public and only when the female is able to conceive. Ape females advertise the time when they are ovulating; human females conceal it even to themselves. While anatomists understand why men's testes are the size that they are, an explanation for men's relatively enormous penis still escapes us. Whatever their explanation, all these features, too, are part of what defines humanity. Certainly, it is hard to picture how fathers and mothers could cooperate harmoniously in rearing their children if human females resembled some primate females in having their genitalia turn bright red at the time of ovulation, becoming sexually receptive only at that time, flaunting their red badge of receptivity, and proceeding to have sex in public with any male in the vicinity.

Human society and child-rearing rest therefore not only on the skeletal changes mentioned in Chapter Two, but also on these remarkable new features of our life-cycle. Unlike the case with our skeletal changes, however, we cannot follow through our evolutionary history the timing of each of these life-cycle changes, because they leave no direct fossil imprint. As a result, they receive only brief attention in paleontology texts despite their importance. Archaeologists have recently discovered a Neanderthal hyoid bone, one of the key pieces of our speech-producing equipment, but as yet no trace of a Neanderthal penis. We do not know whether *Homo erectus* was already on the road to evolving a preference for having sex in private, in addition to having evolved his and her well-documented large brain.

Our sole clues about the dating of these life-cycle changes are that something about longevity can be inferred from skeletons, and that size differences between fossil men and women may be indirect reflections of their mating system (more of that in Chapter Three). We cannot even prove through fossils, as we can for our large brain size, that we rather than living apes are the ones whose life-cycles diverged most from the ancestral condition. Instead, we have to be content with merely inferring that conclusion from the fact that our life-cycles are exceptional compared not just to living apes but also to other primates, suggesting that we were the ones who did more changing.

Darwin established in the mid-Nineteenth Century that the anatomy of animals has evolved through natural selection. Within this century, biochemists have similarly traced how the chemical make-up of animals has evolved through natural selection. But so has the behaviour of animals, including reproductive biology and sexual habits in particular. Life-cycle traits have some genetic basis, as we shall see below, and vary quantitatively among individuals of the same species. For instance, some women are genetically predisposed to give birth to twins, while genes for long lifespan run in some families more than in others. Life-cycle traits affect our success in passing on our genes, through affecting our success in wooing mates, conceiving and rearing babies, and surviving as adults. Just as natural selection tends to adapt an animal's anatomy to its ecological niche and vice versa, so natural selection also tends to mould animals' life-cycles. Those individuals leaving the most numerous surviving offspring promote their genes for life-cycle traits as well as for bones and chemical make-up.

A difficulty with this reasoning is that it seems as if some of our traits, such as menopause and aging, would reduce (rather than enhance) our output of offspring and should not have resulted from natural selection. It often proves profitable to try to understand these paradoxes through the concept of 'trade-offs'. In the animal world there is nothing that is free or pure good. Everything involves costs as well as benefits, by using space, time, or energy that could have been devoted to something else. You might otherwise have thought that women who never underwent menopause would leave more descendants than women who do. But consideration of the hidden costs of foregoing menopause (Chapter

Seven) will help us understand why evolution did not design these strategies into us. The same considerations illuminate such painful questions as why we grow old and die (Chapter Seven), and whether we are better off (even in a narrow evolutionary sense) in being faithful to our spouse or in pursuing extramarital affairs (Chapter Four).

I have been assuming in this discussion that our distinctively human life-cycle traits have some genetic basis. The comments that I made in Chapter One about the function of genes in general apply here as well. Just as our height and most of our observable traits are not influenced by only a single gene, there surely is not a single gene specifying menopause, testis size, or monogamy. In fact, we know little about the genetic bases of human life-cycle traits, though selective breeding experiments in mice and sheep have illuminated the genetic control of their testis size. Enormous cultural influences obviously operate on our motivation for providing child care or seeking extramarital sex, and there is no reason to believe that genes contribute significantly to differences among individual people in these traits. However, genetic differences between humans and the other two chimpanzee species probably do contribute to the consistent differences in many life-cycle traits between all human populations and all chimpanzee populations. There is no human society, regardless of its cultural practices, whose men have chimpanzee-sized testes and whose women forego menopause. Among those 1.6% of our genes that differ between us and chimps and that have any function, a significant fraction is likely to be involved in specifying traits of our life-cycle.

The story of our uniquely human life-cycle occupies the five chapters of Part Two. Chapter Three begins by taking up the distinctive features of human social organization and of sexual anatomy, physiology, and behaviour. As already mentioned, features that make us strange among animals include our societies of nominally monogamous couples, our genital anatomy, and our constant and generally private pursuit of sex. Our sex lives are reflected not only in our genitalia but also in the relative sizes of men's and women's bodies (much more equal than are the bodies of male and female gorillas or orangutans). We shall see how some of these familiar and distinctive features have known functions, while others continue to defy understanding.

No honest discussion of the human life-cycle could get away with noting that we are nominally monogamous and just leaving it at that. Pursuit of extramarital sex is obviously greatly influenced by each individual's particular upbringing and by the norms of the society in which the individual lives. Despite all that cultural influence, we are left with having to explain the facts that *both* the institution of marriage *and* the occurrence of extramarital sex have been reported from all human societies; but that extramarital sex is unknown in gibbons, although they do practise 'marriage' (that is, lasting male/female pairing to rear offspring); and that the question of extramarital sex is meaningless for chimpanzees because they do not practise 'marriage'. Hence an adequate discussion of our uniquely human life-cycle must account for our combination of marriage with extramarital sex. As Chapter Four will show, animal precedents exist to help us make evolutionary sense of our combination: men and women tend to differ in their attitudes towards extramarital sex much as geese and ganders do.

Chapter Five turns to another distinctive human life-cycle trait: how we select our sex partners, marital or otherwise. That problem scarcely arises for baboon troops, in which there is little selection: any male tries to mate with each female as she comes into heat. While common chimpanzees practise some selection of their sex partners, they are still much less selective and much more promiscuously baboon-like than are humans. Mate selection is a decision of major consequence in the human life-cycle, because married couples share parental responsibilities as well as sexual involvement. Precisely because care of human children demands such heavy and prolonged parental investment, we have to select our co-investor much more carefully than does a baboon. Nevertheless, Chapter Five will show that we can find animal precedents for our procedure in choosing sex partners, by going beyond primates to rats and birds.

Our mate selection criteria, explored in Chapter Five, are relevant to human racial variation, as will be discussed in Chapter Six. Humans native to different parts of the globe vary conspicuously in external appearance, as do gorillas, orangutans, and most other animal species occupying a sufficiently extensive geographic range. Our visible geographic variability has often been taken as a pretext for exercising a human hallmark to be discussed in Chapter Sixteen: genocidal killings. Some of the geographic variation in our appearance surely reflects natural selection moulding us to local climate, just as weasels in areas with winter snow develop white fur in winter for better camouflage and survival. But I shall argue in Chapter Six, as Darwin maintained, that our visible geographic variability arose mainly through sexual selection, as a result of those mate-choice procedures of ours discussed in Chapter Five.

Chapter Seven brings the discussion of our life-cycle to an end, by asking why our lives have to come to an end. Aging is another feature of our life-cycle so familiar that we take it for granted: of course we shall all grow old and eventually die. So will all individuals of all animal species, but different species age at very different rates. Among animals we are relatively long-lived and became even more so around the time that Cro-Magnons replaced Neanderthals. Our longevity has been important for our humanity, by permitting effective transmission of learned skills between generations. But even humans grow old. Why is aging inevitable, given our extensive capacity for biological self-repair?

Here, more than in any other chapter, the importance of thinking in terms of evolutionary trade-offs becomes clear. If measured by the ability to leave increased numbers of offspring, it just would not pay us to make the increased investment in self-repair mechanisms required to live longer. We shall see that the trade-off concept also illuminates the puzzle of menopause: a shutdown of child-bearing, paradoxically programmed by natural selection so that women can leave more surviving children.

THREE

THE EVOLUTION OF HUMAN SEXUALITY

Human sexuality seems normal to us but is bizarre by the standards of other animals. Our bizarre sex lives were as crucial to our rise to human status as were our large brains.

*

No week passes without publication of yet another book about sex. Our desire to read about sex is surpassed only by our desire to practise it. You might suppose that the basic facts of human sexuality must be familiar to lay people and understood by scientists. Just test your own grasp of sex by trying to answer these five easy questions:

Among the various ape species and man, which has by far the biggest penis, and what for?

Why should men be bigger than women?

How can men get away with having much smaller testes than chimpanzees?

Why do humans copulate in private, while all other social animals do it in public?

Why don't women resemble almost all other female mammals in having easily recognized days of fertility, with sexual receptivity confined to those days?

If your answer to the first question was 'the gorilla', put on a dunce's cap; the correct answer is man. If you gave any intelligent answers to the next four questions, publish them; scientists are still debating rival theories.

These five questions illustrate how hard it is to explain the most obvious facts of our sexual anatomy and physiology. Part of the problem is our hang-ups about sex: scientists did not even begin to study the subject seriously until recently, and they still have trouble being objective. Another difficulty is that scientists cannot do controlled experiments on the sexual practices of us humans, as they can on our cholesterol intake or tooth-brushing habits. Finally, sex organs do not exist in isolation: they are adapted to their owners' social habits and life-cycle, which are in turn adapted to food-gathering habits. In our own case that means, among other things, that evolution of human sex organs has been intertwined with that of human tool use, large brains, and child-rearing practices. Thus, our progress from being just another species of big mammal to being uniquely human depended on the remodelling not only of our pelvises and skulls, but also of our sexuality.

*

Given knowledge of how an animal feeds, a biologist can often predict that animal's mating system and genital anatomy. If we want to understand how human sexuality came to be the way it is, we have to begin by understanding the evolution of our diet and our society. From the vegetarian diet of our ape ancestors, we diverged within the last several million years to become social carnivores as well as vegetarians. Yet our teeth and claws remained those of apes, not of tigers. Our hunting prowess depended instead on large brains: by using tools and operating in coordinated groups, our ancestors were able to hunt successfully despite their deficient anatomical equipment, and they regularly shared food with each other. Our ability to gather roots and berries also came to depend on tools and thus to require large brains.

As a result, human children took years to acquire the information and the practice needed to be an efficient hunter-gatherer, just as they still take years to learn how to be a farmer or computer programmer today. During those many years after weaning, our children are still too dumb and helpless to acquire their own food; they depend entirely on their parents to bring food to them. These habits are so natural to us that we forget that baby apes gather food as soon as they are weaned.

The reasons why human infants are totally incompetent at food-gathering are actually two-fold – mechanical and mental. Firstly, making and wielding the tools used to obtain food requires fine finger co-ordination that children take years to develop. Just as my three-year-old sons still cannot tie their own shoelaces, three-year-old hunter-gatherer children cannot sharpen a stone axe, weave a net, or build a dugout canoe. Secondly, we depend on much more brainpower than do other animals in acquiring food, because we have a much more varied diet and more varied and complicated food-gathering techniques. For instance, New Guineans with whom I work typically have separate names for about a thousand different species of plants and animals living in the vicinity. For each of those species they know something about its distribution and life history, how to recognize it, whether it is edible or otherwise useful, and how best to capture or harvest it. All this information takes years to acquire.

Weaned human infants cannot support themselves because they lack these mechanical and mental skills. They need adults to teach them, and they also need adults to feed them for the decade or two that they are being taught. As is true of so many other human hallmarks, these problems of ours have animal precedents. In lions and many other species, the young must be trained to hunt by their parents. Chimpanzees too have a varied diet, employ varied foraging techniques, and assist their young in obtaining food, while common (but not pygmy) chimps make some use of tools. Our distinction is not absolute but one of degree: for us the necessary skills and hence the parental burden are far greater than for lions or chimpanzees.

The resulting parental burden makes care by the father as well as the mother important for a child's survival. Orangutan fathers provide their offspring with nothing beyond their initial donation of semen; gorilla, chimpanzee, and gibbon fathers go beyond that to offer protection; but hunter–gatherer human fathers provide some food and much teaching as well. Hence human food-gathering habits required a social system in which a male retained his relationship with a female after fertilizing her, in order to assist in rearing the resulting child. Otherwise, the child would be less likely to survive, and the father less likely to pass on his genes. The orangutan system, in which the father departs after copulation, would not work for us.

The chimpanzee system, in which several adult males are likely to copulate with the same oestrus female, also would not work for us. The result of that system is that a chimpanzee father has no idea which infants in the troop he has sired. For the chimp father that is no loss, as his exertions on behalf of troop infants are modest. The human father, however, who will contribute significantly to the care of what he thinks is his child, had better have some confidence in his paternity – for example, through having been the exclusive sexual partner of the child's mother. Otherwise, his child-care contribution may help pass on some other man's genes.

Confidence in paternity would be no problem if humans, like gibbons, were scattered over the landscape as separate couples, so that each female would only rarely encounter a male other than her consort. But there are compelling reasons why almost all human populations have consisted of groups of adults, despite the paranoia about paternity that this causes. Among the reasons: much human hunting and gathering involves cooperative group efforts among men, women, or both; much of our wild food occurs in scattered but concentrated patches, able to sustain many people; and groups offer better protection against predators and aggressors, especially against other humans.

In short, the social system we evolved to accommodate our un-apelike food habits seems utterly normal to us, but is bizarre by ape standards and is virtually unique among mammals. Adult orangutans are solitary; adult gibbons live as separate monogamous male/female pairs; gorillas live in polygamous harems, each consisting of several adult females and usually one dominant adult male; common chimpanzees live in fairly promiscuous communities consisting of scattered females plus a group of males; and pygmy chimpanzees form even more promiscuous communities of both sexes. Our societies, like our food habits, resemble those of lions and wolves: we live in bands containing many adult males *and* many adult females. Furthermore, we diverge from even lions and wolves in how those societies are organized: our males and females are paired off with each other. In contrast, any male lion within a lion pride can and regularly does mate with any of the pride's lionesses, making paternity unidentifiable. Our peculiar societies instead have their closest parallels in colonies of seabirds, like gulls and penguins, which also consist of male/female pairs.

At least officially, human pairing is more or less monogamous in most modern political states, but is 'mildly polygynous' among most surviving hunter-gatherer bands, which are better models for how mankind lived over the last million years. (This description omits consideration of extramarital sex, through which we become effectively more polygamous and whose scientifically fascinating aspects I shall discuss in Chapter Four.) By 'mildly polygynous', I mean that most hunter-gatherer men can support only a single family, but a few powerful men have several wives. Polygyny on the scale of elephant seals, among which powerful males have dozens of wives, is impossible for hunter-gatherer men, because they differ from elephant seals in having to provide child care. The big harems for which some human potentates are famous didn't become possible until the rise of agriculture and centralized government let a few princes tax everyone else in order to feed the royal harem's babies.

*

Now let's see how this social organization shapes the bodies of men and women. Take first the fact that adult men are slightly bigger than similarly aged women (about eight per cent taller and twenty per cent heavier, on the average). A zoologist from outer space would take one look at my 5-foot 8-inch wife next to me (5 foot 10 inches), and would instantly guess that we belonged to a mildly polygynous species. How,

MALES, AS FEMALES SEE THEM

Humans and great apes differ with respect to the relative body size of males and females, penis length, and testis size. The main circles represent the body size of the male of each species, relative to that of the female of the same species. Female body size is arbitrarily shown as the same for all species at upper right. Thus, chimps of both sexes weigh about the same; men are slightly larger than women; but male orangutans and gorillas are much bigger than females. The arrows on the male symbols are proportional to the length of the erect penis, while the twin circles represent testis weight relative to that of the body. Men have the longest penis, chimps the largest testes, and orangutans and gorillas the shortest penis and smallest testes.

you may ask, can one possibly guess mating practices from relative body size?

It turns out that, among polygynous mammals, average harem size increases with the ratio of the male's body size to the female's body size. That is, the biggest harems are typical of species in which males are much larger than females. For example, males and females are the same size in gibbons, which are monogamous; male gorillas, with a typical harem of three to six females, weigh nearly double the weight of each female; but

FEMALES, AS MALES SEE THEM

Human females are unique in their breasts, which are considerably larger than those of apes even before the first pregnancy. The main circles represent female body size relative to male body size of the same species.

the average harem is forty-eight wives for the southern elephant seal, whose 3-ton male dwarfs his 700-pound wives. The explanation is that, in a monogamous species, every male can win a female, but in a very polygynous species most males languish without any mate, because a few dominant males have succeeded in rounding up all the females into their harems. Hence, the bigger the harem, the fiercer is the competition among males and the more important it is for a male to be big, since the bigger male usually wins the fights. We humans, with our slightly bigger males and slight polygyny, fit this pattern. (However, at some point in human evolution, male intelligence and personality came to count for more than size: male basketball players and sumo wrestlers don't tend to have more wives than male jockeys or coxswains.)

Because competition for mates is fiercer in polygynous than in monogamous species, the polygynous species also tend to have more marked differences between males and females in other respects besides body size. These differences are the secondary sexual characteristics that play a role in attracting mates. For instance, males and females of the monogamous gibbons look identical at a distance, while male gorillas (befitting their polygyny) are easily recognized by their crested heads and silver-haired backs. Here too, our anatomy reflects our mild polygyny. The external differences between men and women are not nearly as marked as sex-related differences in gorillas or orangutans, but the zoologist from outer space could probably still distinguish men and women by the body and facial hair of men, men's unusually large penis, and the large breasts of women even before first pregnancy (in this we are unique among primates).

*

Proceeding now to the genitalia themselves, the combined weight of the testes in the average man is about 1½ ounces. This may boost the macho man's ego when he reflects on the slightly lower testis weight in a 450-pound male gorilla. But wait – our testes are dwarfed by the 4-ounce testes of a 100-pound male chimpanzee. Why is the gorilla so economical, and the chimp so well-endowed, compared to us?

The Theory of Testis Size is one of the triumphs of modern physical anthropology. By weighing the testes of thirty-three primate species, British scientists identified two trends: species that copulate more often need bigger testes; and promiscuous species in which several males routinely copulate in quick sequence with one female need especially big testes (because the male that injects the most semen has the best chance of being the one to fertilize the egg). When fertilization is a competitive lottery, large testes enable a male to enter more sperm-tickets in the lottery.

Here is how these considerations account for the differences in testis size among the great apes and humans. A female gorilla does not resume sexual activity until three or four years after giving birth, and she is receptive for only a couple of days a month until she becomes pregnant again. So even the successful male gorilla with a harem of several females experiences sex as a rare treat – if he is lucky, a few times a year. His relatively tiny testes are quite adequate for those modest demands. The sex life of a male orangutan may be somewhat more demanding, but not much. However, each male chimp in a promiscuous troop of many females lives in sexual nirvana, with nearly daily opportunities to copulate for a common chimp and several daily copulations for the average pygmy chimp. That, plus his need to outdo other male chimps in semen output if he is to fertilize the promiscuous female, explains his need for gigantic testes. We humans make do with medium-sized testes because the average man copulates more often than gorillas or orangutans but less often than chimps. In addition, the typical woman in a typical menstrual cycle does not force several men into sperm competition to fertilize her.

Thus, primate testis design well illustrates the principles of trade-offs and evolutionary cost/benefit analyses explained on page 52. Each species has testes big enough to do their job, but not unnecessarily larger ones. Bigger testes would just entail more costs without proportional benefits, by diverting space and energy from other tissues and increasing the risk of testicular cancer.

From this triumph of scientific explanation we descend to a glaring failure: the inability of twentieth-century science to formulate an adequate Theory of Penis Length. The length of the erect penis averages 1¼ inches in a gorilla, 1½ inches in an orangutan, 3 inches in a chimp, and 5 inches in a man. Visual conspicuousness varies in the same sequence: a gorilla's penis is inconspicuous even when erect because of its black colour, while the chimp's pink erect penis stands out against the bare white skin behind it. The flaccid penis is not even visible in apes. Why does the human male need his relatively enormous, attention-getting penis, which is larger than that of any other primate? Since the male ape successfully propagates his kind with much less, does not the human penis represent largely wasted protoplasm that would be more valuable if devoted, say, to cerebral cortex or improved fingers?

Biologist friends to whom I pose this conundrum usually think of distinctive features of human coitus where they suppose a long penis might somehow be useful: our frequent use of the face-to-face position, our acrobatic variety of coital positions, and the supposedly long duration of our coital bouts. None of these explanations survives close scrutiny. The face-to-face position is also a preferred one for orangutans and pygmy chimps, and is used occasionally by gorillas. Orangutans vary face-to-face copulation with dorso-ventral and sideways positions, and do it while hanging from branches of trees – surely that demands more penile acrobatics than our comfortable boudoir exercises. Our mean duration of coitus (about four minutes for Americans) is much longer than for gorillas (one minute), pygmy chimps (fifteen seconds), or common chimps (seven seconds), but shorter than for orangutans (fifteen minutes) and lightning-fast compared to the twelve-hour-long copulations of marsupial mice. (Are you listening, ghosts of Errol Flynn and Don Juan?)

Since it thus seems unlikely that special features of human coitus demand a large penis, a popular alternative theory is that the human penis has also become an organ of display, like a peacock's tail or a lion's mane. This theory is reasonable but begs the question, what type of display, and to whom?

Proud male anthropologists unhesitatingly answer, an attractive display, to women, but this represents mere wishful thinking. Many women say that they are turned on by a man's voice, legs, and shoulders more than by the sight of his penis. A telling point is that the women's magazine *Viva* initially published photos of nude men but dropped them after surveys showed lack of female interest. When *Viva*'s nude men disappeared, the number of female readers increased, and the number of male readers decreased. Evidently, the male readers were the ones buying *Viva* for its nude photos. While we can agree that the human penis is an organ of display, the display is intended not for women but for fellow men.

Other facts confirm the role of a large penis as a threat or status display towards other men. Recall all the phallic art created by men for men, and the widespread obsession of men with their penis size. Evolution of the human penis was effectively limited by the length of the female vagina: a man's penis would damage a woman if it were significantly larger. However, I can guess what the penis would look like if this practical constraint were removed and if men could design it themselves. It would resemble the penis sheaths (phallocarps) used as male attire in some areas of New Guinea where I do field work. Phallocarps vary in length (up to 2 feet), diameter (up to 4 inches), shape (curved or straight), angle made with the wearer's body, colour (yellow or red), and decoration (such as a tuft of fur at the end). Each man has a wardrobe of several sizes and shapes from which to choose each day, depending on his mood that morning. Embarrassed male anthropologists interpret the phallocarp as something used for modesty or concealment, to which my wife had a succinct answer on seeing a phallocarp: 'The most immodest display of modesty I've ever seen!'

Thus, astonishing as it seems, important functions of the human penis remain obscure. Here is a rich field for research.

*

Passing now from anatomy to physiology, we are immediately confronted by our sexual activity pattern, which must be considered freakish by the standards of other mammal species. Most mammals are sexually inactive most of the time. They copulate only when the female is in oestrus — that is, when she is ovulating and capable of being fertilized. Female mammals apparently 'know' when they are ovulating, for they solicit copulation then by presenting their genitals towards males. Lest a male miss the point, many female primates go further; the area around the vagina, plus in some species the buttocks and breasts, swells up and turns red, pink, or blue. This visual advertisement of female availability affects male monkeys in the same way that the sight of a seductively dressed woman affects male humans. In the presence of females with brightly swollen genitals, male monkeys stare much more often at the female's genitals, develop higher testosterone levels, attempt to copulate more often, and penetrate more quickly and after fewer pelvic thrusts than in the presence of females not displaying their wares.

Human sexual cycles are quite different. The human female maintains her sexual receptivity more or less constantly, instead of having it sharply confined to a short oestrus phase. Indeed, despite numerous studies aimed at settling whether a woman's receptivity varies at all through her cycle, there is still no agreement about the answer – nor about the cycle phase when receptivity is maximal if it does vary.

So well concealed is human ovulation that we did not have accurate scientific information on its timing until around 1930. Before that, many physicians thought that women could conceive at any point in their cycle, or even that conception was most likely at the time of menstruation. In contrast to the male monkey who has only to scan his surroundings for brightly swollen lady monkeys, the unfortunate human male has not the faintest idea which ladies around him are ovulating and capable of being fertilized. A woman herself may learn to recognize sensations associated with ovulation, but it is often tricky, even with the help of thermometers and ratings of vaginal mucus quality. Furthermore, today's would-be mother, who tries in such ways to sense ovulation in order to achieve (or avoid) fertilization, is responding by cold-blooded calculation to hard-won, modern book knowledge. She has no other choice; she lacks the innate, hot-blooded sense of sexual receptivity that drives other female mammals.

Our concealed ovulation, constant receptivity, and brief fertile period in each menstrual cycle ensure that most copulations by humans are at the wrong time for conception. To make things worse, menstrual cycle length varies more between women, or from cycle to cycle in a given woman, than for other female mammals. As a result, even young newlyweds who omit contraception and make love at maximum frequency have only a twenty-eight per cent probability of conception in each menstrual cycle. Animal breeders would be in despair if a prize cow had such low fertility, but in fact they can schedule a *single* artificial insemination so that the cow has a seventy-five per cent chance of being fertilized!

Whatever the main biological function of human copulation, it is not conception, which is just an occasional by-product. In these days of growing human overpopulation, one of the most ironic tragedies is the Catholic Church's claim that human copulation has conception as its natural purpose, and that the rhythm method is the only proper means of birth control. The rhythm method would be terrific for gorillas and most other mammal species, but not for us. In no species besides humans has the purpose of copulation become so unrelated to conception, or the rhythm method so unsuited for contraception.

For animals, copulation is a dangerous luxury. While occupied *in acto flagrante*, an animal is burning up valuable calories, neglecting opportunities to gather food, vulnerable to predators eager to eat it, and vulnerable to rivals eager to usurp its territory. Copulation is something to be accomplished in the minimum time required to do the job of fertilization. In contrast, human sex, as a device to achieve fertilization, would have to be rated a huge waste of time and energy, an evolutionary failure. Had we retained a proper oestrus cycle like other mammals, the wasted time could have been diverted by our hunter-gatherer ancestors to butchering more mastodons. By this results-oriented view of sex, any hunter-gatherer band whose females advertised their oestrus period could thereby have fed more babies and out-competed neighbouring bands.

Thus, the most hotly debated problem in the evolution of human reproduction is to explain why we nevertheless ended up with concealed ovulation, and what good all our mistimed copulations do us. For scientists, it is no answer just to say that sex is fun. Sure, it's fun, but evolution made it that way. If we were not getting big benefits from our mistimed copulations, mutant humans who had evolved not to enjoy sex would have taken over the world.

Related to this paradox of concealed ovulation is the paradox of concealed copulation. All other group-living animals have sex in public, whether they are promiscuous or monogamous. Paired seagulls mate in the middle of the colony; an ovulating female chimpanzee may mate consecutively with five males in each other's presence. Why are we unique in our strong preference for copulating in private?

Biologists are currently arguing over at least six different theories to explain the origin of concealed ovulation and concealed copulation in humans. Interestingly, the debate proves to be a Rorschach test for the gender and outlook of the scientists involved. Here are the theories and their proponents:

1. *Theory preferred by many traditional male anthropologists.*

According to this view, concealed ovulation and copulation evolved in order to enhance cooperation and reduce aggression among male hunters. How could cavemen bring off the precise teamwork needed to spear a mammoth, if they had been fighting that morning for the public favours of a cavewoman in oestrus? The implicit message of this theory is that women's physiology is important chiefly for its effect on bonds between men, the real movers of society. However, one can broaden this theory to make it less blatantly sexist. Visible oestrus and sex would disrupt human society by affecting female/female and male/female as well as male/male bonds.

To illustrate this broadened version of the prevalent theory, consider the following scene from an imaginary soap opera, showing what life would be like for us modern hunter-gatherers if we did not have concealed ovulation and private copulation. Our soap opera stars Bob and Carol and Ted and Alice and Ralph and Jane. Bob, Alice, Ralph, and Jane work together in an office where the men hunt contracts and the women gather accounts payable. Ralph is married to Jane. Bob's wife is Carol, and Alice's husband is Ted. Carol and Ted work elsewhere.

One morning, Alice and Jane both discover on awakening that they have turned bright red in order to advertise impending ovulation and sexual receptivity. Alice and Ted make love at home before they go off in their separate directions to work. Jane and Ralph go together to work, where they copulate occasionally on the office sofa in the presence of their co-workers.

Bob cannot help lusting for Alice and Jane when he sees them bright red and sees Jane and Ralph copulating. He is unable to concentrate on his work. He repeatedly propositions Jane and Alice.

Ralph drives Bob away from Jane.

Alice is faithful to Ted and rejects Bob, but the hassle also interferes with her work.

All day, Carol in her office elsewhere is seething with jealousy at the thought of Alice and Jane, because Carol knows that Alice and Jane are bright red and attractive to Bob, while she (Carol) is not.

As a result, the office succeeds in bagging few contracts and accounts. In the meantime, other offices, where ovulation is concealed and where copulation is private, prosper. Eventually, Bob's, Alice's, Ralph's, and Jane's office goes extinct. The only offices that survive are those with concealed ovulation and copulation.

This parable suggests that the traditional theory, by which concealed ovulation and copulation evolved to promote cooperation within human societies, is plausible. Unfortunately, there are other, equally plausible theories that I will now explain more briefly.

2. *Theory preferred by many other traditional male anthropologists.*
Concealed ovulation and copulation cement the bonds between a particular man and woman, thereby laying the foundations of the human family. A woman remains sexually attractive and receptive so that she can satisfy a man sexually all the time, bind him to her, and reward him for his help in rearing her baby. The sexist message: women evolved to make

men happy. Left unexplained by this theory is the question of why pairs of gibbons, whose unflinching devotion to monogamy should make them role models for the Moral Majority, remain constantly together despite having sex only every few years.

3.   *Theory of a more modern male anthropologist (Donald Symons).*
Symons noted that a male chimpanzee who kills a small animal is more likely to share the meat with an oestrus female than with a non-oestrus female. This suggested to Symons that human females might have evolved a constant state of oestrus, in order to ensure a frequent meat supply from male hunters by rewarding them with sex. As an alternative theory, Symons noted that women in most hunter–gatherer societies have little say in selection of a husband. The societies are male-dominated, and male clans just suit themselves by exchanging daughters in marriage. However, by being constantly attractive, even a woman wed to an inferior male could privately seduce a superior male and secure his genes for her children. Symons' theories, while still male-orientated, at least represent a step forward in that he views women as cleverly pursuing their own goals.

4.   *Theory produced jointly by a male biologist and a female biologist (Richard Alexander and Katherine Noonan).*
If a man could recognize signs of ovulation, he could use that knowledge to fertilize his wife by copulating with her only while she is ovulating. He could then safely neglect her the rest of the time and go off and philander, secure in the knowledge that the wife he left behind was unreceptive, if not already fertilized. Hence women evolved concealed ovulation to force men into a permanent marriage bond, by exploiting male paranoia about fatherhood. Not knowing the time of ovulation, a man must copulate often with his wife to have a chance of fertilizing her, and that leaves him less time to develop dalliances with other women. The wife benefits, but so does the husband. He gains confidence in his paternity of his children, and he need not worry that his wife will suddenly attract many competing men by turning bright red on a particular day. At last, we have a theory seemingly grounded in sexual equality.

5.   *Theory of a female sociobiologist (Sarah Hrdy).*
Hrdy was impressed by the frequency with which many primates – including not only monkeys but also baboons, gorillas, and common chimps – kill infants not their own. The bereaved mother is thereby induced to come into oestrus again and often mates with the murderer, thus increasing his output of progeny. (Such violence has been common in human history: male conquerors kill the vanquished men and children

but spare the women.) As a counter-measure, Hrdy reasoned, women evolved concealed ovulation in order to manipulate men by confusing the issue of paternity. A woman who distributed her favours widely would thereby enlist many men to help feed (or at least not to kill) her infant, since many men could suppose themselves to be the infant's father. Whether this theory is right or wrong, we must applaud Hrdy's overturning of conventional masculine sexism and transferring sexual power to women.

6. *Theory of another female sociobiologist (Nancy Burley).*

The average 7-pound newborn human weighs double a newborn gorilla, but the 200-pound gorilla mother dwarfs the average human mother. Because the newborn human is so much larger in relation to its mother than are newborn apes, birth is exceptionally painful and dangerous in humans. Until the advent of modern medicine, women often died in childbirth, whereas I have never heard of such a fate befalling a female gorilla or chimpanzee. Once humans had evolved enough intelligence to associate conception with copulation, oestrous women could have chosen to avoid copulating at the time of ovulation, and could have thereby spared themselves the pain and peril of childbirth, but such women would have left fewer descendants than women who could not detect their ovulation. Thus, where male anthropologists saw concealed ovulation as something evolved by women for men (Theories 1 and 2), Nancy Burley sees it as a trick that women evolved to deceive themselves.

*

Which of these six theories for the evolution of concealed ovulation is correct? Not only are biologists uncertain; it is only in recent years that the question has begun to receive serious attention. This dilemma exemplifies a pervasive problem in establishing causation in evolutionary biology, as well as in history, psychology, and many other fields where one cannot manipulate variables to perform controlled experiments. Such experiments would afford the most convincing way to demonstrate cause or function. If we could remodel one tribe of people so that all women advertised their day of ovulation, we could then see whether cooperation within or between couples broke down, or whether the women used their new knowledge to avoid becoming pregnant. In the absence of such experiments, we can never be certain what human society would really be like today without concealed ovulation.

If it is hard to determine the function of things happening today under our eyes, how much harder must it be to determine functions in the vanished past! We know that human bones and tools were different hundreds of thousands of years ago, when concealed ovulation may have been evolving. Probably human sexuality, including the function of concealed ovulation, may also have been different then, in ways now hard for us to picture. Interpretation of our past runs the constant risk of degenerating into mere 'paleopoetry' stories that we spin today, stimulated by a few bits of fossil bone, and expressing like Rohrschach tests our own personal prejudices, but devoid of any claim to validity about the past.

Nevertheless, having mentioned six plausible theories, I cannot just walk away from the problem without attempting some synthesis. Here again, we come up against another pervasive problem in establishing causation. It is rare for complex phenomena such as concealed ovulation to be influenced by only a single factor. It would be as silly to seek a single cause of concealed ovulation as to claim that there was a single root cause of the First World War. Instead, there were many independent factors in the period 1900–1914 pushing towards war, others pushing towards peace. War finally broke out when the net weight of factors tipped towards war. Yet that does not excuse going to the opposite extreme of 'explaining' complex phenomena by an unweighted laundry list encompassing every conceivable factor.

As a first step towards pruning down our laundry list of six theories, let's realize that, whatever factors caused our distinctive sexual habits to evolve in the distant past, they would not be persisting today if there were not some factors still sustaining them. But the factors responsible for their initial appearance need not have been the same as the ones now operative. In particular, although the factors behind Theories 3, 5, and 6 may have been major ones long ago, they do not seem to be so now. Only a minority of modern women uses sex either to obtain food or other resources from a number of men, or to confound paternity and induce many men simultaneously to support a woman's child. Postulates of their former role are paleopoetry, albeit plausible paleopoetry. Let's just content ourselves with trying to understand why concealed ovulation and frequent private copulation might make sense now. At least, our guesses can be guided by introspection about ourselves plus observations of others.

The factors behind Theories 1, 2, and 4 seem to me still operative today, and to be facets of the same paradox, the most distinctive feature of human social organization. That paradox is that a man and woman desirous for their child (and genes) to survive must cooperate with each other for a long time to rear their child, *and* must also cooperate economically with many other couples living close by. It is obvious that regular sexual relations between a man and woman intensifies their connection, compared to their connections with other women and men whom they see daily but with whom they are not involved sexually. Concealed ovulation and constant receptivity advance this 'new' function of sex (new by the standards of most mammals) as a social cement, not just a device for fertilization. This function is not, as in the traditional male chauvinist version of Theories 1 and 2, a sop thrown by a cold, calculating woman to a sex-starved man, but instead an inducement for both sexes. Not only have all signs of female ovulation vanished, but the act of sex itself takes place privately, to emphasize the distinction between sexual and non-sexual partners within the same close group. As for the objection that gibbons remain monogamously involved without the reward of constant sex, that is easy to explain: each gibbon couple has minimal social – and no economic – involvement with other gibbon couples.

Human testis size also seems to me an outcome of that same basic paradox of human social organization. While our testes are larger than a gorilla's, because we have frequent sex for fun, they are still smaller than a chimpanzee's, because we are more monogamous. The oversized human penis may have evolved as an arbitrary sexual display symbol, as arbitrary as a lion's mane or a woman's enlarged breasts. Why were lionesses not the ones to develop enlarged breasts, lions an oversized penis, and men a mane? If they had, those permuted signals could have functioned equally well. That it did not come out that way may be just an accident of evolution, a result of each species' and sex's relative ease of evolving those various structures.

But there is still something basic missing from our discussion so far. I have talked about an idealized form of human sexuality: monogamous couples (plus a few polygynous households), husbands confident in their paternity of their wives' children, and husbands helping their wives with child-rearing rather than neglecting the kids in order to philander. As justification for discussing this fictitious ideal, I maintain that actual human practice is much closer to this ideal than to baboon or chimpanzee practice. But the ideal is still fictitious. Any social system with rules of conduct is open to the risk of individuals cheating when they find the advantages of cheating to outweigh the burden of sanctions. The question is thus a quantitative one. Does cheating become so regular that the whole system collapses, or does cheating occur but not so often as to destroy the system, or is cheating vanishingly rare? As translated for human sexuality, that question becomes one of whether ninety, thirty or one per cent of human babies are fathered extramaritally. That question and its consequences will be the subject of the next chapter.

FOUR

THE SCIENCE OF ADULTERY

*Cold-blooded analysis of adultery views life as an evolutionary
contest whose winners are those individuals leaving the largest
number of surviving offspring. This view helps one understand why
humans reinvented adultery after the other two chimps had bypassed
it.*

*

People have many reasons to lie when asked whether they have
committed adultery. Consequently, it is notoriously difficult to get
accurate scientific information about this important subject. One of the
few existing sets of hard facts emerged as a totally unexpected by-product
of a medical study, performed nearly half-a-century ago for a different
reason. That study's findings have never been revealed until now.

I recently learned those facts from the distinguished medical scientist
who ran the study. (Since he does not wish to be identified in this
connection, I shall refer to him as Dr X.) In the late 1940s Dr X was
studying the genetics of human blood groups, which are molecules that
we acquire only by inheritance. Each of us has dozens of blood-group
substances on our red blood cells, and we inherit each substance either
from our mother or from our father. The study's research plan was
straightforward: go to the obstetrics ward of a highly respectable US
hospital; collect blood samples from 1,000 newborn babies and their
mothers and fathers; identify the blood groups in all the samples; and then
use standard genetic reasoning to deduce the inheritance patterns.

To Dr X's shock, the blood groups revealed nearly ten per cent of those
babies to be the fruits of adultery! Proof of the babies' illegitimate origin
was that they had one or more blood groups lacking in both alleged
parents. There could be no question of mistaken maternity – the blood
samples were drawn from an infant and its mother soon after the infant
emerged from the mother. A blood group present in a baby but absent in
its undoubted mother could only have come from its father. Absence of
the blood group from the mother's husband as well showed conclusively
that the baby had been sired by some other man, extramaritally. The true

incidence of extramarital sex must have been considerably higher than ten per cent, since many other blood-group substances now used in paternity tests were not yet known in the 1940s, and since most bouts of intercourse do not result in conception.

At the time that Dr X made his discovery, research on American sexual habits was virtually taboo. He decided to maintain a prudent silence, never published his findings, and it was only with difficulty that I got his permission to mention his results without betraying his name. However, his results have more recently been confirmed by several similar genetic studies whose results did get published. Those studies variously showed between about five and thirty per cent of American and British babies to have been adulterously conceived. Again, the proportion of the tested couples of whom at least one practised adultery must have been higher, for the same two reasons as in Dr X's study.

We can now answer the question posed at the end of the last chapter: whether extramarital sex is for humans a rare aberration, a frequent exception to a 'normal' pattern of marital sex, or so frequent as to make a sham of marriage. The middle alternative proves to be the correct one. Most fathers really are raising their own children, and human marriage is not a sham. We are not just promiscuous chimpanzees pretending to be otherwise. Yet it is also clear that extramarital sex is an integral, albeit unofficial, part of the human mating system. Adultery has also been observed in many other animal species whose societies resemble ours in being based on male and female co-parents with a lasting bond. Since such lasting bonds do not characterize common chimpanzee or pygmy chimpanzee society, it is meaningless to talk of adultery in chimps. We must have reinvented it after our chimp-like ancestors had rendered it obsolete. Therefore, we cannot discuss human sexuality, and its role in our rise to humanity, without carefully considering the science of adultery.

Most of our information about adultery's incidence has come from researchers asking people about their sex lives, rather than from blood-grouping their babies. Since the 1940s, the myth that marital infidelity is rare in the US has been publicly exploded by a long succession of surveys, beginning with the Kinsey report. Nevertheless, even though this is the supposedly liberated 1990s, we are still profoundly ambivalent about adultery. It is thought of as exciting; no television soap opera could attract many viewers without it. It has few rivals as a basis of humour. Yet, as Freud pointed out, we often use humour to deal with things that are intensely painful. Thus, throughout history, adultery has also had few rivals as a cause of murder and human misery. In writing about this subject, it is impossible to remain completely serious, but it is also impossible not to be revolted at the sadistic institutions by which societies have attempted to deal with extramarital sex.

*

What makes a married person decide to seek or avoid adultery? Scientists have theories to explain many other things, so it should not be surprising that there is also a theory of extramarital sex (abbreviated to EMS, and not to be confused with premarital sex or PMS, in turn not to be confused with premenstrual syndrome, also PMS). With many species of animals the problem of EMS never arises, because they do not opt for marriage in the first place. For instance, a female Barbary macaque in heat copulates promiscuously with every adult male in her troop and averages one copulation per seventeen minutes. However, some mammals and most bird species do opt for 'marriage'. That is, a male and a female form a lasting pair-bond to devote care or protection to their joint offspring. Once there is marriage, there is also the possibility of what socio-biologists euphemistically term 'the pursuit of a mixed reproductive strategy' (abbreviated to MRS). In plain English, that means being married while simultaneously seeking extramarital sex.

Married animals vary enormously in the degree to which they mix their reproductive strategies. There appears to be no recorded instance of EMS in the little apes called gibbons, while snow geese indulge regularly. Human societies similarly vary, but I suspect that none comes close to the faithful gibbons. To explain all this variation, sociobiologists have found it useful to apply the reasoning of game theory. That is, life is considered an evolutionary contest whose winners are those individuals leaving the largest number of surviving offspring.

Contest rules are set by the ecology and reproductive biology of the particular species. The problem is then to figure out which strategy is most likely to win the contest: rigid fidelity, pure promiscuity, or a mixed strategy. But I must make one thing clear right at the outset. While this sociobiological approach certainly proves useful for understanding adultery in animals, its relevance for human adultery is an explosive issue and one to which I shall return.

The first thing one realizes is that the best game strategy differs between males and females of the same species. This is because of two profound differences between the reproductive biology of males and females, in the minimum necessary reproductive effort, and in the risk of being cuckolded. Let's consider these differences, which are painfully familiar to humans.

For men, the minimum effort needed to sire an offspring is the act of copulation, a brief expenditure of time and energy. The man who sires a baby by one woman one day is biologically capable of siring a baby by another woman the same day. For women, however, the minimum effort consists of copulation plus pregnancy plus (throughout most of human history) several years spent nursing – a huge commitment of time and energy. Thus, a man potentially can sire far more offspring than can a woman. A nineteenth-century visitor who spent a week at the court of the Nizam of Hyderabad, a polygamous Indian potentate, reported that four of the Nizam's wives gave birth within eight days, and that nine more births were anticipated for the following week. The record lifetime number of offspring for a man is 888, sired by Emperor Moulay Ismail the Bloodthirsty of Morocco, while the corresponding record for a woman is only sixty-nine (a nineteenth-century Moscow woman specializing in triplets). Few women have topped twenty children, whereas some men easily do so in polygynous societies.

As a result of this biological difference, a man stands to gain much more from EMS or polygamy than does a woman – if one's sole criterion is the number of offspring born. (To female readers about to stop reading in outrage, or to male readers about to cheer, I warn you now – keep reading, there is much more to the question of EMS.) For human EMS the statistical evidence is naturally hard to come by, but for human polygamy it is readily available. In the sole polyandrous society for which I could find data, the Tre-ba of Tibet, women with two husbands average *fewer* children, not more children, than women with one husband. In contrast, nineteenth-century American Mormon men realized big benefits from polygyny: men with one wife averaged only seven children, but men with two wives averaged sixteen children, and those with three wives averaged twenty. Polygynous Mormon men as a group averaged 2.4 wives and fifteen children, while polygynous Mormon church leaders in particular averaged five wives and twenty-five children. Similarly, among the polygynous Temne people of Sierra Leone, a man's average number of children increases from 1.7 to seven as his number of wives increases from one to five.

The other sexual asymmetry relevant to the best game strategy involves confidence that one really is the biological parent of one's putative offspring. A cuckolded animal, deceived into rearing offspring not its own, has thereby lost the evolutionary game while advancing the victory of another player, the real parent. Barring a switch of babies in the hospital nursery, women cannot be cuckolded; they see their baby emerge from their bodies. Nor can there be cuckoldry of males in animal species practising external fertilization (that is, fertilization of eggs outside the female's body). For instance, some male fish watch a female shed eggs, then immediately deposit sperm on the eggs and scoop them up to care for them, secure in their paternity. However, men and other male animals practising internal fertilization – fertilization of eggs inside the female's body – can readily be cuckolded. All that the putative father knows for sure is that his sperm went into the mother, and eventually an offspring came out. Only observation of the female throughout her whole fertile period can absolutely exclude the possibility that some other male's sperm also entered and did the actual fertilizing.

An extreme solution to this simple asymmetry is the one formerly adopted by southern India's Nayar society. Among the Nayar, women freely took many lovers simultaneously or in sequence, and husbands accordingly had no confidence in paternity. To make the best of a bad situation, a Nayar man did not live with his wife or care for his supposed children, but he instead lived with his sisters and cared for his sisters' children. At least, those nieces and nephews were sure to share one-quarter of his genes.

Bearing in mind these two basic facts of sexual asymmetry, we can now examine what is the best game strategy, and when EMS pays. Let's examine three game plans of increasing complexity:

Game Plan 1.

A man should always seek EMS, because he has so little to lose and so much to gain. Consider the hunter-gatherer conditions prevailing throughout most of human evolution, under which a woman could at best rear about four children in the course of her life. Through one dalliance, her otherwise faithful husband could increase his lifetime reproductive output from four to five: an enormous increase of twenty-five per cent, for only a few minutes' work. What is wrong with this dazzlingly naive reasoning?

Game Plan 2.

A moment's reflection should expose a basic flaw of Game Plan 1; it considers only the potential benefits of EMS to a man and ignores his potential costs. Obvious costs would include the risk of detection and injury or murder by the husband of the woman sought as EMS partner; the risk that one's own wife will desert; the risk of being cuckolded by one's wife while one is off seeking EMS; and the risk that one's legitimate children will suffer through one's neglect of them. Thus, according to Game Plan 2 the would-be Casanova, like a sophisticated investor, should seek to maximize his gains while minimizing his losses. What reasoning could be more impeccably judicious?

Game Plan 3.

The man silly enough to be satisfied with Game Plan 2 has obviously never approached a lady with an offer of EMS or PMS. Worse yet, the silly man has never even thought about the statistics of human heterosexual intercourse, which dictate that, for every bout of EMS by a man, there must be one bout of EMS (or at least PMS) by a woman.

Game Plans 1 and 2 share the flaw that they ignore considerations of the woman's strategy, without which any male strategy is doomed to failure. Hence Game Plan 3 must combine a male strategy and a female strategy. But, since one husband suffices to realize a woman's maximum reproductive potential, what could possibly attract a woman to EMS or PMS? This question puzzles the current generation of theoretical sociobiologists with a purely intellectual interest in EMS, just as it has taxed the ingenuity of would-be male adulterers throughout human history.

*

To proceed further with our theoretical exploration of Game Plan 3, we need rigorous empirical data on EMS. As surveys of people's sexual habits are notoriously unreliable, let's first turn to some recently published studies of birds that nest as mated pairs in large colonies. These, rather than our closest relatives the apes, are the animals whose mating system most closely resembles our own. Compared to us, birds have the disadvantage that one cannot ask them about their motives for EMS, but this is no great loss, as our answers are often lies anyway. The great virtue of colonial birds for EMS research is that one can band the birds in a colony, then sit nearby for hundreds of hours and determine exactly who does what with whom. I am unaware of equivalent information for a large human population.

Important recent observations of adultery among birds were made on five species of herons, gulls, and geese. All five nest in dense colonies composed of nominally monogamous male/female pairs. One parent alone is incapable of rearing a chick, as an unguarded nest is likely to be destroyed while the parent is off gathering food, nor is a male capable of feeding or guarding two families simultaneously. Consequently, among the ground rules of sexual strategy for these colonial birds are the following: polygamy is forbidden; copulation with or by an unmated female is pointless, unless she soon acquires a mate to care for the resulting offspring; but surreptitious fertilization by one male of another male's mate is a viable strategy.

The first study involved great blue herons and great egrets at Hog Island, Texas. In these species the male builds a nest and stays there to court visiting females. Eventually a male and female accept each other and copulate about twenty times. The female then lays eggs and goes off to spend most of the daylight hours feeding, while the male remains to guard the nest and eggs. During the first day or two after pairing, the male often resumes courting any passing female as soon as his mate leaves to feed, but EMS does not result. Instead, the male's halfway-unfaithful behaviour seems to constitute 'divorce insurance' that reserves a back-up mate for him in case his own mate deserts (she does desert him in up to twenty per cent of the pairings reported). The passing 'back-up' females pursue the courtship out of ignorance. They are seeking a mate and have no way of knowing that the male is already mated, until his spouse returns (which she does at frequent intervals) and drives them off. Eventually, the male gains complete confidence that he will not be deserted, and he ceases to court any passing females.

In the second study, of little blue herons in Mississippi, behaviour that might have originated as divorce insurance took a more serious turn. Sixty-two cases of EMS were documented, mostly between a female on her nest and a male from the neighbouring nest while the female's mate was busy finding food. Most females initially resisted but then ceased resisting, and some females engaged in more EMS than marital sex. To reduce his own risk of being cuckolded, the adulterous male did his feeding as quickly as possible, returned often to his own nest to guard his mate, and travelled no further than neighbouring nests to seek EMS. EMS was usually timed to occur when the chosen female had not yet completed egg-laying and could still be fertilized. However, adulterous copulations were quicker than marital copulations (eight versus twelve seconds), hence possibly less effective at fertilizing, and nearly half of all nests involved in EMS were subsequently abandoned.

Among herring gulls in Lake Michigan, thirty-five per cent of mated males were observed to engage in EMS. This percentage is nearly the same as the value of thirty-two per cent reported for young American husbands in a study published by Playboy Press in 1974, but there is a big difference between gulls and humans in female behaviour. Whereas Playboy Press reported EMS for twenty-four per cent of young American wives, every mated female gull virtuously rejected adulterous male advances and never solicited the neighbouring male in her own mate's absence. Instead, all cases of male EMS involved unmated female gulls practising PMS. To decrease his own risk of being cuckolded, the male spent more time chasing intruders away from his nest when his mate was fertile than when she was not fertile. As for how the male induced his mate to remain faithful during the time that he was off seeking EMS, his secret – like that of some married men similarly pursuing a mixed reproductive strategy – consisted of feeding her diligently and copulating often whenever she was receptive.

Our final set of rigorous data involves snow geese breeding in Manitoba. Just as I explained in the case of little blue herons, EMS in snow geese mainly involves a male approaching an initially resisting female on a neighbouring nest in the absence of her mate. The mate's absence is usually due to the fact that he himself is off seeking EMS. It may seem as if the male thereby loses as much as he gains, but a male goose is not so dumb. As long as the female is still laying eggs, her mate remains to guard her. (A nesting female is propositioned fifty times less often in her mate's presence than in his absence.) Only after the female has finished laying does her mate go off on EMS quests, with his paternity assured at home.

Such bird studies illustrate the value of a scientific approach to adultery. They have revealed a series of sophisticated strategies by which adulterous male birds try to have it both ways, so as to obtain confidence of paternity at home while sowing their seed abroad. The strategies include wooing unmated females for 'divorce insurance', as long as one feels unsure of one's wife's fidelity; guarding one's fertile spouse; feeding her copiously and copulating with her often, to induce her to remain faithful in one's absence; and coveting one's neighbour's spouse at a time when she is fertile and one's own spouse is no longer fertile. However, not even these applications of the scientific method in all its power sufficed to clarify what, if anything, female birds gain from EMS. One possible answer is that female herons weighing desertion of their mates may use EMS to shop around for a new mate. Another is that some unmated female gulls in colonies with a deficit of males may get fertilized by PMS, and then try to rear the chicks with the help of another, similar female.

The chief limitation of these colonial bird studies is that the females often seem to be unwilling participants in EMS. For further understanding of a more active female role, we have no choice but to turn to human studies, riddled as they are with problems of cultural variation, observer bias, and dubiously reliable survey responses.

*

Surveys comparing men with women in various cultures scattered around the world typically purport to find the following differences: men are more interested in EMS than are women; men are more interested than women in seeking a variety of sexual partners for the sake of variety itself; women's motives for EMS are more likely to be marital dissatisfaction and/or a desire for a lasting new relationship; and men are less selective in taking on a casual female sexual partner than vice versa. For example, among the New Guinea highlanders with whom I work, the men say they seek EMS because sex with their own wife (or even wives, in the case of polygynous men) inevitably becomes boring, while the women who seek EMS do so mainly because their husband cannot satisfy them sexually (for example, because of old age). In the questionnaires that several hundred young Americans filled out for a computer dating service, women expressed stronger partner preferences than men did in almost every respect: intelligence, status, dancing ability, religion, race, etc. The only category in which men were more selective than women was physical attractiveness. After a date the men and women then filled out 'debriefing' questionnaires, with the result that two-and-a-half times as many men as women expressed a strong romantic attraction to their computer-selected partner. Thus, the women were choosier, the men more undiscriminating, in their reactions to partners.

Obviously, we are on shaky ground if we expect an honest answer when we ask people their attitudes about EMS. However, people also express their attitudes in laws and in their behaviour. In particular, some widespread hypocritical and sadistic features of human societies stem from two fundamental difficulties that men face in seeking EMS. Firstly, a man who pursues an MRS is trying to have it both ways: he wishes to obtain sex with other men's wives, while denying sex with his own wife (or wives) to other men. Some men therefore inevitably gain at the expense of other men. Secondly, as we have discussed, there is a realistic biological basis for men's widespread paranoia about being cuckolded.

Adultery laws provide a clear example of how men have dealt with these dilemmas. Until recently, essentially all such laws – Hebraic, Egyptian, Roman, Aztec, Moslem, African, Chinese, Japanese, and others – were asymmetrical. They existed to secure a married man's confidence in his paternity of his children, and for no other purpose. Consequently these laws define adultery by the marital status of the participating woman; that of the participating man is irrelevant. EMS by a married woman is considered an offence against her husband, who is commonly entitled to damages, often including violent revenge or else divorce with refund of the bride price. EMS by a married man is not considered an offence against his wife. Instead, if his partner in adultery is married, the offence is against her husband; if she is unmarried, the offence is against her father or brothers (because her value as a prospective bride is reduced).

No criminal law against male infidelity even existed until a French law of 1810, and that law only forbade a married man to keep a concubine in his conjugal house against his wife's wishes. Viewed from the perspective of human history, the absence or near-symmetry of modern Western adultery laws is a novelty that only appeared in the last 150 years. Even today, prosecutors, judges, and juries in the US and England often reduce a homicide charge to manslaughter of the lowest degree, or else acquit altogether, when a husband kills an adulterous wife or her lover caught in the act.

Perhaps the most elaborate system to uphold confidence of paternity was that maintained by Chinese emperors of the T'ang Dynasty. For each of the emperor's hundreds of wives and concubines, a team of court ladies kept records on dates of menstruation, so that the emperor could copulate with that wife on a date likely to result in fertilization. Dates of copulation were also recorded, and as an auxiliary form of record-keeping, were commemorated by an indelible tattoo on the woman's arm and by a silver ring on her left leg. It goes without saying that equal thoroughness was applied to excluding men other than the emperor from the harem.

Men of other cultures have resorted to less complicated but even more repulsive means of ensuring paternity. These measures limit sexual access to wives, or else to daughters or sisters who would command a high bride price if delivered as proven virgin goods. Relatively mild measures include close chaperoning or virtual imprisonment of women. Similar purposes are served by the code of 'honour and shame' widespread in Mediterranean countries. (Translation: EMS for me but not for you; only the latter is a shame to *my* honour.) Stronger measures include the barbaric mutilations euphemistically and misleadingly termed 'female circumcision'. These consist of removal of the clitoris or most of the external female genitalia to reduce female interest in sex, marital or extramarital. Men bent on total certainty invented infibulation, suturing a woman's *labia majora* nearly shut, so as to make intercourse impossible. An infibulated wife can be de-infibulated for childbirth or for re-insemination after each child is weaned, and can be re-infibulated when the husband takes a long trip. Female circumcision and infibulation are still practised in twenty-three countries today, from Africa through Saudi Arabia to Indonesia.

When adultery laws, imperial records, and coercive restraint still fail to ensure paternity, murder is available as a last resort. The role of sexual jealousy as one of the commonest causes of homicide emerges from studies in many American cities and in many other countries. Usually, the murderer is a husband while the victim is his adulterous wife or her lover, or else the lover kills the husband. The table on the following page gives some actual numbers for murders committed in Detroit in 1972. Until the formation of centralized political states provided soldiers with loftier motives, sexual jealousy also loomed large in human history as a cause of war. It was the seduction (abduction, rape) by Paris of Menelaus's wife Helen that provoked the Trojan War. In the modern New Guinean highlands, only disputes over ownership of pigs rival disputes over sex in triggering war.

Asymmetric adultery laws, tattooing of wives after insemination, virtual imprisonment of women, genital mutilation of women – these behavioural habits are unique to the human species, defining humanity as much as does invention of the alphabet. More exactly, they are new means to the old evolutionary goal of males promoting their genes. Some of our other means to this goal are ancient ones shared with many animals, including jealous murder, infanticide, rape, inter-group warfare, and adultery itself. Human male infibulators stitch the vagina closed; some male animals achieve the same result by cementing a female's vagina after copulating with her.

BREAKDOWN OF MURDERS CAUSED BY SEXUAL JEALOUSY IN THE US CITY OF DETROIT IN 1972

	No. of cases
Murders precipitated by jealous man	
jealous man killed the unfaithful woman	16
jealous man killed the rival man	17
jealous man was killed by the accused woman	9
jealous man was killed by the accused woman's relatives	2
jealous man killed unfaithful homosexual male lover	2
jealous man killed innocent bystander accidentally	1
	47
Murders precipitated by jealous woman	
jealous woman killed the unfaithful man	6
jealous woman killed the rival woman	3
jealous woman was killed by the accused man	2
	11
Total murders	58

Sociobiologists have had considerable success at understanding the marked differences among animal species in the details of these practices. As a result of recent research, it is no longer controversial to conclude that natural selection caused animals to evolve behavioural patterns, as well as anatomical structures, that tend to maximize the number of their descendants. Few scientists doubt that natural selection moulded human anatomy. However, no theory has caused such bitter divisions among my fellow biologists today as the claim that natural selection likewise moulded our social behaviour. Most of the human behaviour discussed in this chapter is considered barbaric by modern Western society. Some biologists are outraged not only by the behaviours themselves, but also by sociobiological explanations for the evolution of the behaviours. To 'explain' a behaviour seems uncomfortably close to defending it.

Like nuclear physics and all other knowledge, sociobiology is available for abuse. People have never lacked pretexts to justify the abuse or killing of other people, but ever since Darwin formulated his theory of evolution, evolutionary reasoning has also been abused as such a pretext. Sociobiological discussions of human sexuality can be seen as seeking to justify men's abuse of women, analogous to the biological justifications advanced for whites' treatment of blacks or Nazis' treatment of Jews. In the critiques that some biologists have directed at sociobiology, two fears recur: that a demonstrated evolutionary basis for a barbaric behaviour would seem to justify it; and that a demonstrated genetic basis for the behaviour would imply the futility of attempts at change.

In my view, neither fear is warranted. As for the first, one can seek to understand how something arose, regardless of whether one considers that something admirable or abominable. Most books analysing the motives of murderers are not written in an effort to justify murder, but instead to understand its causes as a way of preventing it. As for the second fear, we are not mere slaves to our evolved characteristics, not even to our genetically acquired ones. Modern civilization is fairly successful at thwarting ancient behaviours like infanticide, and one of the main objectives of modern medicine is to thwart the effects of our harmful genes and microbes, despite our having come to understand why it is natural for those genes and microbes to tend to kill us. The case against infibulation does not collapse even if the practice can be shown to be genetically advantageous to male infibulators. Instead, we condemn it because we hold the mutilation of one person by another to be ethically loathsome.

While sociobiology is therefore useful for understanding the evolutionary context of human social behaviour, this approach still should not be pushed too far. The goal of all human activity cannot be reduced to the leaving of descendants. Once human culture was firmly in place, it acquired new goals. Many people debate today whether to have children, and many decide that they prefer to devote their time and energy to other activities. We shall reach a similar perspective in later chapters for other attributes as uniquely human as our sexuality, including our art and our abuse of drugs. For these activities too, one can identify animal precursors and discern original roles in promoting survival and gene transmission, but these activities also proceeded to take on a life of their own. Hence I claim only that evolutionary reasoning is valuable for understanding the origin of such human practices, and not that it is necessarily the only way to understand their current forms.

In short, we evolved, like other animals, to win the reproduction game. That contest has a single aim, to leave as many descendants as possible. Much of the legacy of that game strategy is still with us. But we have also chosen to pursue ethical goals, which can conflict with the goals and methods of the sexual contest. Having that choice among goals represents one of our most radical departures from other animals.

FIVE

How We Pick Our Mates and Sex Partners

Most humans are choosier about their sex partners than are the (other two) chimpanzees. By what criteria do we select our spouse or bedmate, and how does each of us develop our individual standard of beauty?

*

One evening, while I was camping with some New Guinea men of the Foré tribe, the conversation turned to women and sex, and my Foré friends proceeded to explain to me their tastes:

The most beautiful women are Foré women. They have gorgeous black skin, thick, dark frizzy hair, full lips, broad noses, small eyes, a nice smell, and perfectly shaped breasts and nipples. Women of other New Guinea tribes are less attractive, and white women are unspeakably hideous. Just compare your white women with our women to see why – white skin like a sick albino's, straight hair like strings, sometimes even hair coloured yellow like dead grass or red like a poisonous snail, thin lips and narrow noses like axe blades, big eyes like a cow's, a repulsive smell when they sweat, and breasts and nipples of the wrong shape. When you get ready to buy a wife, find a Foré if you want someone beautiful.

Among the reasons I did not follow that advice was that I happen to find those 'unspeakably hideous' women attractive. But then I was conditioned by my own society's ideals, just as my Foré friends were by theirs. Darwin commented that every people he knew about – Chinese, Hottentots, black Africans, Fijians, and others – measure beauty by their own appearance. Are there really no universal rules of human beauty and sex appeal? If not, do we inherit our particular taste in marriage partner through our genes, or do we learn it by looking at other members of our society? How, really, do we pick our sex partners and spouse?

It may be surprising to realize that this problem is one that arose anew during the evolution of the human species – or at least became much more important for us than for the other two chimpanzees. As we saw in Chapter Three, our familiar human mating-system, based ideally on couples maintaining on-going involvement, is a human innovation. Pygmy chimps are the opposite of sexually selective; females mate in sequence with many males, and there is much sexual activity between females and between males as well. Common chimps are not so completely promiscuous – a male and female may sometimes go off and 'consort' with each other for a few days – but they still rank as promiscuous by human standards. However, humans are much more selective sexually, since rearing a human child is difficult (at least for hunter-gatherers) without a father's help, and since sex becomes part of the cement that differentiates co-parents from other men and women frequently encountered. Choosing a mate or sex partner is not so much a human invention as a reinvention of something practised by many other (nominally) monogamous animals with lasting pair-bonds, and lost by our chimpanzee-like ancestors. Those choosy animals include many bird species, plus our distant ape relatives, the gibbons.

We saw in Chapter Four that this ideal depiction of a human society based on monogamous couples coexists with a good deal of extramarital sex. That activity also involves selection of sex partners, with adulterous women tending to be more selective than adulterous men. Thus, selection of spouses and sex partners is another important piece of what defines humanity. It is as basic to our rise from chimpanzee status as is the remodelled pelvis discussed in detail in physical anthropology texts. We shall see in the next chapter that our sexual choosiness may be central to the origin of the most conspicuous visible variability in modern humans. That is, much of what we think of as human racial variation may have arisen as a by-product of the beauty standards by which we choose our sex partners.

*

In addition to this theoretical interest, the question of how we select our mates and sex partners is of much personal interest. It preoccupies most of us for much of our lives. Those of us who are still unattached spend daily hours dreaming about whom we will consort with or marry. The question becomes more intriguing when we compare what turns on different people within the same culture. Think of the men or women that you find sexually attractive. If you are a man, for instance, do you prefer women who are blonde or brunette, flat-chested or buxom, and with big or small eyes? If you are a woman, do you like men who are bearded or smooth-shaven, tall or short, and smiling or scowling? Probably you do not go for just anyone, only certain types attract you. Everyone can name friends who got divorced, then chose a second spouse who was the spitting image of the first. A colleague of mine went through a long series of plain, slim, brown-haired, round-faced girlfriends, until he finally found one he got along with and married her. Whatever your own preference, you will have noticed that some of your friends have completely different tastes.

The particular ideal that each of us pursues is an example of what are called 'search images'. (A search image is a mental picture against which we compare objects and people around us in order to be able to recognize something quickly, like a Perrier bottle amidst all the other bottled waters on the supermarket shelf, or one's child at a playground with other kids.) How do we develop our private search image for a mate? Do we seek someone familiar and similar to us, or are we more turned on by someone exotic? Would most European men really marry a Polynesian woman if given the chance? Do we seek someone complementary to us so as to fulfil our needs? For instance, there undoubtedly are some dependent men who marry a mothering woman, but how typical are such pairings?

Psychologists have tackled this question by examining many married couples, measuring everything conceivable about their physical appearance and other characteristics, and then trying to make sense out of who married whom. A simple numerical way of describing the result is by means of a statistical index called the correlation coefficient. If you line up 100 husbands in order of their ranking for some characteristic (say, their height), and if you also line up their 100 wives with respect to the same characteristic, the correlation coefficient describes whether a man tends to be at the same position in the husbands' line-up as his wife is in the line-up of wives. A correlation coefficient of plus one would mean perfect correspondence: the tallest man marries the tallest woman, the thirty-seventh tallest man marries the thirty-seventh tallest woman, and so on. A correlation coefficient of minus one would mean perfect matching by opposites: the tallest man marries the shortest woman, the thirty-seventh tallest man marries the thirty-seventh shortest woman, and so on. Finally, a correlation coefficient of zero would mean that husbands and wives assort completely randomly by height: a tall man is as likely to marry a short woman as a tall woman. These examples are for height, but correlation coefficients can also be calculated for anything else, such as income and IQ.

If you measure enough things about enough couples, here is what you will find. Not surprisingly, the highest correlation coefficients – typically around +0.9 – are for religion, ethnic background, race, socioeconomic status, age, and political views. That is, most husbands and wives prove to be of the same religion, ethnic background, and so on. Perhaps you also will not be surprised that the next highest correlation coefficients, usually around +0.4, are for measures of personality and intelligence, such as extroversion, neatness, and IQ. Slobs tend to marry slobs, though the chances of a slob marrying a compulsively neat person are not as low as the chances of a political reactionary marrying a left-winger.

What about matching of husbands and wives for physical characteristics? The answer is not one that would leap out at you immediately if you just looked at a few married couples. That is because we do not select our own mates for their bodies as carefully as we select the mates of our show dogs, racehorses, and beef cattle. But we select nevertheless. If you measure enough couples, the answer that finally emerges is unexpectedly simple. *On the average*, spouses resemble each other slightly but significantly in almost every physical feature examined.

That is true of all the obvious traits you would first think of when asked to design your ideal beloved – his or her height, weight, hair colour, eye colour, and skin colour – but it is also true of an astonishing variety of other traits that you probably would not have mentioned in your description of the perfect sex partner. Those other traits include ones as diverse as breadth of nose, length of ear lobe or middle finger, circumference of wrist, distance between eyes, and lung volume! Experimenters have made this finding for people as diverse as Poles in Poland, Americans in Michigan, and Africans in Chad. If you do not believe it, try noting eye colours (or measuring ear lobes) the next time you are at a dinner party with many couples, and then get your pocket calculator to give you the correlation coefficient.

Coefficients for physical traits are on the average +0.2 – not so high as for personality traits (+0.4) or religion (+0.9), but still significantly higher than zero. For a few physical traits the correlation is even higher than 0.2 – for instance, an astonishing 0.61 for length of middle finger. At least unconsciously people care more about their spouse's middle finger length than about his or her hair colour and intelligence!

*

Thus, like tends to marry like. Among the obvious explanations that contribute to these results is propinquity: we tend to live in neighbourhoods defined by socioeconomic status, religion, and ethnic background. For instance, in large American cities one can point to the rich neighbourhoods and the poor neighbourhoods, and also to the Jewish section, Chinese section, Italian section, black section, and so on. We meet people of the same religion when we go to church, and we tend to meet people of similar socioeconomic status or political views in many of our daily activities. Since we thus have far more opportunities to meet people like us than unlike us in these respects, of course we are more likely to marry someone of our religion, socioeconomic status, and so on. But we don't live in neighbourhoods grouped by length of ear lobe, so there must be some other reason why spouses tend to be matched in that respect as well.

Another obvious reason why like tends to marry like is that marriage is not just a choice; it is a negotiation. We do not go out searching until we find a person with the right eye colour and length of middle finger, then announce to that person, 'You are marrying me'. For most of us, marriage results from a proposal rather than a unilateral announcement, and the proposal is the culmination of some sort of negotiation. The more similar a man and woman are in political views, religion, and personality, the smoother will be the negotiation. Hence the match in personality traits is on the average closer for married couples than for dating couples, closer for happily than unhappily married couples, and closer for couples who stay married than for those who get divorced. But this still does not explain spousal resemblance in ear lobe length, which is only rarely cited as a factor in divorce.

The remaining factor deciding whom you will marry, besides propinquity and smoothness of negotiation, is surely sexual attraction based on physical appearance. That in itself is no surprise. Most of us are aware of our preferences in obvious visible features like height, build, and hair colour. What is initially surprising is the importance of so many other physical traits that we usually do not consciously notice, such as ear lobes, middle fingers, and interocular distances. Nevertheless, all those other traits contribute unconsciously to the snap decisions we make when we are introduced to someone and a voice inside tells us, 'She's my type!'

Here is an example. When my wife and I were introduced to each other, I instantly found Marie attractive and vice versa. In retrospect, I can understand why: we are both brown-eyed, similar in height and build and hair colour, and so on. But, on the other hand, I also had a sense that there was something about Marie that did not quite match my ideal, even though I could not figure out what exactly it was. Not until Marie and I first went to a ballet together did I solve the puzzle. I lent Marie my opera glasses, and when she passed them back to me, I found that she had pushed the eye-pieces so close together that I could not see through them until I had spread them apart again. I then realized that Marie has more close-set eyes than I do, and that most women whom I had pursued before had wide-set eyes like my own. Thanks to Marie's ear lobes and other merits, I have been able to make peace with my and her mismatched

interocular distances. Nevertheless, the episode with the opera glasses
made me appreciate for the first time that I have always found wide-set
eyes a turn-on, even though I had not been explicitly aware of it.

So, we tend to marry someone who looks like us. But – wait a minute.
The men who look most similar to a woman are the men who share half
of her genes – her father or brother! Similarly, the best-matched mate for
a man would be his mother or sister! Yet most of us obey the incest taboo
and certainly do not marry our parent or sibling of the opposite sex.
Instead, I am saying that people tend to marry a person who *looks like* the
parent or sibling of the opposite sex. Our actual behaviour is summed up
by a popular song of the 1920s.

> I want a girl
> Just like the girl
> That married dear old Dad . . .

The reason we tend to resemble our mates is that many of us are
looking for someone who reminds us of our parent or sibling of the
opposite sex, who in turn resembles us. As children, we already begin to
develop our search image of a future sex partner, and that image is heavily
influenced by the people of the opposite sex whom we see most often. For
most of us that is our mother (or father) and sister (or brother), plus close
childhood friends.

*

At this point, you are probably turning to your spouse or Significant
Other, pulling out your tape measure, and discovering a gross mismatch
between your and his (or her) ear lobes. Or perhaps you have pulled out a
photo of your mother or sister, and you detect not the faintest
resemblance when you hold it beside your spouse. You may be about to
throw away this book as patent nonsense. But if your wife is not a dead
ringer for your mother, don't stop reading, and conversely don't get
worried that you should see a psychiatrist about your pathological search
image. After all, remember:

1.   Studies consistently show that factors like religion and personality influence our choice of spouse much more strongly than physical
appearance. All I am making is the obvious point that physical traits have
some influence. In fact, I would predict much higher correlation
coefficients for physical traits between casual sex partners than between
spouses. That is because we can select casual sex partners solely on the

basis of physical attraction, without regard to religion or political views. This prediction awaits testing.

2. Remember also that your search image could have been influenced by any of the people of the opposite sex that you regularly saw around you as you were growing up. That includes playmates and siblings as well as parents. Perhaps your spouse resembles the little girl next door, rather than Mother.

3. Finally, remember that lots of independent physical traits enter into our search image, so most of us end up with a mild average resemblance to our spouses in many traits, rather than with a very close resemblance in a few traits. This idea is known as the 'buxom redhead theory'. If a man's mother and sister were both buxom redheads, he might grow up to consider buxom redheads very exciting, but redheads are relatively rare, and buxom redheads still rarer. Furthermore, the man's preference even in a casual sex partner is likely to depend on some other physical traits as well, and his preference in a wife will certainly depend on her views about children, politics, and money. Hence, in a group of sons of buxom redheads, a few lucky ones will find a girl like Mother in those two respects, some will have to settle for buxom non-redheads, others for non-buxom redheads, and most for run-of-the-mill non-buxom brunettes.

You may also be objecting at this point that my argument applies only to societies where spouses pick each other. As friends from India and China are quick to remind me, that is a peculiar custom of the twentieth-century US and Europe. It was not true of the US and Europe in the past, and it is still not true of most of the world today, where marriages are instead arranged by the families involved. The bride and groom often are not even introduced until the wedding day. How could my argument possibly apply to such marriages?

Of course it couldn't, if one is talking just about legal marriages. But my argument would still apply to the choice of extramarital sex partners, who may father a non-trivial fraction of children, just as blood-group studies proved for American and British children (Chapter Four). In fact, I would expect that if extramarital fathering is frequent even in societies where a woman already exercises her sexual preferences in choosing a husband, it may be even more frequent in societies with arranged marriages, where a woman's choice can only be expressed extramaritally.

*

It is not just the case, then, that Foré men prefer Foré women over Californian women, and vice versa: our search images are much more specific. However, these insights still leave questions unanswered. Did I inherit or learn my search image for someone like Mother? If I were offered the choice of sex with my sister or a strange woman, I would certainly reject the offer of my sister and probably my first cousin, but would I prefer my second cousin over a strange woman (because the cousin probably resembles me more)? There are some crucial experiments that would settle these questions – for instance, keeping a man in a large cage with his female first, second, third, fourth, and fifth cousins, counting how many times he had sex with each, and repeating the experiment with many men (or women) and their cousins. Alas, such experiments are hard to do with humans, but they have been done for several animal species, with instructive results. I shall give just three examples, the cousin-loving quail, and the perfumed mice and rats. (We cannot use our closest relatives the chimpanzees for these examples, since they are so unselective.)

Consider first the case of Japanese quail, which are either brown or white. Quail normally grow up with their biological parents and siblings. However, it is also possible to 'cross-foster' quail by switching eggs between quail mothers and their nests before the eggs hatch. In that way, a baby quail may be reared by foster-parents and grow up with 'pseudo-siblings' – that is, littermates among whom the baby hatched but to whom the baby is not genetically related.

The preferences of male quail have been tested by putting a male in a cage with two females and observing with which female the male spent more time or copulated. It turns out that males preferred whichever colour of female they grew up with. Furthermore, when a brown-loving male was given a choice between brown females that he had never seen before (although some were his relatives from whom he had been separated before hatching), he preferred his first cousin to his third cousin or an unrelated female, but he also preferred his first cousin to his sister. Evidently, male quail as they grow up learn the appearance of their sisters (or mother) with whom they are reared, then seek a mate that is very similar but not *too* similar. In fancy technical language, biologists term this the Principle of Optimal Intermediate Similarity. Like other things in life, inbreeding seems to be good in moderation – a little inbreeding, but not too much. For instance, among unrelated brown females a male prefers an unfamiliar one over a familiar one with whom he grew up (a 'pseudo-sister', who pushes the male's not-too-much-incest button).

Mice and rats similarly learn in childhood what to look for in a mate, but they choose by smell more than by appearance. When infant female mice were reared by parents sprayed repeatedly with Parma Violet perfume, the females on reaching adulthood sought out Parma-Violet-scented males in preference to unscented males. ('I want a boy, just like the boy, that smells like dear old Dad'.) In another experiment, infant male rats were reared by mother rats whose nipples and vagina were sprayed with lemon odour, then the male on reaching adulthood was put in a cage with a lemon-smelling or unscented female rat. Each such encounter was videotaped and played back to note the times of key events. It turned out that males with scented mothers mounted and ejaculated more quickly when placed with a scented female than with an unscented one, while the reverse was true for males with unscented mothers. For example, sons of scented mother rats were so excited by a scented sex partner that they ejaculated in only eleven-and-a-half minutes, while they took over seventeen minutes to ejaculate with an unscented female. But sons of unscented mother rats took over seventeen minutes with the *scented* partner and only twelve minutes with the *unscented* partner. Obviously, the males had learned to be sexually excited by their mother's smell (or lack of smell); they did not inherit the knowledge.

*

What do these experiments on quail, mice, and rats show? The message is clear. Animals of those species learn to recognize their parents and siblings as they grow up, then are programmed to seek out an individual fairly similar to the parent or sibling of the opposite sex – but not Mother or Sister herself. They may *inherit* some search image of what constitutes a rat, but they evidently *learn* their search image of who in particular is a beautiful, eligible rat.

We can immediately appreciate what experiments are needed to get unequivocal proof of this theory for humans. We should take an average happy family, spray Father every day with Parma Violet, spray Mother's nipples daily with lemon oil while she is nursing, and then wait twenty years to see whom the sons and daughters marry. Alas, we would be frustrated by the many obstacles to establishing Scientific Truth for humans. But some observations and accidental experiments still let us tip-toe towards the truth.

Take the incest taboo. Scientists debate whether the taboo itself in humans is instinctive or learned. However, this chapter is concerned with a separate question: given that we somehow acquire an incest taboo, do we learn to whom to apply it, or do we inherit that information in our genes? Normally we grow up with our closest relatives (parents and siblings), so our subsequent avoidance of them as sex partners could equally well be genetic or learned, but adoptive brothers and sisters also tend to avoid incest, suggesting learned avoidance.

This conclusion is strengthened by an interesting set of observations made in Israeli kibbutzim – the collective settlements whose members house, school, and care for all their children together as a large group. Thus, kibbutz children live from birth until young adulthood in intimate association with each other, like a gigantic family of brothers and sisters. If propinquity were the main factor influencing whom we marry, most kibbutz children should marry within the kibbutz. In fact, a study of 2,769 marriages contracted by kibbutz-reared children turned up only thirteen between children from the same kibbutz. All the other children married outside the kibbutz on reaching maturity.

Even those thirteen cases turned out to be the exceptions that proved the rule: all involved couples of whom one had moved into that kibbutz only after the age of six! Among children reared in the same peer group since birth, there were not only no marriages, but also no adolescent or adult heterosexual activity at all. This is astonishing restraint on the part of nearly 3,000 young men and women who enjoyed daily opportunities for sexual involvement with each other, and who had far fewer opportunities for involvement with outsiders. It illustrates dramatically that the period between birth and the age of six is a critical time for formation of our sexual preferences. We *learn*, however unconsciously, that our intimate associates from that period are ineligible as sex partners when we become mature.

We also appear to learn the part of our search image that tells us whom to seek, not just the part that tells us whom to avoid. For instance, a friend of mine who is 100% Chinese herself happened to grow up in a community in which every other family was white. Eventually she moved as an adult to an area with many Chinese men, and for some time she dated both Chinese and white men, but came to realize that it was the whites who attracted her. She has been married twice, both times to white men. Her own experiences led her to ask her Chinese women friends about their backgrounds. It turned out that most of her friends reared in white enclaves also ended up marrying white men, while those reared in Chinese neighbourhoods married Chinese men – although all had plenty of men of both types from whom to choose during their young adult years. Hence those who surround us as we grow up, though ineligible themselves as eventual mates, nevertheless shape our standards of beauty and search image.

Think to yourself: what sort of men or women do you find physically attractive, and where did you develop that taste? I would guess that most people, like myself, can trace their preference to the appearance of parents or siblings or childhood friends. So do not be discouraged by all those old generalizations about sex appeal – 'Gentlemen prefer blondes,' 'Men seldom make passes at girls who wear glasses,' etc. Each such 'rule' applies only to some of us, and there are plenty of men out there whose mothers were myopic brunettes. Fortunately for my wife and me – both of us brunettes requiring glasses, born of brunette glass-wearing parents – beauty is in the eye of the beholder.

SIX

SEXUAL SELECTION, AND THE ORIGIN OF HUMAN RACES

People from different parts of the world can be distinguished at a glance by so-called racial characteristics. But those same traits – ones such as the colour of our skin and hair and eyes, or the shapes of our breasts and genitals – play a big role in how we select our mates and sex partners. Thus, our outward appearances and our beauty standards have evolved in tandem to different local end points.

*

'White man! Lookim this-feller line three-feller man. This-feller number-one he belong Buka Island, na 'nother-feller number-two he belong Makira Island, na this-feller number-three he belong Sikaiana Island. Yu no savvy? Yu no enough lookim straight? I think, eye-belong-yu he bugger-up finish?'

No, damn it, my eyes-belong-me were not ruined beyond repair. It was my first visit to the Solomon Islands in the Southwest Pacific, and I told my scornful guide through the medium of pidgin English that I saw perfectly well the differences between those three men in a row over there. The first one had jet-black skin and frizzy hair, the second had much lighter skin and frizzy hair, and the third had straighter hair and more slanty eyes. The only thing the matter with me was that I had no experience of what people from each particular Solomon island looked like. By the end of my first trip through the Solomons, I too could match people to their islands by their skin and hair and eyes.

In those variable features, the Solomons are a microcosm of humanity.

Simply by looking at a person, even laymen can often tell what part of the world that person comes from, and trained anthropologists may be able to 'place' him or her in the right part of the right country. For example, given one person each from Sweden, Nigeria, and Japan, none of us would have any trouble deciding at a glance which person was from which country. The most visibly variable features in clothed people are of course skin colour, the colour and form of the eyes and hair, body shape, and (in men) the amount of facial hair. If the people to be identified were undressed, we might also notice differences in amount of body hair, the size and shape and colour of a woman's breasts and nipples, the form of her labia and buttocks, and the size and angle of a man's penis. All those variable features contribute to what we know as human racial variation.

Those geographic differences among humans have long fascinated travellers, anthropologists, bigots, and politicians, as well as the rest of us. Since scientists have solved so many arcane questions about obscure unimportant species, surely you might expect them to have answered one of the most obvious questions about ourselves: 'Why do people from different areas look different?' Our understanding of how humans came to differ from other animals would remain incomplete if we did not also consider how, in the process, human populations acquired their most visible differences from each other. Nevertheless, the subject of human races is so explosive that Darwin excised all discussion of it from his famous 1859 book *On the Origin of Species*. Even today, few scientists dare to study racial origins, lest they be branded racists simply for being interested in the problem.

There is another reason why we do not understand the significance of human racial variation: it proves to be an unexpectedly difficult problem. Twelve years after Darwin wrote his book attributing the origin of species to natural selection, he wrote another book 898 pages long, attributing the origin of human races to our sexual preferences which I described in the last chapter, and entirely rejecting a role of natural selection. Despite that verbal overkill, many readers were unconvinced. To this day, Darwin's theory of sexual selection (as he called it) remains controversial. Instead, modern biologists generally invoke natural selection to explain the visible differences among human races – especially the differences in skin colour, whose relation to sun exposure seems obvious. However, biologists cannot even agree on why natural selection led to dark skin in the tropics. I shall explain why I believe natural selection to have played only a secondary role in our racial origins, and why Darwin's preference for sexual selection seems to me correct. I therefore consider visible human racial variation to be largely a by-product of the remodelled human life-cycle that forms the subject of Part Two of this book.

*

Firstly, to place matters in perspective, let's realize that racial variation is not at all confined to humans. Most animal and plant species with sufficiently wide distributions, including all higher ape species except the geographically localized pgymy chimp, also vary geographically. So marked is variation in some bird species, such as North America's white-crowned sparrow and Eurasia's yellow wagtail, that experienced bird-watchers can identify an individual bird's approximate birthplace by its plumage pattern.

Variation in apes encompasses many of the same characteristics that vary geographically in humans. For example, among the three recognized races of gorillas, western lowland gorillas have the smallest bodies and rather grey or brown hair, while mountain gorillas have the longest hair, and eastern lowland gorillas share black hair with mountain gorillas. Races of white-handed gibbons similarly vary in hair colour (variously black, brown, reddish, or grey), hair length, tooth size, protrusion of the jaws, and protrusion of the bony ridges over the eyes. All these traits that I have just mentioned as varying among gorilla or gibbon populations also differ among human populations.

How does one decide whether recognizably distinct animal populations from different localities constitute different species, or belong instead to the same species and just constitute different races (also known as subspecies)? As explained in Chapter Two, the distinction is based on interbreeding under normal circumstances: members of the same species may interbreed normally if given the opportunity, while members of different species do not. (But closely related species that would not normally interbreed in the wild, like lions and tigers, may do so if a male of one is caged with a female of the other and given no other choice.) By this criterion, all living human populations belong to the same species, since some interbreeding has occurred whenever humans from different regions have come into contact – even people as dissimilar in appearance as African Bantus and Pygmies. With humans as with other species, populations may intergrade into each other, and it becomes arbitrary to decide which populations to group as races. By the same criterion of interbreeding, the large gibbons known as siamangs are a distinct species from the smaller gibbons, since both occur together in the wild without hybridizing. This is also the criterion for considering Neanderthals possibly as a species distinct from *Homo sapiens*, since hybrid skeletons have not been identified despite apparent Cro-Magnon/Neanderthal contact (see Chapter Two).

Racial variation has characterized humans for at least the past several thousand years, and possibly much longer. Already around 450 BC, the

Greek historian Herodotus described the Pygmies of West Africa, the black-skinned Ethiopians, and a blue-eyed red-haired tribe in Russia. Ancient paintings, mummies from Egypt and Peru, and bodies of people preserved in European peat bogs confirm that people several thousand years ago differed in their hair and facial features much as they do today. Origins of modern races can be pushed back still further, to at least ten thousand years ago, since fossil skulls of that age from various parts of the world differ in many of the same respects that modern skulls from the same regions differ. More controversial are the studies of some anthropologists, contested by others, reporting continuity of racial skull characteristics for hundreds of thousands of years. If those studies are correct, then some of the human racial variation that we see today may predate the Great Leap Forward, and may have gone back to the times of *Homo erectus*.

*

Now let's turn to the question of whether natural selection or sexual selection has made the larger contribution to those visible geographic differences of ours. Take first the arguments about natural selection, the selection of traits that enhance survival. No scientist denies today that natural selection does account for many of the differences between species, such as why lions have paws with claws while we have grasping fingers. No one denies either that natural selection explains some geographic variation ('racial variation') within some animal species. For instance, Arctic weasels that live in areas covered by winter snow change colour from brown in summer to white in winter, while more southerly weasels stay brown all year. That racial difference enhances survival, because white weasels against a brown background would be glaringly conspicuous to their prey if they were not camouflaged against snow.

By the same token, natural selection surely explains *some* geographic variation in humans. Many black Africans but no Swedes have the sickle-cell haemoglobin gene, because the gene protects against malaria, a tropical disease that would otherwise kill many Africans. Other localized human traits that surely evolved through natural selection include the big chests of Andean Indians (good for extracting oxygen from thin air at high altitudes), the compact shapes of Eskimos (good for conserving heat), the slender shapes of southern Sudanese (good for losing heat), and the slit-like eyes of northern Asians (good for protecting eyes against cold and against sun glare off the snow). All these examples are easy to understand.

Can natural selection similarly explain the racial differences that we think of first, those in skin colour and eye colour and hair? If so, one might expect that the same trait (for instance, blue eyes) would reappear in different parts of the world with similar climates, and that scientists would agree on what the trait is good for.

Seemingly the simplest trait to understand is skin colour. Our skins run the spectrum from various shades of black, brown, copper, and yellowish to pink with or without freckles. The usual story to explain this variation by natural selection goes as follows. People from sunny Africa have blackish skins. So too (supposedly) do people from other sunny places, like southern India and New Guinea. Skins are said to get paler as one moves north or south from the equator, until one reaches northern Europe, with the palest skins of all. Obviously, dark skins evolved in those people who were exposed to much sunlight. That is just like the skins of whites tanning under the summer sun (or in tanning salons!), except that tanning is a reversible response to sun rather than a permanent genetic one. It is equally obvious what good a dark skin does in sunny areas: it protects against sunburn and skin cancer. Whites who spend lots of time outdoors in the sun tend to get skin cancer, and they get it on exposed parts of their body like their head and hands. Does that not all make sense?

Yes, but . . . it is really not so simple at all. To begin with, skin cancer and sunburn cause little debilitation and few deaths. As agents of natural selection, they have an utterly trivial impact compared to infectious diseases of childhood. Hence many other theories have been proposed to explain the supposed pole-to-equator gradient in skin colour.

One favourite competing theory notes that the sun's ultraviolet rays promote vitamin D formation in a layer of our skin beneath the main pigmented layer. Thus, people in sunny tropical areas might have evolved dark skin to protect them against the risk of kidney disease caused by too much vitamin D, while people in Scandinavia with its long dark winters evolved pale skins to protect them against the risk of rickets caused by too little vitamin D. Two other popular theories are that dark skins are to protect our internal organs against overheating by the tropical sun's infrared rays, or – just the opposite – dark skins help keep tropical people warm when the temperature drops. And if those four theories are not enough for you, consider four more: that dark skins provide camouflage in the jungle, or that pale skins are less sensitive to frostbite, or that dark skins protect against beryllium poisoning in the tropics, or that pale skins cause deficiency of another vitamin (folic acid) in the tropics.

With at least eight theories in the running, we can hardly claim to understand why people from sunny climates have dark skins. That in itself does not refute the idea that, somehow, natural selection caused the evolution of dark skins in sunny climates. After all, dark skins could have multiple advantages, which scientists may sort out some day. Instead, the heaviest objection to any theory based on natural selection is that the association between dark skins and sunny climates is a very imperfect one. Native peoples had very dark skins in some areas receiving relatively little sunlight, like Tasmania, while skin colour is only medium in sunny areas of tropical Southeast Asia. No American Indians have black skins, not even in the sunniest parts of the New World. When one takes cloud cover into account, the world's most dimly lit areas, receiving a daily average of under three-and-a-half hours of sunlight, include parts of equatorial West Africa, southern China, and Scandinavia, inhabited respectively by some of the world's blackest, yellowest, and palest peoples! Among the Solomon Islands, all of which share a similar climate, jet-black people and lighter people replace each other over short distances. Evidently, sunlight has not been the sole selective factor that moulded skin colour.

The first response of anthropologists to these objections is to raise a counter-objection, the time factor. This argument tries to explain away the cases of pale-skinned people in the tropics by claiming that those particular peoples migrated to the tropics too recently to have evolved black skins. For example, the ancestors of American Indians may have reached the New World only 11,000 years ago (Chapter Eighteen): perhaps that has not been long enough to evolve black skins in the tropical Americas. But if you are going to evoke the time factor to explain away objections to the climate theory of skin colour, then you also have to consider the time factor for peoples who supposedly support that theory.

One of the prime supports of the climate theory is the pale skin of Scandinavians, living in the cold, dark, foggy North. Unfortunately, Scandinavians have been in Scandinavia for an even shorter time than American Indians have been in the Amazon. Until about 9,000 years ago, Scandinavia was covered by an ice-sheet and could hardly have supported any people, pale-skinned or dark-skinned. Modern Scandinavians reached Scandinavia only around 4,000 or 5,000 years ago, as a result of the expansion of farmers from the Near East (Chapter Ten) and of Indo-European speakers from southern Russia (Chapter Fifteen). Either Scandinavians acquired their pale skins long ago in some other area with a different climate, or else they acquired them in Scandinavia within half the time that Indians have spent in the Amazon without becoming dark-skinned.

The sole people in the world about whom we can be certain that they spent the last 10,000 years in the same location were the natives of Tasmania. Lying south of Australia, at the temperate latitude of Chicago or Vladivostok, Tasmania used to be connected to Australia until it was cut off by rising sea levels 10,000 years ago and became an island. Since modern Tasmanian natives did not have boats capable of going more than a few miles, we know that they were derived from colonists who walked out to Tasmania at the time of its connection to Australia, and who remained there continuously until they were exterminated by British colonists in the Nineteenth Century (Chapter Sixteen). If any people had enough time for natural selection to match their skin colour to their local temperate-zone climate, it was the Tasmanians. Yet they had blackish skins, supposedly adapted to the Equator.

If the case for natural selection of skin colour seems weak, that for hair colour and eye colour is virtually non-existent. There are no consistent correlations with climate, and not even any half-plausible theories for the supposed advantage lent by each colour type. Blonde hair is common in cold, wet, dimly lit Scandinavia and also among Aborigines of the hot, dry, sunny desert of central Australia. What do those two areas have in common, and how does being blonde help both Swedes and Aborigines to survive? Do freckles and red hair help Irishmen catch leprechauns? Blue eyes are common in Scandinavia and supposedly help their owners see farther in dim, misty light, but that speculation is unproven, and all my friends in the even dimmer, mistier mountains of New Guinea see just fine with their dark eyes.

The racial traits for which it seems most absurd to seek an explanation based on natural selection are our variable genitalia and secondary sex characteristics. Are hemispherical breasts an adaptation to summer rainfall and conical breasts an adaptation to winter fog, or vice versa? Do the protruding *labia minora* of Bushmen women protect them against pursuing lions, or reduce their water losses in the Kalahari Desert? You surely don't think that men with hairy chests can thereby keep warm while going shirtless in the Arctic, do you? If you do think so, then please explain why women do not share hairy chests with men, since women also have to keep warm.

*

Facts such as these were what made Darwin despair of imputing human racial variation to his own concept of natural selection. He finally dismissed the attempt with a succinct statement: 'Not one of the external differences between the races of man are of any direct or special service to him.' When Darwin came up with a theory that he preferred, he termed it 'sexual selection' to contrast with natural selection, and he devoted an entire book to explaining it.

The basic notion behind this theory is easily grasped. Darwin noted many animal features that had no obvious survival value but that did play an obvious role in securing mates, either by attracting an individual of the opposite sex or by intimidating a rival of the same sex. Familiar examples are the tails of male peacocks, the manes of male lions, and the bright red buttocks of female baboons in oestrus. If an individual male is especially successful at attracting females or intimidating rival males, that male will leave more descendants and will tend to pass on his genes and traits – as a result of sexual selection, not natural selection. The same argument applies to female traits as well.

For sexual selection to work, evolution must produce two changes simultaneously: one sex must evolve some trait, and the other sex must evolve in tandem a liking for that trait. Female baboons could hardly afford to flash red buttocks if the sight revolted male baboons to the point of their becoming impotent. As long as the female has it and the male likes it, sexual selection could lead to any arbitrary trait, just as long as it does not impair survival too much. In fact, many traits produced by sexual selection do seem quite arbitrary. A visitor from outer space who had yet to see humans could have no way of predicting that men rather than women would have beards, that the beards would be on the face rather than above the navel, and that women would not have red and blue buttocks.

That sexual selection really can work, at least in birds, was proved by an elegant experiment carried out by the Swedish biologist Malte Andersson on the long-tailed widowbird of Africa. In this species the male's tail in the breeding season grows to 20 inches long, while the female's tail is only 3 inches. Some males are polygamous and acquire up to six mates, at the expense of other males who get none. Biologists had guessed that a long tail served as an arbitrary signal by which males attracted females to join their harem. Andersson's test was to cut off part of the tail from nine males until their tails were only 6 inches long. He then glued those cut segments to the tails of nine other males to give them 30-inch tails, and he waited to see where the females built their nests. It turned out that the males with the artificially lengthened tails attracted on the average over four times as many mates as the males with artificially shortened tails.

Perhaps our first reaction to Andersson's experiment is: those dumb birds! Imagine a female selecting a particular male to father her offspring merely because his tail is longer than other males' tails! But before we get too smug, let's consider again what we learned in the last chapter about how we humans select our own mates. Are our criteria such good indicators of genetic worth? Do not some men and women set disproportionate value on the size or form of certain body parts, which are really nothing more than arbitrary signals for sexual selection? Why did we evolve to pay any attention at all to a beautiful face, which is useless to its owner in the struggle for survival?

In animals some of the traits that vary racially are ones produced by sexual selection. For instance, lions' manes vary in length and in colour. Males of the astrapia birds of paradise in New Guinea have fancy tails to display to females, but different populations evolved tails of different shapes and colours. From west to east, the tails are broad and purple, short and white-based, very long and white, long and purple, and broad and purple again. Similarly, snow geese occur in two colour phases, a blue phase commoner in the western Arctic and a white phase commoner in the eastern Arctic. Birds of each phase prefer a mate of the same phase. Could human breast shape and skin colour similarly be the outcome of sexual preferences that vary arbitrarily from area to area?

After 898 pages Darwin convinced himself that the answer to this question was a resounding 'yes'. He noted that we pay inordinate attention to breasts, hair, eyes, and skin colour in selecting our mates and sex partners. He noted also that people in different parts of the world define beautiful breasts, hair, eyes, and skin by what is familiar to them. Thus, Fijians, Hottentots, and Swedes each grow up with their own learned, arbitrary beauty standards, which tend to maintain each population in conformity with those standards, since individuals deviating too far from the standards would find it harder to obtain a mate.

Darwin died before his theory could be tested against rigorous studies of how people actually do select their mates. Such studies have proliferated in recent decades, and I summarized the results in Chapter Five. There I showed that people tend to marry individuals who resemble themselves in every conceivable character, including hair and eye and skin colour. To explain that seeming narcissism of ours, I reasoned that we develop our beauty standards by imprinting on the people we see around us in childhood – especially on our parents and siblings, the people of which we see the most. But our parents and siblings are also the people to whom we bear the strongest physical resemblance, since we share their genes. Thus, if you are a fair-skinned, blue-eyed blonde who grew up in a family of fair-skinned, blue-eyed blondes, that is the sort of person whom you will consider most beautiful and will seek as a mate. In the meantime, my dark-skinned, dark-haired New Guinea friends were growing up with other New Guineans and learning to regard fair-skinned, blue-eyed blondes as grotesquely revolting.

To test that imprinting theory of human mate choice rigorously, one would have to do experiments like shipping some Swedish babies to adoptive parents in New Guinea, or painting some Swedish parents permanently black. Then, after waiting twenty years for the babies to grow up, one could study whether they preferred Swedes or New

Guineans as sex partners. Alas, once again, the Search for Truth about humans founders on practical problems, but such tests can be performed with full experimental rigour on animals.

Take snow geese, for example, with their blue or white colour phases. Do white geese learn or inherit their preference in the wild for white geese over blue ones? Canadian biologists hatched gosling eggs in an incubator, then put the goslings into a nest of goose 'foster-parents'. When those goslings grew up, they chose a mate with the colour of the foster-parents. When goslings were reared in a large mixed flock of both blue and white birds, they showed no preference between blue and white prospective mates on reaching adulthood. Finally, when the biologists dyed some white parents pink, their offspring came to prefer pink-dyed geese. Thus, geese do not inherit but learn a colour preference, by imprinting on their parents (and on their siblings and playmates).

*

How, then, do I think that people in different parts of the world evolved their differences? Our insides remained invisible to us and were moulded only by natural selection, with results such as that of tropical Africans but not Swedes evolving the anti-malarial defence of a sickle-cell haemoglobin gene. Many visible features of our outsides also got moulded by natural selection. But, just as in animals, sexual selection had a big effect in moulding the external traits by which we pick our mates.

For us humans those traits are especially the skin, eyes, hair, breasts, and genitals. In each part of the world those traits evolved in tandem with our imprinted aesthetic preferences to reach different, somewhat arbitrary results. Which particular human population ended up with any given eye or hair colour may have been partly an accident of what biologists term the 'founder effect'. That is to say, if a few individuals colonize an empty land and their descendants then multiply to fill the land, the genes of those few founding individuals may still dominate the resulting population many generations later. Just as some birds of paradise ended up with yellow plumes and others with black plumes, so some human populations ended up with yellow hair and others with black hair, some with blue eyes and others with green eyes, some with orange nipples and others with brown nipples.

I do not mean thereby to claim that climate has nothing whatsoever to do with skin colour. I acknowledge that tropical peoples tend on the average to have darker skins than temperate-zone peoples, though there are many exceptions, and that this is probably due to natural selection, though we are unsure of the exact mechanism. Instead, I am saying that sexual selection has been strong enough to render the correlation between skin colour and sun exposure quite imperfect.

If you are still sceptical about how traits and aesthetic preferences can evolve together to different and arbitrary end points, just think about our changing fashion preferences. When I was a schoolboy in the early 1950s, women rated men with crew-cuts and clean-shaven faces as handsome. Since then, we have seen a parade of men's fashions, including beards, long hair, earrings, purple-dyed hair, and the Mohawk hair style. A man daring to flaunt any of those fashions in the 1950s would have revolted the girls and enjoyed zero mating success. That is not because crew-cuts were better adapted to atmospheric conditions of Stalin's last years, while a purple Mohawk has higher survival value in our post-Chernobyl era. Instead, men's appearances and women's tastes changed in tandem, and the changes occurred far more rapidly than evolutionary changes in skin colour, since no gene mutations were required. Either women came to like crew-cuts because good men had them, or men adopted crew-cuts because good women liked them, or something of both happened. The same goes for women's appearances and men's tastes.

To a zoologist, the visible geographic variability that sexual selection produced in humans is impressive. I have argued in this chapter that much of our variability is a by-product of a distinctive feature of the human life-cycle, our choosiness with respect to our spouses and sex partners. I do not know of any other wild animal species in which eye colour of different populations can be green, blue, grey, brown, or black, while skin colour varies geographically from pale to black and hair is either red, yellow, brown, black, grey, or white. There may be no limits, except those imposed by evolutionary time, on the colours with which sexual selection can adorn us. If humanity survives another 20,000 years, I predict that there will be women with naturally green hair and red eyes – plus men who think such women are the sexiest.

SEVEN

WHY DO WE GROW OLD AND DIE?

We constantly invest resources in the repair of our bodies, just as we do with our cars. Unfortunately for us and for all other animals, there is a limit to the resources that natural selection found it worthwhile to programme into our self-repair. As a result, we eventually grow old and die, but at least we age more slowly than our ape relatives.

*

'Mother, why did Grandpa have to die? Will you die some day? Will I die too? Why?'

Death and aging constitute a mystery that we often ask about as children, deny in youth, and reluctantly come to accept as adults. I scarcely reflected on aging when I was a college student. Now that I am fifty-three years old, I find it decidedly more interesting. Life expectancy among US white adults is presently about seventy-eight years for men, eighty-three for women. But few of us will survive to 100. Why is it so easy to live to eighty, so hard to live to 100, and almost impossible to live to 120? Why do humans with access to the best medical care, and animals kept in a cage with plenty of food and no predators, inevitably grow infirm and die? It is the most obvious fact of life, but there is nothing obvious about what causes it.

In the bare fact of our aging and dying, we resemble all other animals. In the details, however, we have improved considerably over the course of our evolutionary history. Not a single individual of any ape species has been recorded as achieving the current life expectancy of US whites, and only exceptional apes reach their fifties. Hence we age more slowly than do our closest relatives. Some of that slowdown may have developed recently, around the time of the Great Leap Forward, since quite a few Cro-Magnons lived into their sixties while few Neanderthals passed forty.

Slow aging is crucial to the human lifestyle because the latter depends on transmitted information. As language evolved, far more information became available to us to pass on than previously. Until the invention of writing, old people acted as the repositories of that transmitted information and experience, just as they continue to do in tribal societies today. Under hunter-gatherer conditions, the knowledge possessed by even one person over the age of seventy could spell the difference between survival and starvation or defeat for a whole clan. Thus, our long lifespan was important for our rise from animal to human status.

Obviously, our ability to survive to a ripe old age depended ultimately upon advances in culture and technology. It is easier to defend yourself against a lion if you are carrying a spear than just a hand-held stone, and easier yet with a high-powered rifle. However, advances in culture and technology alone would not have been enough, unless our bodies had also become redesigned to last longer. No caged ape in a zoo, enjoying all the benefits of modern human technology and veterinary care, reaches eighty. We shall see in this chapter that our biology became remoulded to the increased life expectancy that our cultural advances made possible. In particular, I would guess that Cro-Magnon tools were not the sole reason why Cro-Magnons lived on the average longer than Neanderthals. Instead, around the time of the Great Leap Forward our biology must have changed so that we aged more slowly. That may even have been the time when menopause, the concomitant of aging that paradoxically functions to let women live longer, evolved.

In short, cultural and biological change had to develop hand-in-hand to permit our long lives. Along with the changes in our sexual anatomy, physiology, behaviour, and preferences discussed in Chapters Three to Six, retarded aging is the last of the life-cycle changes that made possible the third chimpanzee's rise.

*

The way in which scientists think about aging depends on whether they are interested in so-called proximate explanations or ultimate explanations. To appreciate this difference, consider the question, 'Why do skunks smell bad?' A chemist or molecular biologist would answer,

'It's because skunks secrete chemical compounds with certain particular molecular structures. Due to the principles of quantum mechanics, those structures result in bad smells. Those particular chemicals would smell bad no matter what the biological function of their bad smell was.'

But an evolutionary biologist would instead reason,

'It's because skunks would be easy victims for predators if they didn't defend themselves with bad smells. Natural selection made skunks evolve to secrete bad-smelling chemicals; those skunks with the worst smells survived to produce the most baby skunks. The molecular structure of those chemicals is a mere incidental detail; any other bad-smelling chemicals would suit skunks equally well.'

The chemist has offered a proximate explanation: that is, the mechanism immediately responsible for the observation that was to be explained. The evolutionary biologist has instead offered an ultimate explanation: the function or chain of events that caused that mechanism to be present. The chemist and the evolutionary biologist would each dismiss the other's answer as not being 'the real explanation'.

Similarly, studies of aging are pursued independently by two groups of scientists who scarcely communicate with each other. One group seeks a proximate explanation, the other an ultimate explanation. Evolutionary biologists try to understand how natural selection could ever permit aging to occur, and they think that they have found an answer to this question. Physiologists inquire instead into the cellular mechanisms underlying aging, and admit that they do not yet have an answer. But I shall argue that aging cannot be understood unless we seek both explanations simultaneously. In particular, I expect that the evolutionary (ultimate) explanation will help us find the physiological (proximate) explanation of aging that has so far eluded scientists.

*

Before I can pursue this reasoning, I must anticipate objections of my physiologist friends. They tend to believe that something about our physiology somehow makes aging inevitable, and that evolutionary considerations are irrelevant. For instance, one such theory attributes aging to the progressive difficulties that our immune system is said to face in distinguishing our own cells from foreign cells. Physiologists subscribing to this view make an implicit assumption that natural selection could not lead to an immune system without that fatal defect. Is this belief warranted?

To evaluate this objection, let's consider biological repair mechanisms, because aging may be thought of simply as unrepaired damage or deterioration. Our first association with the word 'repair' is likely to be to those repairs that cause us the most frustration, car repairs. Our cars tend to grow old and die, but we spend money to postpone their inevitable fate. Similarly, we are unconsciously but constantly repairing ourselves too, at every level from that of molecules to that of tissues or whole organs. Our own self-repair mechanisms, like those we lavish on our cars, are of two sorts – damage control, and regular replacement.

An automotive example of damage control is that we replace a car's bumper only if it is bashed in; we do not routinely replace the bumper at every regular oil change. The most visible example of damage control applied to our bodies is wound healing, by which we repair damage to our skin. Many animals can achieve more spectacular results: lizards regenerate severed tails, starfish and crabs their limbs, sea cucumbers their intestines, and ribbon worms their poison stylets. At the invisible molecular level our genetic material, DNA, is repaired exclusively by damage control. We have enzymes that recognize and fix damaged sites in the DNA helix while ignoring intact DNA.

The other type of repair, regular replacement, is also familiar to every car-owner. We periodically change the oil, air filter, and ball-bearings to eliminate slight wear, without waiting for the car to break down first. In the biological world, teeth are similarly replaced on a pre-scheduled basis: humans go through two sets, elephants six sets, and sharks an indefinite number, during their lifetimes. Though we humans go through life with the same skeleton with which we were born, lobsters and other arthropods regularly replace their exoskeleton by moulting it and growing a new one. Still another highly visible example of scheduled repair is the continual growth of our hair: no matter how short we cut it, its growth will replace the cut portion.

Regular replacement also goes on at a microscopic or submicroscopic level. We constantly replace many of our cells about once every few days for the cells lining our intestine, once every two months for the cells lining the urinary bladder, and once every four months for our red blood cells. At the molecular level, our protein molecules are subject to continuous turnover at a rate characteristic of each particular protein; we thereby avoid the accumulation of damaged molecules. If you compare your beloved's appearance today with a photograph taken a month ago, he (or she) may look the same, but many of the individual molecules forming that beloved body are different. While all the king's horses and men couldn't put Humpty Dumpty together again, Nature is taking us apart and putting us back together every day.

Thus, much of an animal's body can be repaired as needed, or is regularly replaced anyhow, but the details of how much is replaceable vary greatly with the part and with the species. There is nothing physiologically inevitable about the limited repair capabilities of us humans. Since starfish can regrow amputated limbs, why can't we? What prevents us from having six sequential sets of teeth like an elephant, rather than just baby teeth and adult teeth? With four more natural sets, we would not need fillings, crowns, and dentures as we got older. Why don't we protect ourselves against arthritis? – all we would need is to replace our joints periodically, as crabs do. Why don't we guard against heart disease by periodically replacing our hearts, as ribbon worms replace their poison stylets? One might suppose that natural selection would favour the man or woman who did not die of heart disease around the age of eighty but continued to live and produce babies at least until the age of 200. Why, for that matter, cannot we repair or replace everything in our bodies?

The answer surely has something to do with the expense of repair. Here again, the analogy of car repair is helpful. If the boasts of the Mercedes-Benz company are to be believed, their cars are so well built that, even should you do no maintenance whatsoever – not even lubrication or oil changes – your Mercedes will still run for years. At the end of that time, of course, it will fall apart from accumulated irreversible damage. So Mercedes-owners generally do choose to service their cars regularly. My Mercedes-owning friends tell me that Mercedes service is very expensive, hundreds of dollars every time they drive into the workshop. Nevertheless, they consider the expense worth it. A serviced Mercedes lasts much longer than an unserviced Mercedes, and it is much cheaper to service your old Mercedes regularly than to discard it and buy a new one every few years.

That is how Mercedes-owners reason in Germany and the US. But suppose you were living in Port Moresby, the capital of Papua New Guinea, automobile accident capital of the world, where any car is likely to be written off within a year no matter how you maintain it. Many car-owners in New Guinea do not go to the expense of maintaining their car; they use the saved money to help buy the inevitable next car.

By analogy, how much an animal 'should' invest in biological repair depends on the expense of the repairs, and on a comparison of the animal's expected lifespan with and without the repairs. But such 'should' questions belong to the realm of evolutionary biology, not physiology. Natural selection tends to maximize one's rate of producing offspring that survive to leave offspring of their own. Evolution can thus be regarded as a strategy game, in which the individual whose strategy leaves the most descendants wins. Hence the type of reasoning used in game theory is helpful in understanding how we came to be the way we are.

*

This problem of lifespan, and of investment in biological repair, is in turn one of an even broader class of evolutionary problems addressed by game theory: the mystery of what sets the maximum limit on any advantageous trait. There are lots of other biological traits, besides lifespan, that beg the question why natural selection has not made them longer or bigger or faster or made more of them. For instance, people who are big or smart or can run fast have obvious advantages over small, dumb, slow people – especially throughout most of human evolution, when we were still fending off lions and hyenas. Why did we not evolve to become on the average even bigger, smarter, and faster than we now are?

The complication that makes these evolutionary design problems less simple than they might at first seem is this: natural selection acts on whole individuals, not on single parts of an individual. It is you, not your big brain or fast legs, that does or does not survive and leave offspring. Increasing one part of an animal's body may be beneficial in some obvious respect but harmful in other respects. For instance, that one larger part might not fit in well with other parts of the same animal, or it might drain off energy from other parts.

To evolutionary biologists, the magic word that expresses this complication is 'optimize'. Natural selection tends to mould each trait to the size, speed, or number that maximizes the survival and reproductive success of the whole animal, given the animal's basic design. Hence each trait in itself does not tend towards a maximal value. Instead, each trait converges on some optimal intermediate value, neither too big nor too small. The whole animal is thereby more successful than it would be if that trait were bigger or smaller.

Should this reasoning about animals seem abstract, think instead of our everyday machines. Essentially the same principles apply to engineering design, of machines by humans, as to evolutionary design, of animals by natural selection. For example, consider my pride and joy among my machines, my 1962 Volkswagen Beetle, the only car I have ever owned. (Car buffs will remember 1962 as the year that Volkswagen introduced the big rear window in the Beetle.) On a smooth, level road with an assisting tailwind, my VW can go at 65 mph. To BMW owners, that may sound distinctly submaximal. Why don't I junk my puny 4–cylinder, 40–horsepower engine, install instead the 12–cylinder, 296–horsepower engine from my neighbour's BMW 750 IL, and roar off at 180 mph down the freeway?

Well, even I, dodo about cars that I am, know that that would not work. To begin with, that huge BMW engine would not fit into my VW's engine compartment, which would need enlarging. Then, the BMW engine is meant to go in front, but the VW engine compartment is in the back, so I would have to change the gearbox and transmission and other things. I would also have to change the shock absorbers and brakes, designed to smooth the ride and stop a car at 65 mph but not at 180 mph. By the time I had finished modifying my VW to take the BMW engine, there would not be much remaining from my original Beetle, and the modifications would have cost me a big pile of money. I suspect that my puny 40–horsepower engine is optimal, in the sense that I could not increase my cruising speed without sacrificing other performance features of my car – as well as sacrificing other money-requiring features of my lifestyle.

While the marketplace *eventually* eliminates engineering monstrosities like a VW with a BMW engine, all of us can think of monstrosities that took quite a while to eliminate. To those of you who share my fascination with naval warfare, British battle-cruisers are a good example. Before and during the First World War, the British navy launched thirteen warships called battle-cruisers, designed to be as large and with as many big guns as battleships but much faster. By maximizing speed and firepower, the battle-cruisers immediately caught the public imagination and became a propaganda sensation. However, if you take a 28,000-ton battleship, keep the weight of the big guns nearly constant, and greatly increase the weight of the engines while still maintaining total weight around 28,000 tons, you have to skimp on the weight of some other parts. The battle-cruisers skimped especially on weight of armour, but also on weight of small guns, internal compartments, and anti-aircraft defence.

The results of this suboptimal overall design were inevitable. In 1916 H.M.S. *Indefatigable*, *Queen Mary*, and *Invincible* all blew up almost as soon as they were hit by shells at the Battle of Jutland. H.M.S. *Hood* blew up in 1941, a mere eight minutes after entering battle with the German battleship *Bismarck*. H.M.S. *Repulse* was sunk by Japanese bombers a few days after the Japanese attack on Pearl Harbor, thereby acquiring the dubious distinction of being the first large warship to be destroyed from the air while in combat at sea. Faced with this stark evidence that some spectacularly maximal parts do not make an optimal whole, the British navy let its programme of building battle-cruisers become extinct.

In short, engineers cannot tinker with single parts in isolation from the rest of a machine, because each part costs money, space, and weight that might have gone into something else. Engineers instead have to ask what *combination* of parts will optimize a machine's effectiveness. By the same reasoning, evolution cannot tinker with single traits in isolation from the rest of an animal, because every structure, enzyme, or piece of DNA consumes energy and space that might have gone into something else. Instead, natural selection favoured that combination of traits that maximizes the animal's reproductive output. Thus, both engineers and evolutionary biologists have to evaluate the trade-offs involved in increasing anything; that is, its costs, as well as the benefits that it would bring.

*

An obvious difficulty in applying this reasoning to our life-cycles is that they have many features seeming to reduce, not to maximize, our ability to produce offspring. Growing old and dying is just one example; other examples are human female menopause, bearing one baby at a time, producing babies only once every year or so at most, and not even starting to produce babies until the age of twelve to sixteen. Would not natural selection favour the woman who reached puberty at age five, completed gestation in three weeks, regularly bore quintuplets, never underwent menopause, put lots of biological energy into repair of her body, lived to 200, and thereby left hundreds of offspring?

But posing the question in that form pretends that evolution can change our bodies one piece at a time, and ignores the hidden costs. For example, a woman certainly could not reduce the length of pregnancy to three weeks without changing anything else about herself or her baby. Remember that we only have a finite amount of energy available to us. Even people doing hard exercise and eating rich food – lumberjacks, or marathon runners in training – cannot metabolize much more than about 5,000 calories per day. How should we allocate those calories between repairing ourselves and rearing babies, if our goal is to raise as many babies as possible?

At the one extreme, if we put all our energy into babies and devoted no energy to biological repair, our bodies would age and disintegrate before we could rear our first baby. At the other extreme, if we lavished all our available energy on keeping our bodies in shape, we might live a long time but would have no energy left for the exhausting process of making and rearing babies. What natural selection must do is to adjust an animal's relative expenditures of energy on repair and on reproduction, so as to maximize its reproductive output, averaged over its lifetime. The answer to that problem varies among animal species, depending on factors such as their risk of accidental death, their reproductive biology, and the costs of various types of repair.

This perspective can be employed to make testable predictions about how animals should differ in their repair mechanisms and rates of aging. In 1957 the evolutionary biologist George Williams cited some striking facts about aging that become comprehensible only from an evolutionary perspective. Let's consider several of Williams's examples and re-express them in the physiological language of biological repair, by taking slow aging as an indication of good repair mechanisms.

The first example concerns the age at which an animal first breeds and produces offspring. That age varies enormously among species: few humans are so precocious as to produce babies before the age of twelve years, while any self-respecting mouse a mere two months old can already make baby mice. Animals belonging to a species whose age of first breeding is late, like us, need to devote much energy to repair, in order to ensure that they survive to that reproductive age. Hence we expect investment in repair to increase with age at first reproduction.

For instance, correlated with our having a much later age of first reproduction than do mice, we humans age far more slowly than mice and are thus presumed to repair our bodies much more effectively. Even with plenty of food and the best medical care, a mouse is lucky to reach its second birthday, while we would be unlucky not to reach our seventy-second birthday. The evolutionary reason: a human who invested no more of his/her energy in repair than does a mouse would be dead long before reaching puberty. Hence it is more worthwhile to repair a human than a mouse.

What might that postulated extra energy expenditure of ours actually consist of? At first, our human repair capabilities seem unimpressive. We cannot regrow an amputated arm, and we do not regularly replace our skeleton, in the way that some short-lived invertebrates do. However, such spectacular but infrequent replacements of a whole structure probably are not the biggest items in an animal's repair budget. Instead, the biggest expense is all that invisible replacement of so many of your cells and molecules, day after day. Even if you spend all day every day just lying in bed, you need to eat about 1,640 calories per day if you are a man (1,430 for a woman) just to maintain your body. Much of that maintenance metabolism goes to our invisible scheduled replacement. And so I would guess that we cost more than a mouse in the respect of putting a bigger fraction of our energy into self-repair, and a smaller fraction into other purposes like keeping warm or caring for babies.

The second example I shall discuss involves the risk of irreparable injury. Some biological damage is potentially reparable, but there is also damage that is guaranteed to be fatal (for example, being eaten by a lion). If you are likely to be eaten by a lion tomorrow, there is no point paying a dentist to start expensive orthodontic work on your teeth today. You would do better to let your teeth rot and start having babies immediately. But if an animal's risk of death from irreparable accidents is low, then there is a potential payoff, in the form of increased lifespan, from putting energy into expensive repair mechanisms that retard aging. This is the reasoning by which Mercedes-owners decide to pay for lubrication of their cars in Germany and the US but not in New Guinea.

Biological analogies are that the risk of death from predators is lower for birds than for mammals (because birds can escape by flying), and lower for turtles than for most other reptiles (because turtles are protected by a shell). Thus, birds and turtles stand to gain a lot from expensive repair mechanisms, compared to flightless mammals and shell-less reptiles that will soon be eaten by predators anyway. Indeed, if one compares longevities of well-fed pets protected from predators, birds do live longer (that is, do age more slowly) than similarly sized mammals, and turtles live longer than similarly sized shell-less reptiles. The bird species best protected from predators are seabirds like petrels and albatrosses that nest on remote oceanic islands free of predators. Their leisurely life-cycles rival our own. Some albatrosses do not even breed until they are ten years old, and we still do not know how long they live: the birds themselves last longer than the metal rings that biologists began putting on their legs a few decades ago in order to age them. In the ten years that it takes an albatross to start breeding, a mouse population could have gone through sixty generations, most of which would already have succumbed to predators or old age.

As our third example, let's compare males and females of the same species. We expect more potential payoff from repair mechanisms, and lower rates of aging, in that sex with the lower accidental mortality rate. For many or most species, males suffer greater accidental mortality than females, partly because males put themselves at greater risk by fighting and bold displays. This is certainly true of human males today and has probably been so throughout our history as a species – men are the sex most likely to die in wars against men of other groups, and in individual fights within a group. Also, in many species the males are bigger than the females, but studies of red deer and of New World blackbirds show that males are thereby more likely than females to die when food becomes scarce.

Correlated with this greater accidental death rate of men, men also age faster and have a higher non-accidental death rate than women. At present, women's life expectancy is about six years greater than that of men; some of this difference is because more men than women are smokers, but there is a sex-linked difference in life expectancy even among non-smokers. These differences suggest that evolution has programmed us so that women put more energy into self-repair, while men put more energy into fighting. Expressed another way, it just is not worth as much to repair a man as it is to repair a woman. I do not thereby mean to denigrate male fighting, which serves a useful evolutionary purpose for a man: to gain wives and to secure resources for his children and his tribe, at the expense of other men and their children and tribe.

*

My remaining example of how some striking facts of aging become comprehensible only from an evolutionary perspective concerns the distinctively human phenomenon of survival past reproductive age, especially past female menopause. Since transmitting one's genes to the next generation is what drives evolution, other animal species rarely survive past reproductive age. Instead, Nature programmes death to coincide with the end of fertility, because there is then no longer an evolutionary benefit to gain from keeping one's body in good repair. It is an exception in need of explanation to realize that women are programmed to live for decades after menopause, and that men are programmed to live to an age when most men are no longer busy siring babies.

But the explanation becomes apparent on reflection. The intense phase of parental care is unusually protracted in the human species and lasts nearly two decades. Even those older people whose own children have reached adulthood are tremendously important to the survival of not just their children but of their whole tribe. Especially in the days before writing, they acted as the carriers of essential knowledge. Thus, Nature has programmed us with the capacity to keep the rest of our bodies in reasonable repair even at an age when the female reproductive system itself has fallen into disrepair.

Conversely, though, we have to wonder why natural selection programmed female menopause into us in the first place. It too, like aging, cannot be explained away as something physiologically inevitable. Most mammals, including human males plus chimps and gorillas of both sexes, merely experience a gradual decline and eventual cessation of fertility with age, rather than the abrupt shutdown of women's fertility. Why did that peculiar, seemingly counter-productive feature of ours evolve? Would not natural selection favour the woman who remained fertile until the bitter end?

Human female menopause probably resulted from two other distinctively human characteristics: the exceptional danger that childbirth poses to the mother, and the danger that a mother's death poses to her offspring. Recall from Chapter Three the enormous size of the human infant at birth relative to its mother: our big 7-pound babies emerging from 100-pound mothers, compared to little 4-pound gorilla babies emerging from a 200-pound gorilla mother. As a result, childbirth is dangerous to women. Especially before the advent of modern obstetrics, women often died in childbirth, whereas mother gorillas and chimps virtually never do.

Now recall also from Chapter Three the extreme dependence of human infants on their parents, especially on their mother. Because human infants develop so slowly and cannot even feed themselves after weaning (unlike young apes), the death of a hunter-gatherer mother would have been likely to be fatal to her offspring up to a later age in childhood than for any other primate. Hence a hunter-gatherer mother with several children was gambling the lives of those children at every subsequent childbirth. Since her investment in those prior children increased with their age, and since her own risk of death in childbirth also increased with her age, the odds of her gamble paying off got worse and worse as she got older. When you already have three children alive but still dependent on you, why risk those three for a fourth?

Those worsening odds probably led through natural selection to menopausal shutdown of human female fertility, in order to protect a mother's prior investment in children. Since childbirth carries no risk of death for fathers, men did not evolve menopause. Like aging, menopause illustrates how an evolutionary approach illuminates features of our life-cycle that would otherwise be counter-intuitive. It is even possible that menopause evolved only within the past 40,000 years, when Cro-Magnons and other anatomically modern humans began frequently to survive to the age of sixty or more. Neanderthals and earlier humans usually died before the age of forty anyway, so that menopause would have brought their women no benefits if it were to occur at the same age as in modern *Femina sapiens*.

Thus, the longer lifespan of modern humans than of apes rests not only on cultural adaptations, such as tools to acquire food and deter predators. It also rests on the biological adaptations of menopause and increased investment in self-repair. Whether those biological adaptations developed especially at the time of the Great Leap Forward or earlier, they rank among the life history changes that permitted the rise of the third chimpanzee to humanity.

*

The last conclusion that I wish to draw from an evolutionary approach to aging is that it undermines the approach which has long dominated the physiological study of aging. The gerontological literature is obsessed with a search for The Cause of Aging – preferably a single cause, certainly not more than a few major causes. Within my own lifetime as a biologist, hormonal changes, deterioration in the immune system, and neural degeneration have vied in popularity for the title of The Cause, without compelling support having been adduced to date for any of the candidates. But evolutionary reasoning suggests that this search will remain futile. There *should* not be just one, or even a few, dominant physiological mechanisms of aging. Instead, natural selection should act to match rates of aging in all physiological systems, with the result that aging involves innumerable simultaneous changes.

The basis of this prediction is as follows. There is no point doing expensive maintenance on one piece of the body if other pieces are deteriorating more rapidly. Conversely, natural selection should not permit a few systems to deteriorate long before all the others, as the cost of extra repairs on just those few systems would then buy a big increase in life expectancy and would be worth it. By analogy, Mercedes-owners should not install cheap ball-bearings when they are lavishing expense on all other parts of the car. Had they been so foolish, they could have doubled the lifetime of their costly car just by spending a few more dollars for better ball-bearings. But it would not pay either to go to the expense of installing diamond ball-bearings, when all the rest of the car would have rusted away before those ball-bearings wore out. Thus, the optimal strategy for Mercedes-owners, and for us, is to repair all parts of our cars or bodies at such rates that everything finally collapses all at once.

It seems to me that this depressing prediction is borne out, and that we come closer to this evolutionary ideal than to the physiologists' long-sought Cause of Aging. Signs of aging can be found wherever one looks for them. Already I am conscious in myself of tooth wear, considerable decreases in muscle performance, and significant losses in hearing, vision, smell, and taste. For all these senses, the acuity of women is greater than that of men of equal age, whatever the age group compared. Ahead of me lies the familiar litany: weakening of the heart, hardening of the arteries, increasing brittleness of bones, decreases in kidney filtration rates, lower resistance of the immune system, and loss of memory. The list could be extended almost indefinitely. Evolution seems indeed to have arranged things so that all our systems deteriorate, and that we invest in repair only as much as we are worth.

From a practical standpoint, this conclusion is disappointing. If there had been one dominant cause of aging, curing that cause would have provided us with a fountain of youth. This thought, operating at a time when aging was thought to be largely a hormonal phenomenon, inspired some attempts at miraculous rejuvenation of old people by hormonal injections or implantation of young gonads. Such an attempt was the subject of Sir Arthur Conan Doyle's story, *The Adventure of the Creeping Man*, in which the aged Professor Presbury becomes infatuated with a young woman, desperately wants to rejuvenate himself, and instead is found creeping around like a monkey after midnight. The great Sherlock Holmes discovers the reason: the Professor has been seeking youth by injecting himself with the serum of langur monkeys.

I could have warned Professor Presbury that his myopic obsession with proximate causation would lead him astray. Had he thought of ultimate evolutionary causation, he would have realized that natural selection would never permit us to deteriorate through a single mechanism with one simple cure. Perhaps it is just as well. Sherlock Holmes worried greatly about what would happen if such an elixir of life were found:

There is danger there – a very real danger to humanity. Consider, Watson, that the material, the sensual, the worldly would all prolong their worthless lives . . . It would be the survival of the least fit. What sort of cesspool may not our poor world become?

Holmes would be relieved to know that his worries now appear unlikely to materialize.

PART THREE

UNIQUELY HUMAN

PARTS ONE AND TWO DISCUSSED THE GENETICALLY SPECIFIED FOUNDATIONS of our unique cultural traits. We saw that those foundations include our familiar skeletal hallmarks, such as our large braincase and our adaptations for upright gait. They also include features of our soft tissues, behaviour, and endocrinology concerned with reproduction and social organization.

But if those genetically specified features were our sole distinctions, we would not stand out among animals, and we would not now be threatening the survival of ourselves and other species. Other animals, such as ostriches, walk erect on two legs. Others have relatively large brains, though not as large as ours. Others live monogamously in colonies (many seabirds), or are very long-lived (albatrosses and tortoises).

Instead, our uniqueness lies in the cultural traits that rest on those genetic foundations and that in turn give us our power. Our cultural hallmarks include spoken language, art, tool-based technology, and agriculture. But if we stopped there, we would have a one-sided and self-congratulatory view of our uniqueness. The hallmarks I just mentioned are ones that we are proud of. Yet the archaeological record shows the introduction of agriculture to have been a mixed blessing, seriously harming many people while benefitting others. Chemical abuse is a wholly ugly human hallmark. At least it does not threaten our survival, as do two of our other cultural practices: genocide, and mass exterminations of other species. We are uncomfortable about whether to regard these as occasional pathological aberrations, or as features no less basic to humanity than the traits we are proudest of.

All of these cultural traits that define humanity are seemingly absent in animals, even in our closest relatives. They must have arisen some time after our ancestors parted company from the other chimpanzees around seven million years ago. Furthermore, while we have no way of knowing whether Neanderthals spoke or indulged in drug abuse and genocide, they certainly did not have agriculture, art, or the capacity to build radios. These latter traits must therefore be very recent human innovations of the last few tens of thousands of years. However they could not have arisen from nothing. There had to have been animal precursors, if we could only recognize them.

For each of our defining cultural traits, we need to ask, what were those precursors? When in our ancestry did the trait approach its modern form? What were the early stages of its evolution like, and can those stages be traced archaeologically? We are unique on Earth, but how unique are we in the universe?

Our two most dangerous traits, genocide and environmental destruction, will be reserved for discussion in Parts Four and Five. Here we will consider some of the above-mentioned questions for our noble, two-edged, or only mildly destructive characteristics. Chapter Eight takes up the origin of spoken language, which I suggested in Chapter Two might have triggered the Great Leap Forward, and which anyone would list among our most important distinctions from animals. On first reflection, the task of tracing the development of human language appears plainly impossible. Language before the dawn of writing left no archaeological remains, unlike our first experiments in art, agriculture, and tools. There seems to be no surviving simple human language, no animal language, that could exemplify the early stages.

In fact, there are innumerable animal precursors: the vocal communication systems evolved by many species. We are just beginning to appreciate the sophistication of some of these systems. We shall also see that there really are some simple languages that modern humans have unconsciously invented and that prove unexpectedly instructive. Taken together, these complex animal 'languages' and simple human languages begin to bridge, from both sides, the apparent chasm with respect to speech between animals and ourselves.

Chapter Nine turns to the origin of art, the noblest human invention. There seems to be a gulf separating human art, supposedly created just for pleasure and doing nothing to perpetuate our genes, from any animal behaviour. Yet paintings and drawings created by captive apes and elephants, whatever the motives of those animal artists, look so similar to work of human artists that they have fooled experts and have been bought by art collectors. If one nevertheless dismisses those animal artworks as unnatural productions, what is one to say about the carefully arranged coloured displays of normal male bowerbirds? Those bowers play an unquestioned crucial role in passing on genes. I shall argue that human art also had that role originally, and often still does today. Since art, unlike language, does show up in archaeological deposits, we know that human art did not proliferate until the time of the Great Leap Forward.

Agriculture, the subject of Chapter Ten, has an animal precedent, but not precursor, in the gardens of leaf-hopper ants, which lie far off from our direct lineage. The archaeological record lets us date our 'reinvention' of agriculture to a time long after the Great Leap Forward, within the last

10,000 years. That transition from hunting and gathering to agriculture is generally considered a decisive step in our progress, when we at last acquired the stable food supply and leisure time prerequisite to the great accomplishments of modern civilization. In fact, careful examination of that transition suggests another conclusion: for most people the transition brought infectious diseases, malnutrition, and a shorter lifespan. For human society in general it worsened the relative lot of women and introduced class-based inequality. More than any other milestone along the path from chimpanzeehood to humanity, agriculture inextricably combines causes of our rise and our fall.

Abuse of toxic chemicals is a widespread human hallmark documented only within the last 5,000 years, though it may well go back much earlier into pre-agricultural times. Unlike agriculture, it does not even rank as a mixed blessing but as a pure evil threatening the survival of individuals, though not of our species. Like art, drug abuse seems at first to lack animal precedents or biological functions. I shall argue in Chapter Eleven, however, that it fits into a broad class of animal structures or behaviours that are dangerous to their owners or practitioners, and whose function depends paradoxically on that danger.

While animal precursors can thus be identified for all of our hallmarks, they still rank as human hallmarks because we are unique on Earth in the extreme degree to which we have developed them. How unique are we in the universe? Once conditions suitable for life exist on a planet, how likely are intelligent, technologically advanced life forms to evolve? Was their emergence on Earth practically inevitable, and do they now exist on innumerable planets circling other stars?

There is no direct way to prove whether creatures capable of language, art, agriculture, or drug abuse exist elsewhere in the universe, because from Earth we could not detect the existence of those traits on planets of other stars. However, we might be able to detect high technology elsewhere in the universe if it included our own capacity to send out space probes and interstellar electromagnetic signals. In Chapter Twelve I shall examine the on-going search for extraterrestrial intelligent life. I shall argue that evidence from a quite different field – studies of woodpecker evolution on Earth – instructs us about the inevitability of evolving intelligent life, and hence about our uniqueness, not only on Earth but also in the accessible universe.

EIGHT

BRIDGES TO HUMAN LANGUAGE

The gulf between animal vocal communication and human speech has traditionally been viewed as unbridgeable. In fact, recent studies of animal vocalizations show some of them to be far more sophisticated than we had previously suspected. On the other hand, there are dozens of cases in which humans have been forced by exceptional social circumstances to create simplified languages, possibly illustrating two primitive stages in the evolution of human language. Thus, we are beginning to understand how our most unique and important distinction from animals nevertheless arose from animal precursors.

*

The mystery of human language origins is the most crucial in understanding how we became uniquely human. After all, it is language that allows us to communicate with each other far more precisely than any animal can. Language enables us to formulate joint plans, to teach one another, and to learn from what others have experienced elsewhere or in the past. With it, we can store precise representations of the world in our minds, and hence encode and process information far more efficiently than any animal can. Without language we could never have conceived and built Chartres Cathedral – or V-2 rockets. For these reasons, I speculated in Chapter Two that the Great Leap Forward (the stage in human history when innovation and art at last emerged) was made possible by the emergence of spoken language as we know it.

Between human language and the vocalizations of any animal lies a seemingly unbridgeable gulf. It has been clear since the time of Darwin that the mystery of human language origins is an *evolutionary* problem: how was this unbridgeable gulf nevertheless bridged? If we accept that we evolved from animals lacking human speech, then our language must have evolved and become perfected with time, along with the human pelvis, skull, tools, and art. There must once have been intermediate language-like stages linking monkey grunts to Shakespeare's sonnets. Darwin diligently kept notebooks on his children's linguistic development, and reflected on the languages of 'primitive' peoples, in the hope of solving this evolutionary mystery.

Unfortunately, the origins of language prove harder to trace than the origins of the human pelvis, skull, tools, and art. All of the latter may persist as fossils that we can recover and date, but the spoken word vanishes in an instant. In frustration, I often dream of a time machine that would let me place tape-recorders in ancient hominoid camps. Perhaps I would discover that australopithecines uttered grunts little different from those of chimpanzees; that early *Homo erectus* used recognizable single words, progressing after a million years to two-word sentences; that *Homo sapiens* before the Great Leap Forward became capable of strings of words that were longer but still without much grammar; and that syntax and the full range of modern speech sounds arrived only with the Great Leap.

Alas, we have no such retrospective tape-recorder, and no prospects for ever getting one. How can we hope to trace speech origins without such a magic time machine? Until recently, I would have said that it was hopeless to do more than speculate. In this chapter, however, I shall try to draw on two exploding bodies of knowledge that may allow us to begin building bridges across the gulf between animal and human sounds, by starting from each of its opposite shores.

Sophisticated new studies of wild animal vocalizations, especially those of our primate relatives, constitute the bridgehead on the animal shore of the gulf. It has always been obvious that animal sounds must have been precursors of human speech, but only now are we beginning to sense how far animals have come towards inventing their own 'languages'. In contrast, it has not been clear where to locate the bridgehead on the human shore, since all existing human languages seem infinitely advanced over animal sounds. Recently, though, it has been argued that a numerous set of human languages neglected by most linguists truly exemplifies two primitive stages on the human side of the causeway.

*

Many wild animals communicate with each other by sounds, of which bird-songs and the barking of dogs are especially familiar to us. Most of us are within earshot of some calling animal on most days of our lives. Scientists have been studying animal sounds for centuries. Despite this long history of intimate association, our understanding of these ubiquitous and familiar sounds has suddenly expanded because of the application of new techniques: use of modern tape-recorders to record animal calls, electronic analysis of the calls to detect subtle variations imperceptible to the unaided human ear, broadcasting recorded calls back to animals to observe how they react, and observing their reactions to electronically reshuffled calls. These methods are revealing animal vocal communication to be much more like language than anyone would have guessed thirty years ago.

The most sophisticated 'animal language' studied to date is that of a common, cat-sized African monkey known as the vervet. Equally at home in trees and on the ground in savannah and rainforest, vervets are among the monkey species that visitors to East African game parks are most likely to see. They must have been familiar to Africans for the hundreds of thousands of years that we have existed as the species *Homo sapiens*. They may have reached Europe as pets over 3,000 years ago, and they certainly have been familiar to European biologists exploring Africa since the Nineteenth Century. Many laypeople who have never visited Africa are still acquainted with vervets from visits to the zoo.

Like other animals, wild vervets regularly face situations in which efficient communication and representation would help them to survive. About three-quarters of wild vervet deaths are caused by predators. If you are a vervet, it is essential to know the differences between a martial eagle, one of the leading killers of vervets, and a white-backed vulture, an equally large soaring bird that eats carrion and is no danger to live monkeys. It is vital to act appropriately when the eagle appears, and to tell your relatives. If you fail to recognize the eagle, you die; if you fail to tell your relatives, they die, carrying your genes with them; and if you think it is an eagle when it is really just a vulture, you are wasting time on defensive measures while other monkeys are safely out there gathering food.

Besides these problems posed by predators, vervets have complex social relationships with each other. They live in groups and compete for territory with other groups. Hence it is also essential to know the difference between a monkey intruding from another group, an unrelated member of your own group likely to push you, and a close relative in your own group on whose support you can count. Vervets that get into trouble need ways of telling their relatives that they, and not some other monkey, are in trouble. They also need to know and communicate about sources of food: for instance, which of the thousand plant and animal species in the environment are good to eat, which are poisonous, and where and when the edible ones are likely to be found. For all these reasons, vervets would profit from efficient ways of communicating about and representing their world.

Despite these reasons, and despite the long and close association between vervets and humans, we had no appreciation of their complex world knowledge and vocalizations until the mid-1960s. Since then, observations of vervet behaviour have revealed that they make finely graded discriminations among types of predators, and among each other. They adopt quite different defensive measures when threatened by leopards, eagles, and snakes. They respond differently to dominant and subordinate members of their own troop, differently again to dominant and subordinate members of rival troops, differently to members of different rival troops, and differently to their mother, maternal grandmother, sibling, and unrelated members of their own troop. They know who is related to whom: if an infant monkey calls, its mother turns towards it, but other vervet mothers turn instead towards that infant's mother to see what she will do. It is as if vervets had names for several predator species and several dozen individual monkeys.

The first clue to how vervets communicate this information came from observations that the biologist Thomas Struhsaker made on vervets in Kenya's Amboseli National Park. He noted that three types of predator triggered different defensive measures by vervets, and also triggered alarm calls sufficiently distinct that Struhsaker could hear the differences even without making any sophisticated electronic analysis. When vervets encounter a leopard or any other species of large wild cat, male monkeys give a loud series of barks, females give a high-pitched chirp, and all monkeys within earshot may run up a tree. The sight of a martial or crowned eagle soaring overhead causes vervets to give a short cough of two syllables, whereupon listening monkeys look up into the air or run into a bush. A monkey who spots a python or other dangerous snake gives a 'chuttering' call, and that stimulates other vervets in the vicinity to stand erect on their hind legs and look down (to see where the snake is).

Beginning in 1977, a husband-and-wife team named Robert Seyfarth and Dorothy Cheney proved that these calls really had the different functions suggested by Struhsaker's observations. Their experimental procedure was as follows. Firstly, they made a tape-recording of a monkey giving a call whose apparent function Struhsaker had observed (say, the 'leopard call'). Then, on a later day, after locating the same troop of monkeys, either Cheney or Seyfarth hid the tape and loudspeaker equipment in a bush nearby, while the other started filming the monkeys with a cine or video camera. After fifteen seconds, one of the two scientists broadcast the tape while the other kept filming the monkeys for one minute to see whether the monkeys behaved appropriately for the call's suspected function (for example, whether the monkeys ran up a tree on hearing a broadcast of the supposed 'leopard' call). It turned out that playback of the 'leopard call' really did stimulate the monkey to run up a tree, while the 'eagle call' and 'snake call' similarly stimulated monkeys into behaviour that seemed to be associated with these calls under natural conditions. Thus, the apparent association between the observed behaviour and the calls was not coincidental, and the calls did have the functions suggested by observation.

The three calls that I have mentioned by no means exhaust a vervet's vocabulary. Besides those loud and frequently given alarm calls, there appear to be at least three fainter alarms that are given less frequently. One, triggered by baboons, causes listening vervets to become more alert. A second, given in response to mammals like jackals and hyenas that prey on vervets only infrequently, causes the monkeys to watch the animal and perhaps move slowly towards a tree. The third faint alarm call is a response to unfamiliar humans and results in the vervets quietly moving towards a bush or the top of a tree. However, the postulated functions of these three fainter alarm calls remain unproven because they have not yet been tested by playback experiments.

Vervets also utter grunt-like calls when interacting with each other. Even to scientists who have spent years listening to vervets, all these social grunts sound the same. When the grunts are recorded and displayed as a frequency spectrum on the screen of a sound-analysing instrument, they look the same. Only when the spectra were measured in elaborate detail could Cheney and Seyfarth detect (sometimes but not always!) average differences between the grunts given in four social contexts: when a monkey approaches a dominant monkey, when it approaches a subordinate monkey, when it watches another monkey, or when it sees a rival troop.

Broadcasts of grunts recorded in these four different contexts caused monkeys to behave in subtly different ways. For example, they looked towards the loudspeaker if the grunt had originally been recorded in the 'approach dominant monkey' context, while they looked in the direction towards which the call was being broadcast if it had originally been recorded in the 'see rival troop' context. Further observations of the monkeys under natural conditions showed that the natural calls had also been eliciting this subtly different behaviour.

Vervets are much more finely attuned than we are to their calls. Merely listening to and watching vervets, without recording and playing back their calls, gave no hint that they had at least four distinct grunts – and may have many more. As Seyfarth writes, 'Watching vervets grunt to each other is really very much like watching humans engaged in conversation without being able to hear what they're saying. There aren't any obvious reactions or replies to grunts, so the whole system seems very mysterious – mysterious, that is, until you start doing playbacks.' These discoveries illustrate how easy it is to underestimate the size of an animal's vocal repertoire.

*

The vervets of Amboseli have *at least* ten putative 'words': their words for 'leopard', 'eagle', 'snake', 'baboon', 'other predatory mammal', 'unfamiliar human', 'dominant monkey', 'subordinate monkey', 'watch other monkey', and 'see rival troop'. However, virtually every claim of animal behaviour suggesting elements of human language is greeted with scepticism by many scientists, who are convinced of the linguistic gulf separating us from animals. Such sceptics consider it simpler to assume that humans are unique, and that the burden of proof should be borne by anyone who thinks otherwise. Any claim of language-like elements for animals is considered a more complicated hypothesis, to be dismissed as unnecessary in the absence of positive proof. Yet the alternative hypotheses by which the sceptics instead attempt to explain animal behaviour sometimes strike me as more complicated than the simple, and often plausible, explanation that humans are not unique.

It seems a modest claim to propose that the different calls which vervets give in response to leopards, eagles, and snakes actually refer to these animals or are intended as communications to other monkeys. However, sceptics were disposed to believe that only humans could emit voluntary signals referring to external objects or events. The sceptics proposed that the vervet alarm calls were merely an involuntary expression of the monkey's emotional state ('I'm scared out of my wits!') or of its intent ('I'm going to run up a tree'). After all, those explanations apply to some of our own 'calls'. If I saw a leopard coming at me, I too might emit a reflex scream even though there was no one around with whom to communicate. We grunt as a reflex when we throw ourselves into some physical activities, such as lifting a heavy object.

Suppose that zoologists from an advanced civilization in outer space observed me to give a trisyllabic scream, 'argh, leopard', and to climb a tree, when I saw a leopard. The zoologists might well doubt that my lowly species could express anything beyond grunts of emotion or intent – certainly not symbolic communications. To test their hypothesis, the zoologists would resort to experiments and detailed observations. If I screamed regardless of whether any other human was in earshot, that would support the theory of a mere expression of emotion or intent. If I screamed only in the presence of another person, and only when approached by a leopard but not by a lion, that would suggest a communication with a specific external referent. And if I gave the scream to my son but remained silent when I saw the leopard stalk a man with whom I had frequently been seen to fight, the visiting zoologist would feel certain that a purposeful communication was involved.

Similar observations convinced earthling zoologists of the communicative role of vervet alarm calls. A solitary vervet chased by a leopard for nearly an hour remained silent throughout the whole ordeal. Mother vervets give more alarm calls when accompanied by their own offspring than by unrelated monkeys. Vervets occasionally give the 'leopard call' when no leopard is present but when their troop is fighting with another troop and losing the fight. The fake alarm sends all combatants scrambling for the nearest tree and thereby serves as a deceptive 'time out'. The call is clearly a voluntary communication, not an automatic expression of fear at the sight of a leopard. Nor is the call a mere reflex grunt given in the act of climbing a tree, since a calling monkey may either climb a tree, jump out of a tree, or do nothing, depending on the circumstances.

The supposition that the call has a well-defined external referent is especially well illustrated by the 'eagle call'. Among large, broad-winged, soaring hawks, vervets usually respond with the eagle call to the martial eagle and the crowned eagle, their two most dangerous avian predators. They usually do not respond to the tawny eagle, and almost never to the black-chested snake eagle and white-backed vulture, which do not prey on vervets. Seen from below, black-chested snake eagles look rather similar to martial eagles in their shared pale underparts, banded tail, and black head and throat. Hence vervets rate as good bird-watchers. Their lives depend on it!

Vervet alarm calls are not an involuntary expression of either fear or intent. They have an external referent that may be quite exact. They are finely targeted communications which are more likely to be given honestly if the caller cares about the listener, and which may also be given dishonestly to enemies.

Sceptics dispute proposed analogies between animal sounds and human speech on the additional grounds that human speech is learned, but that many animals are born with the instinctive ability to utter the sounds characteristic of their species. However, young vervets appear to learn how to utter and respond to sounds appropriately, just as human infants. The grunts of an infant vervet sound different from those of an adult. 'Pronunciation' gradually improves with age until it becomes virtually adult at about the age of two years, somewhat less than half the age for vervet puberty. That is like human children attaining adult pronunciation at the age of five years; my sons, who are almost four years old, are still sometimes difficult to understand. Infant vervets do not learn to give reliably the correct response to an adult's call until the age of six or seven months. Until then, an adult's snake alarm call may send the infant jumping into a bush, the correct response to an eagle but a suicidal response to a snake. Not until the age of two years does the infant consistently emit each alarm call in the correct context. Before that age, the young vervet may call 'eagle!' not only when a martial or crowned eagle goes overhead, but also when any other bird flies over, and even when a leaf flutters down from a tree. Child psychologists refer to such behaviour in our own children as 'overgeneralizing' – as when a child greets not just dogs but also cats and pigeons with 'bow-wow'.

If vervet calls are indeed partly learned rather than entirely instinctive, one might expect vervet populations in different parts of Africa to have developed different 'dialects' for the same reason that different human populations have. That is, 'word' meanings and pronunciations would gradually change with time, but the changes would develop independently in different areas and would be transmitted by learning, leading first to different dialects and eventually to different languages. This prediction of dialect differences has yet to be tested for vervets, since all the detailed studies of their vocal communication to date have been made in one small area of Kenya. However, song dialects are well developed in some bird species whose young learn the locally correct song from adult birds that they hear around them as they grow up. In a North American songbird called the white-crowned sparrow, such dialect differences are so pronounced that experienced bird-watchers near San Francisco can pinpoint an individual sparrow's home within ten miles.

*

So far, I have loosely applied human concepts such as 'word' and 'language' into vervet vocalization. Let's now compare human vocalizations and those of subhuman primates more closely. In particular, let's ask ourselves three questions. Do vervet sounds really constitute 'words'? How large are animals' 'vocabularies'? Do any animal vocalizations involve 'grammar' and merit the term 'language'?

Firstly, on the question of words, it is clear at least that each vervet alarm call refers to a well-defined class of external dangers. That does not imply, of course, that a vervet's 'leopard call' designates the same animals to a vervet as the word 'leopard' does to a professional zoologist – namely, members of a single animal species, defined as a collection of potentially interbreeding individuals. We already know that vervets give their leopard alarm in response not just to leopards but also to two other medium-sized cat species (caracals and servals). If the 'leopard call' is a word at all, it would not mean 'leopard' but instead 'medium-sized cat that is likely to attack us, hunts in a similar way, and is best avoided by running up a tree'. However, many human words are used in a similar generic sense. For example, most of us other than ichthyologists and ardent fishermen apply the generic word 'fish' to any cold-blooded animal with fins and a backbone that swims in the water and might be worth eating.

Instead, the real question is whether the leopard call constitutes a word ('medium-sized cat that. . . etc.'), a statement ('there goes a medium-sized cat'), an exclamation ('watch out for that medium-sized cat!') or a proposition ('let's run up a tree or take other appropriate action to avoid that medium-sized cat'). At present it is not clear which of those functions the leopard call fills, or whether it fills a combination of them. Similarly, I was excited when my then one-year-old son Max said 'juice', which I proudly took to be one of his first words. To Max, though, the syllable 'juice' was not just his academically correct identification of a external referent with certain properties, but it also served as a proposition: 'Give me some juice!' Only at a later age did Max add more syllables, like 'gimme juice', to distinguish propositions from pure words. Vervets show no evidence of having reached that stage.

On the second question of extent of 'vocabulary', even the most advanced animals seem, on the basis of present knowledge, to be far behind us. The average human has a daily working vocabulary of around a thousand words; my compact desk dictionary claims to contain 142,000 words; but only ten calls have been distinguished even for vervets, the most intensively studied mammal. Animals and humans surely do differ in vocabulary size, yet the difference may not be as great as these numbers suggest. Remember how slow has been our progress in distinguishing vervet calls. Not until 1967 did anyone realize that these common animals had *any* calls with distinct meanings. The most experienced observers of vervets still cannot separate some of their calls without machine analysis, and even with machine analysis the distinctness of some of the suspected ten calls remain unproven. Obviously, vervets (and other animals) could have many other calls whose distinctness we have not yet recognized.

There is nothing surprising about our difficulties in distinguishing animal sounds, when one considers our difficulties in distinguishing human sounds. Children devote much of their time for the first several years of their lives to learning how to recognize and reproduce the distinctions in the utterances of adults around them. As adults, we continue to have difficulty distinguishing sounds in unfamiliar human languages. After four years of high-school French between the ages of twelve and sixteen, my problems with understanding spoken French are embarrassing compared to the abilities of any four-year-old French child. But French is easy compared to the Iyau language of New Guinea's Lakes Plains, in which a single vowel may have eight different meanings depending on its pitch. A slight change in pitch converts the meaning of the Iyau word meaning 'mother-in-law' into 'snake'. Naturally, it would be suicidal for an Iyau man to address his mother-in-law as 'beloved snake', and Iyau children learn infallibly to hear and reproduce pitch distinctions that for years confounded even a professional linguist devoted full-time to the study of the Iyau language. Given the problems we have ourselves with unfamiliar human languages, of course we must still be overlooking distinctions within the vervet vocabulary.

However, it is unlikely that any studies on vervets will reveal to us the limits attained by animal vocal communication, because those limits are probably reached by apes rather than by monkeys. While the sounds made by chimps and gorillas seem to our ears to be unsophisticated grunts and shrieks, so did the sounds made by vervet monkeys until they were studied carefully. Even unfamiliar human languages can sound like undifferentiated gibberish to us.

Unfortunately, vocal communication by wild chimps and other apes has never been studied by the methods applied to vervets, because of logistical problems. The width of a troop's territory is typically less than 2,000 feet for vervets but is several miles for chimps, making it far harder to carry out playback experiments with video cameras and hidden loudspeakers. These logistical problems cannot be overcome by studying groups of apes caught in the wild and held captive in conveniently-sized zoo cages, because the captives generally constitute an artificial community of individuals caught at different African locations and thrown together in a cage. As I will discuss later in this chapter, humans originally speaking different languages, when captured at different African locations and thrown together as slaves, converse in only the crudest shadow of human language, virtually without any grammar. Similarly, captive apes taken from the wild must be virtually useless for studying the degree of sophistication of a vocal community of wild apes. The problem will remain unsolved until someone works out how to do for wild chimps what Cheney and Seyfarth have done for wild vervets.

Several groups of scientists have nevertheless spent years training captive gorillas, common chimps, and pygmy chimps to understand and use artificial languages based on plastic chips of different sizes and colours, or on hand signs similar to those used by deaf people, or on consoles, like a gigantic typewriter with each key bearing a different symbol. The animals have been reported to learn the meanings of up to several hundred symbols, and a pygmy chimp has recently been reported to understand (but not to utter) a good deal of spoken English. At the least, these studies of trained apes reveal that they possess the intellectual capabilities for mastering large vocabularies, begging the obvious question of whether they have evolved such vocabularies in the wild.

It is suspicious that wild gorilla troops may be seen sitting together for a long time, grunting back and forth in seemingly undifferentiated gibberish, until suddenly all the gorillas get up at the same time and head off in the same direction. One wonders whether there really was a transaction concealed within that gibberish. Because the anatomy of apes' vocal tracts restricts their ability to produce the variety of vowels and consonants that we can, the vocabulary of wild apes is unlikely to be anywhere as large as our own. Nevertheless, I would be surprised if wild chimp and gorilla vocabularies did *not* eclipse those reported for vervets and comprise dozens of 'words', possibly including names for individual animals. In this exciting field where new knowledge is being rapidly acquired, we should keep an open mind on the exact size of the vocabulary gap between apes and humans.

The last unanswered question concerns whether animal vocal communication involves anything that could be considered grammar or syntax. Humans do not only have vocabularies of thousands of words with different meanings. We also combine those words and vary their forms in ways prescribed by grammatical rules that determine the meaning of the word combinations. Grammar thereby allows us to construct a potentially infinite number of sentences from a finite number of words. To appreciate this point, consider the different meaning of the following two sentences, composed of the same words and endings but with different word order, which constitutes one set of the grammatical rules that specify sentence meaning in the English language:

'Your hungry dog bit my old mother's leg.'

or

'My hungry mother bit your old dog's leg.'

If human language did not involve grammatical rules, those two sentences would have exactly the same meaning. Most linguists would not dignify an animal's system of vocal communication with the name of language, no matter how large its vocabulary, unless it also involved grammatical rules.

No hint of syntax has been discovered in the studies of vervets to date. Most of their grunts and alarm calls are single utterances. When a vervet gives a sequence of two or more utterances, all analysed cases have proved to consist of the same utterance repeated, as has also been the case when one vervet has been recorded responding to another vervet's call. Capuchin monkeys and gibbons do have calls of several elements used only in certain combinations or sequences, but the meanings of these combinations remain to be deciphered (by us humans, that is).

I doubt that any student of primate vocalizations expects even wild chimps to have evolved a grammar remotely approaching the complexity of human grammar, complete with prepositions, verb tenses, and interrogative particles. However, it remains for the present an open question whether any animal has evolved syntax. The necessary studies on the wild animals most likely to use grammar – pygmy or common chimps – simply have not yet been attempted.

In short, while the gulf between animal and human vocal communication is surely large, scientists are rapidly gaining understanding of the causeway that evolved over that gulf from the animal side. Now let's trace the bridge from the human side. We have already discovered complex animal 'languages'; do any truly primitive human languages still exist?

*

To help us recognize what a primitive human language might sound like if there were any, let's remind ourselves of the ways in which normal human language differs from vervet vocalizations. One difference is that of grammar. Humans, but not vervets, possess grammar, meaning the variations in word order, prefixes, suffixes, and changes in word roots (such as 'they', 'them', 'their') that modulate the sense of the roots. A second difference is that vervet vocalizations, if they constitute words at all, stand only for things that one can point to or act out. One could try to argue that vervet calls do include the equivalents of nouns ('eagle') and verbs or verb phrases ('watch out for the eagle'). Our words clearly include both nouns and verbs that are distinct from each other, as well as adjectives. Those three parts of a speech referring to specific objects, acts, or qualities are termed lexical items. But up to half of the words in typical human speech are purely grammatical items, with no referent that one can point to.

These grammatical words include our prepositions, conjunctions, articles, and auxiliary verbs (words like 'can', 'may', 'do', and 'should'). It is much harder to understand how grammatical items could evolve than it is for lexical items. Given someone who understands no English, you can point to your nose to explain what that noun means. Apes might similarly come to agree on the meanings of grunts functioning as nouns, verbs, or adjectives. How, though, do you explain the meaning of 'by', 'because', 'the', and 'did' to someone who understands no English? How could apes have stumbled on such grammatical terms?

Yet another difference between human and vervet vocalizations is that ours possess a hierarchial structure, such that a modest number of items at each level creates a larger number of items at the next higher level. Our language uses many different syllables, all based on the same set of a few dozen sounds. We assemble those syllables into thousands of words. Those words are not merely strung haphazardly together but are organized into phrases, such as prepositional phrases. Those phrases in turn interlock to form a potentially infinite number of sentences. In contrast, vervet calls cannot be resolved into modular elements and lack even a single stage of hierarchical organization.

As children, we master all of this complex structure of human language without ever learning the explicit rules governing it. We are not forced to formulate the rules unless we study our own language in school or learn a foreign language from books. So complex is our language's structure that many of the underlying rules currently postulated by professional linguists have been proposed only in recent decades. This gulf between human language and animal vocalizations explains why most linguists never discuss how human language might have evolved from animal precursors. They instead regard that question as unanswerable and therefore unworthy even of speculation.

*

The earliest written languages of 5,000 years ago were as complex as those of today. Human language must have achieved its modern complexity long before that. Can we at least recognize linguistic missing links by searching for primitive peoples with simple languages that might represent early stages of language evolution? After all, some tribes of hunter-gatherers retain stone tools as simple as those that characterized the whole world tens of thousands of years ago. Nineteenth-century travel books abound with tales of backward tribes who supposedly used only a few hundred words or who lacked articulated sounds, were reduced to saying 'ugh', and depended on gestures for their communications. That was Darwin's first impression of the speech of the Indians in Tierra del Fuego. But all such tales proved to be pure myth. Darwin and other western travellers merely found it as hard to distinguish the unfamiliar sounds of non-western languages as non-westerners found English sounds, or as zoologists find the sounds of vervet monkeys.

Actually, it turns out that there is no correlation between linguistic and social complexity. Technologically primitive people do not speak primitive languages, as I discovered on my first day among the Foré people in the New Guinea highlands. Foré grammar proved deliciously complex, with postpositions similar to those of the Finnish language, dual as well as singular and plural forms similar to those of Slovenian, and verb tenses and phrase construction unlike any language I had encountered previously. I have already mentioned the eight vowel tones of New Guinea's Iyau people, whose sound distinctions proved imperceptibly subtle to professional linguists for years. Nor could we reverse

Darwin's prejudice by claiming an *inverse* correlation between linguistic and social complexity, citing the advanced civilizations of China and England, whose languages are simple in the sense of having little or no word inflection (verb conjugations and noun declensions). French verbs are much more highly inflected than are modern English verbs (*nous aimons, vous aimez, ils aiment*, etc.), yet the French consider themselves the most highly civilized people.

Thus, while some peoples in the modern world retained primitive tools, none retained primitive languages. Furthermore, Cro-Magnon archaeological sites contain lots of preserved tools but no preserved words. The absence of such linguistic missing links deprives us of what might have been our best evidence about human language origins. We are forced to try more indirect approaches.

*

One indirect approach is to ask whether some people, deprived of the opportunity to hear any of our fully evolved, modern languages, ever spontaneously invented a primitive language. According to the Greek historian Herodotus, the Egyptian king Psammeticus intentionally carried out such an experiment in the hope of identifying the world's oldest language. The king assigned two newborn infants to a solitary shepherd to rear in strict silence, with instructions to listen for their first words. The shepherd duly reported that both children, after mouthing nothing but meaningless babble until the age of two, ran up to him and began constantly repeating the word *becos*. Since that word meant 'bread' in the Phrygian language then spoken in central Turkey, Psammeticus supposedly conceded that the Phrygians were the most ancient people.

Unfortunately, Herodotus's brief account of Psammeticus's experiment fails to convince sceptics that it was carried out as rigorously as described. It illustrates why some scholars prefer to honour Herodotus as the Father of Lies, rather than as the Father of History. Certainly, solitary infants reared in social isolation, like the famous wolf boy of Aveyron, remain virtually speechless and do not invent or discover a language. However, a variant of the Psammeticus experiment has occurred dozens of times in the modern world. In this variant, whole populations of children heard adults around them speaking a grossly simplified and variable form of language, somewhat similar to that which normal children themselves speak at around the age of two years. The children proceeded unconsciously to evolve their own language, far advanced over vervet communication but simpler than normal human languages. The results were the new languages known as pidgins and creoles, which may provide us with models of two missing links in the evolution of normal human language.

My first experience of a creole was with the New Guinea *lingua franca* known either as Neo-Melanesian or pidgin English. (The latter name is a confusing misnomer, since Neo-Melanesian is not a pidgin but rather a creole derived from an advanced pidgin – I shall explain the difference later – and it is only one of many independently evolved languages equally misnamed as pidgin English.) Papua New Guinea boasts about 700 native languages within an area similar to that of Sweden, but no single one of those languages is spoken by more than three per cent of the population. Not surprisingly, a *lingua franca* was needed and it arose after the arrival of English-speaking traders and sailors in the early 1800s. Today, Neo-Melanesian serves in Papua New Guinea as the language not only of much conversation, but also of many schools, newspapers, radios, and parliamentary discussions. The advertisement in the appendix to this chapter (see pages 150–51) gives a sense of this newly evolved language.

When I arrived in Papua New Guinea and first heard Neo-Melanesian, I was scornful of it. It sounded like long-winded, grammarless baby-talk. On speaking a form of English according to my own notion of baby-talk, I was disturbed to discover that New Guineans did not understand me. My assumption that Neo-Melanesian words meant the same as their English cognates led to spectacular disasters, notably when I tried to apologize to a woman in her husband's presence for accidentally jostling her, only to find that Neo-Melanesian *pushim* does not mean 'push' but instead means 'have sexual intercourse with'.

Neo-Melanesian proved to be as strict as English in its grammatical rules. It was a subtle language that let one express anything sayable in English. It even let one make some distinctions that cannot be expressed in English except by means of clumsy circumlocutions. For example, the English pronoun 'we' actually lumps together two quite different concepts: 'I, plus you to whom I am speaking', and 'I, plus one or more other people, but not including you to whom I am speaking'. In Neo-Melanesian these two separate meanings are expressed by the words *yumi* and *mipela* respectively. After I have been using Neo-Melanesian for a few months and then meet an English-speaker who starts talking about 'we', I often find myself wondering, 'am I included or not in your "we"?'

Neo-Melanesian's deceptive simplicity and actual suppleness stem partly from its vocabulary, partly from its grammar. Its vocabulary is based on a modest number of core words whose meaning varies with context and becomes extended metaphorically. For instance, while Neo-Melanesian *gras* can mean English 'grass' (whence *gras bilong solwara* [salt water] means 'seaweed'), it also can mean 'hair' (whence *man i no gat gras long head bilong em* becomes 'bald man').

The derivation of Neo-Melanesian *banis bilong susu* as the word for 'bra' further illustrates the suppleness of the core vocabulary. *Banis*, derived from the English word 'fence', is the Neo-Melanesian word with that meaning, as in the expression *banis pik* for 'pigpen'. *Susu*, taken from Malay as the word for 'milk', is extended to mean 'breast' as well. That sense in turn provides the expressions for 'nipple' (*ai* [eye] *bilong susu*), 'prepubertal girl' (*i no gat susu bilong em*), 'adolescent girl' (*susu i sanap* [stand up]), and 'aging woman' (*susu i pundaun pinis* [fall down finish]). Combining these two roots, *banis bilong susu* denotes a bra as the fence to keep the breasts in, just as *banis pik* denotes pigpen as the fence to keep pigs in.

Neo-Melanesian grammar appears deceptively simple because of what it lacks or else expresses by circumlocutions. These omissions include such seemingly standard grammatical items as plural and case forms of nouns, inflectional endings of verbs, the passive voice of verbs, and most prepositions and verb tenses. Yet Neo-Melanesian has passed far beyond baby-talk and vervet sounds in many other respects, including its conjunctions and auxiliary verbs and pronouns, and its ways of expressing verb moods and aspects. It is a normal complex language in its hierarchical organization of phonemes, syllables, and words. It lends itself so well to hierarchical organization of phrases and sentences that election speeches by New Guinean politicians rival the German prose of Thomas Mann in their convoluted structure.

*

At first, I ignorantly assumed that Neo-Melanesian was a delightful aberration among the world's languages. It had obviously arisen in the 170 years since English ships started visiting New Guinea, but I supposed that it had somehow developed from baby-talk that colonists spoke to the natives they believed incapable of learning English. Only when I began working in Indonesia and learned the Indonesian language did I sense that Neo-Melanesian origins exemplified a much broader phenomenon. On the surface, Indonesian is incomprehensible to an English speaker and totally unrelated to Neo-Melanesian, because its vocabulary is largely Malay. Nevertheless, Indonesian reminded me of Neo-Melanesian in its word use and in the grammatical items that it possessed or lacked.

As it turns out, dozens of other languages resemble Neo-Melanesian and Indonesian in structure. They have arisen independently around the globe, with vocabularies variously derived largely from English, French, Dutch, Spanish, Portuguese, Malay, or Arabic. They appeared especially in and around plantations, forts, and trading posts, where populations speaking different languages came into contact and needed to communicate, but where social circumstances impeded the usual solution of each group learning the other's language. Many cases throughout the tropical Americas and Australia, and on tropical islands of the Caribbean, Pacific, and Indian Oceans, involved European colonists importing workers who came from afar and spoke many different tongues. Other European colonists set up forts or trading posts in already densely populated areas of China, Indonesia, or Africa.

Strong social barriers between the dominant colonists and the imported workers or local populations made the former unwilling, the latter unable, to learn the other's language. Usually the colonists scorned the local people, but in China the scorn was mutual: when English traders set up a post at Canton in 1664, the Chinese would no more abase thenselves by learning the foreign devils' language or teaching them Chinese than would the English learn from or teach the heathen Chinese. Even if those social barriers had not existed, the workers would have had few opportunities to learn the colonists' tongue, because workers so greatly outnumbered colonists. Conversely, the colonists would also have found it difficult to learn 'the' workers' tongue, because so many different languages were often spoken among the workers.

Out of the temporary linguistic chaos that followed the plantations' or forts' founding, simplified but stabilized new languages emerged. Consider the evolution of Neo-Melanesian as an example. After English ships began to visit Melanesian islands just east of New Guinea in about the year 1820, the English also took islanders to work on the sugar plantations of Queensland and Samoa, where workers of many language groups were thrown together. From this Babel sprung the Neo-Melanesian language, of which the vocabulary is eighty per cent English, fifteen per cent Tolai (the Melanesian group that furnished many of the workers), and the rest derived from Malay, German, and other languages. The German element was added when German traders arrived in the 1870s and took over northeastern New Guinea as a colony in 1884. Germans must often have shouted the expletive *heraus!* ('get out!') at New Guineans, thereby inspiring the Neo-Melanesian words *raus* for 'get out' and *rausim* for 'throw out', as well as metaphorical extensions such as *rausim bol* for 'castrate'.

*

Linguists distinguish two stages in the emergence of the new languages: initially, the crude languages (termed pidgins), then later the more complex ones, referred to as creoles. Pidgins arise as a second language for colonists and workers who speak differing native (first) languages and need to communicate with each other. Each group (colonists or workers) retains its native language for use within its own group; each group uses the pidgin to communicate with the other group, and in addition workers on a polyglot plantation may use pidgin to communicate with other groups of workers. An illustration of how quickly pidgins may arise is given by my own experience soon after I first arrived in Indonesia. An Indonesian worker and I were dropped together by helicopter in an uninhabited mountain range to survey birds. We had no Indonesian/English dictionary, knew nothing of each other's language, and could teach each other words only by pointing. Within a week we had evolved a crude pidgin, based solely on Indonesian nouns, to communicate about camp chores: for instance *rice fire* meant 'to cook rice', while *bird binoculars* meant 'to watch birds'.

Compared to normal languages, pidgins are greatly impoverished in their sounds, vocabulary, and syntax. A pidgin's sounds are generally only those common to the two or more native languages thrown together. For example, many New Guineans find it hard to pronounce our consonants *f* and *v*, but I and other native English speakers find it hard to pronounce the vowel tones and nasalized sounds rampant in many New Guinean languages. Such sounds became largely excluded from New Guinean pidgins and then from the Neo-Melanesian creole that developed from them. Words of early-stage pidgins consist largely of nouns, verbs, and adjectives, with few or no articles, auxiliary verbs, conjunctions, prepositions, or pronouns. As for grammar, early-stage pidgin discourse typically consists of short strings of words with little phrase construction, no regularity in word order, no subordinate clauses, and no inflectional endings on words. Together with that impoverishment, variability of speech within and between individuals is a hallmark of early-stage pidgins, which approximate an anarchic linguistic free-for-all.

Pidgins that are used only casually by adults who otherwise retain their own separate native languages persist at this rudimentary level. For example, a pidgin known as Russonorsk grew up to facilitate barter between Russian and Norwegian fishermen who encountered each other in the Arctic. That *lingua franca* persisted throughout the Nineteenth Century but never developed further, as it was used only to transact simple business during brief visits. Both those groups of fishermen spent most of their time speaking Russian or Norwegian with their compatriots. In New Guinea, on the other hand, the pidgin gradually became more regular and complex over many generations because it was used intensively on a daily basis, but most children of New Guinean workers continued to learn their parents' native languages as their first language until after the Second World War.

However, pidgins evolve rapidly into creoles when a generation of one of the groups contributing to a pidgin begins to adopt the pidgin itself as its native language. That generation then finds itself using pidgin for all social purposes, not only for discussing plantation tasks or bartering. Compared to pidgins, creoles have a larger vocabulary, much more complex grammar, and consistency within and between individuals. Creoles can express virtually any thought expressible in a normal language, whereas trying to say anything even slightly complex is a desperate struggle in pidgin. Somehow, without any equivalent of the *Académie Francaise* to lay down explicit rules, a pidgin expands and stabilizes to become a uniform and more sophisticated language.

This process of creolization is a natural experiment in language evolution that has unfolded independently dozens of times in the modern world. The sites for the experiment have ranged from mainland South America and Africa to Pacific islands; the labourers, from Africans and Portuguese to Chinese and New Guineans; the dominant colonists, from English and Spaniards to other Africans and Portuguese; and the century, from at least the Seventeenth to the Twentieth. What is striking is that the linguistic outcomes of all these independent natural experiments share so many similarities, both in what they lack and in what they possess. On the negative side, creoles are simpler than normal languages in that they usually lack conjugations of verbs for tense and person, declensions of nouns for case and number, most prepositions, distinctions between events in the past and present, and agreement of words for gender. On the positive side, creoles are advanced over pidgins in many respects: consistent word order; singular and plural pronouns for the first, second, and third persons; relative clauses; indications of the anterior tense (describing actions occurring before the time under discussion, whether or not that time is the present); and particles or auxiliary verbs preceding the main verb and indicating negation, anterior tense, conditional mood, and continuing as opposed to completed actions. Furthermore, most creoles agree in placing a sentence's subject, verb, and object in that particular order, and also agree in the order of particles or auxiliaries preceding the main verb.

The factors responsible for this remarkable convergence are still controversial among linguists. It is as if you drew a dozen cards fifty times from well-shuffled decks and almost always ended up with no hearts or diamonds, but with one king, a jack, and two aces. The interpretation I find most convincing is that of linguist Derek Bickerton, who views many of the similarities among creoles as a result of a human genetic blueprint for language.

Bickerton derived his view from his studies of creolization in Hawaii, where sugar planters imported workers from China, the Philippines,

Japan, Korea, Portugal, and Puerto Rico in the late Nineteenth Century. Out of that linguistic chaos, and following Hawaii's annexation by the US in 1898, a pidgin based on English developed into a fully fledged creole. The immigrant workers themselves retained their original native language. They also learned the pidgin that they heard, but they did not improve on it, despite its gross deficiencies as a medium of communication. That, however, posed a big problem for the immigrants' Hawaiian-born children. Even if the children were lucky enough to hear a normal language at home because both mother and father were from the same ethnic group, that normal language was useless for communicating with children and adults from other ethnic groups. Many children were less fortunate and heard nothing but pidgin even at home, when mother and father came from different ethnic groups. The children also did not have adequate opportunities to learn English because of the social barriers isolating them and their worker parents from the English-speaking plantation owners. Presented with an inconsistent and impoverished model of human language in the form of pidgin, Hawaiian labourers' children spontaneously 'expanded' pidgin into a consistent and complex creole within a generation.

In the mid-1970s Bickerton was still able to trace the history of this creolization by interviewing working-class people born in Hawaii between 1900 and 1920. Like all of us, those children soaked up language skills in their early years but then became fixed in their ways, so that their speech in their old age continued to reflect the language spoken around them in their youth. (My children too will soon be wondering why their father persists in saying 'icebox' rather than 'refrigerator', decades after the iceboxes of my parents' own childhood disappeared.) Hence the elderly adults of various ages, whom Bickerton interviewed in the 1970s, provided him with virtually frozen snapshots of various stages in Hawaii's pidgin-to-creole transition, depending on the subjects' year of birth. In that way, Bickerton was able to conclude that creolization had begun by 1900, was complete by 1920, and was accomplished by children in the process of acquiring the ability to speak.

In effect, the Hawaiian children lived out a modified version of the Psammeticus experiment. Unlike the Psammeticus children, the Hawaiian children did hear adults speaking and were able to learn words. Unlike normal children, however, the Hawaiian children heard little grammar, and what they did hear was inconsistent and rudimentary. Instead, they created their own grammar. That they did indeed create it, rather than somehow borrowing grammar from the language of Chinese labourers or English plantation owners, is clear from the many features of Hawaiian creole that differ from English or from the workers' languages. The same is true for Neo-Melanesian: its vocabulary is largely English, but its grammar includes many features absent from English.

*

I do not want to exaggerate the grammatical similarities among creoles by implying that they are all essentially the same. Creoles do vary depending on the social history surrounding creolization – especially on the initial ratio between the numbers of plantation owners (or colonists) and workers, how quickly and to what extent that ratio changed, and for how many generations the early-stage pidgin could gradually borrow more complexity from existing languages. Yet many similarities remain, particularly among those creoles that quickly arose from early-stage pidgins. How did each creole's children come so quickly to agree on a grammar, and why did the children of different creoles tend to reinvent the same grammatical features again and again?

It was not because they did it in the easiest or sole way possible to devise a language. For instance, creoles use prepositions (short words preceding nouns), as do English and some other languages, but there are other languages that dispense with prepositions in favour of postpositions following nouns, or else noun case endings. Again, creoles happen to resemble English in placing subject, verb, and object in that order, but the borrowing from English could not account for creole grammar, because creoles derived from languages with a different word order still use the subject-verb-object order.

These similarities among creoles seem likely to stem from a genetic blueprint that the human brain possesses for learning language during childhood. Such a blueprint has been widely assumed ever since the linguist Noam Chomsky argued that the structure of human language is far too complex for a child to learn within just a few years, in the absence of any hard-wired instructions. For example, at the age of two my twin sons were just beginning to use single words. As I write this paragraph a bare twenty months later, still several months short of their fourth birthday, they have already mastered most of the rules of basic English grammar that people who immigrate to English-speaking countries as adults often fail to master after decades. Even before the age of two, my children had learned to make sense of the initially incomprehensible babble of adult sound coming at them, to recognize groupings of syllables into words, and to realize which groupings constituted under-lying words despite variations of pronunciation within and between adult speakers.

Such difficulties convinced Chomsky that children learning their first language would face an impossible task unless much of language's structure were already pre-programmed into them. Hence Chomsky reasoned that we are born with a 'universal grammar' already wired into our brains to give us a spectrum of grammatical models encompassing the range of grammars in actual languages. This pre-wired universal grammar would be like a set of switches, each with various alternative positions. The switch positions would then become fixed to match the grammar of the local language that the growing child hears.

However, Bickerton goes further than Chomsky and concludes that we are pre-programmed not just to a universal grammar with adjustable switches, but to a particular set of switch settings: the settings that surface again and again in creole grammars. The pre-programmed settings can be overridden if they turn out to conflict with what a child hears in the local language around it. But if a child hears no local switch settings at all because it grows up amidst the structureless anarchy of pidgin language, the creole settings can persist.

If Bickerton is correct in that we really are pre-programmed at birth with creole settings that can be overridden by later experience, then one would expect children to learn creole-like features of their local language earlier and more easily than features conflicting with creole grammar. This reasoning might explain the notorious difficulty of English-speaking children in learning how to express negatives: they insist on creole-like double negatives such as 'Nobody don't have this'. The same reasoning could explain the difficulties of English-speaking children with word order in questions.

To pursue the latter example, English happens to be among the languages that uses the creole word order of subject, verb, and object for statements: for instance, 'I want juice'. Many languages, including creoles, preserve this word order in questions, which are merely distinguished by altered tone of voice ('You want juice?'). However, the English language does not treat questions in this way. Instead, our questions deviate from creole word order by inverting the subject and verb ('Where are you?', not 'Where you are?'), or by placing the subject between an auxiliary verb (such as 'do') and the main verb ('Do you want juice?'). My wife and I have been barraging my sons from early infancy onwards with grammatically correct English questions as well as statements. My sons quickly picked up the correct order for statements, but both of them are still persisting in the incorrect creole-like order for questions, despite the hundreds of correct examples that my wife and I utter for them every day. Today's samples from Max and Joshua include 'Where it is?', 'What that letter is?', 'What the handle can do?', and 'What you did with it?'. It is as if they are not yet ready to accept the evidence of their ears, because they are still convinced that their pre-programmed creole-like rules are correct.

I have discussed creoles as if they appeared only with the rise of colonialism in the past 500 years. In fact, the social conditions that produced modern creoles have arisen repeatedly during thousands of years of documented human history, and probably long before that. Hence some of the world's 'normal' languages may have passed through stages of creolization and gradually re-evolved a more complex grammar. The possible example closest to home is the language of these pages. There has been a long controversy among linguists over the history of the Germanic language family that includes English, and that presumably arose in the area of the Baltic Sea. As I shall discuss in Chapter Fifteen, Germanic languages belong to a wider grouping of languages termed Indo-European. All Indo-European languages clearly derived much of their vocabulary and grammar from an ancestral language known as proto-Indo-European, which may have been spoken in southern Russia 5,000 years ago and then spread west across Europe. However, the Germanic languages also include many word roots and grammatical features unique to them, and absent from all other Indo-European families. Familiar examples include the English words 'house', 'wife', and 'hand', close to the modern German words *Haus*, *Weib*, and *Hand*. The shores of the Baltic are the source of prized amber that was traded to southern Europe and Russia thousands of years ago, just as it is still traded around the world today. Could the Germanic languages have arisen as a creole, when proto-Indo-European traders settled among proto-Germanic tribes of the Baltic to buy amber in exchange for pottery, battle-axes, and horses?

*

Now let's pull together all these animal and human studies to try to form a coherent picture of how our ancestors progressed from grunts to Shakespeare's sonnets. A well-studied early stage is represented by vervet monkeys, with at least ten different calls that are under voluntary control, are used for communication, and have external referents. The calls may function as words, explanations, propositions, or as all of those things simultaneously. Scientists' difficulties in identifying those ten calls have been such that more surely await identification, but we still do not know how large the vervet vocabulary really is. We also do not know how far other animals may have progressed beyond vervets, because the vocal communications of the species most likely to have eclipsed vervets, the common and the pygmy chimp, have yet to be studied carefully in the wild. At least in the laboratory, chimps can master hundreds of symbols that we teach them, suggesting that they have the necessary intellectual equipment to master symbols of their own.

The single words of young toddlers, like 'juice' as uttered by my son Max, constitute a next stage beyond animal grunts. Like vervet calls,

Max's 'juice' may have functioned as some combination of a word, an explanation, and a proposition. But Max has made a decisive advance on vervets by assembling his 'juice' word from the smaller units of vowels and consonants, thereby scaling the lowest level of modular linguistic organization. A few dozen such phonetic units can be reshuffled to produce a very large number of words, such as the 142,000 words in my English desk dictionary. That principle of modular organization lets us recognize far more distinctions than can vervets. For example, they name only six types of animals, whereas we name nearly two million.

A further step towards Shakespeare is exemplified by two-year-old children, who in all human societies proceed spontaneously from a one-word to a two-word stage and then to a multi-word one. But those multi-word utterances are still mere word strings with little grammar, and their words are still nouns, verbs, and adjectives with concrete referents. As Bickerton points out, those word strings are rather like the pidgin languages that human adults spontaneously reinvent when necessary. They also resemble the strings of symbols produced by captive apes whom we have instructed in the use of those symbols.

From pidgins to creoles, or from the word strings of two-year-olds to the complete sentences of four-year-olds, is another giant step. In that step were added words lacking external referents and serving purely grammatical functions; elements of grammar such as word order, prefixes and suffixes, and word root variation; and more levels of hierarchical organization to produce phrases and sentences. Perhaps that step is what triggered the Great Leap Forward discussed in Chapter Two. Nevertheless, creole languages reinvented in modern times still give us clues to how these advances arose, through the creoles' circumlocutions to express prepositions and other grammatical elements. As another illustration of how this might have happened, my Indonesian colleague and I were just in the process of reinventing prepositions when the helicopter picked us up and terminated our experiment in pidgin evolution. We had begun to assemble word strings that functioned as locative prepositional phrases but were still composed solely of nouns with concrete referents – strings such as 'spoon top plate' and 'spoon bottom plate', to mean that the spoon was on or under the plate. Many virtual prepositions in Neo-Melanesian, Indonesian, and other creoles are similarly constructed.

If you compare the Neo-Melanesian advertisement on pages 150–51 with a Shakespearean sonnet, you might conclude that a huge gap still exists. In fact, I would argue that, with an advertisement like *Kam insait long stua bilong mipela*, we have come 99.9% of the way from vervet calls to Shakespeare. Creoles already constitute expressive complex languages. For example, Indonesian, which arose as a creole to become the language of conversation and government for the world's fifth most populous country, is also a vehicle for serious literature.

Animal communication and human language once seemed to be separated by an unbridgeable gulf. Now, we have identified not only parts of the bridges starting from both shores, but also a series of islands and bridge segments spaced across the gulf. We are beginning to understand in broad outline how the most unique and important attribute that distinguishes us from animals arose from animal precursors.

Appendix

NEO-MELANESIAN, IN ONE EASY LESSON

Try to understand this Neo-Melanesian advertisement for a department store:

> *Kam insait long stua bilong mipela – stua bilong salim olgeta samting – mipela i-ken helpim yu long kisim wanem samting yu laikim bikpela na liklik long gutpela prais. I-gat gutpela kain kago long baiim na i-gat stap long helpim yu na lukautim yu long taim yu kam insait long dispela stua.*

If some of the words look strangely familiar but do not quite make sense, read the advertisement aloud to yourself, concentrate on the sounds, and ignore the strange spelling. As the next step, here is the same advertisement rewritten with English spelling:

> *Come inside long store belong me-fellow – store belong sellim altogether something – me-fellow can helpim you long catchim what-name something you likim, big-fellow na liklik, long good-fellow price. He-got good-fellow kind cargo long buyim, na he-got staff long helpim you na lookoutim you long time you come inside long this-fellow store.*

A few explanations should help you make sense of the remaining strangenesses. Almost all the words in this sample of Neo-Melanesian are derived from English, except for the word *liklik* for 'little', derived from a New Guinean language (Tolai). Neo-Melanesian has only two pure prepositions: *bilong*, meaning 'of' or 'in order to', and *long*, meaning almost any other English preposition. The English consonant *f* becomes *p* in Neo-Melanesian, as in *stap* for 'staff', and *pela* for 'fellow'. The suffix *-pela* is added to monosyllabic adjectives (hence *gutpela* for 'good', *bikpela* for 'big'), and also makes the singular pronouns 'me' and 'you' into plural ones (for 'we' and 'you' -plural). *Na* means 'and'. So the advertisement means:

Come into our store – a store for selling everything – we can help you get whatever you want, big and small, at a good price. There are good types of goods for sale, and staff to help you and look after you when you visit the store.

NINE

ANIMAL ORIGINS OF ART

Art is often viewed as lacking animal precursors, cultivated solely for pleasure, and serving no biological function. In fact, even art experts have been unable to distinguish human artworks from those produced by apes and elephants. Like the bower decorations of bowerbirds, human art may have evolved as a signal of status and thereby helped us to pass on our genes.

*

Georgia O'Keeffe's drawings were slow to win recognition for her, but Siri's drawings brought her acclaim as soon as other knowledgeable artists saw them. 'They had a kind of flair and decisiveness and originality' – that was the first reaction of the famous abstract-expressionist painter Willem de Kooning. Jerome Witkin, an authority on abstract expressionism who teaches art at Syracuse University, was even more effusive: 'These drawings are very lyrical, very, very beautiful. They are so positive and affirmative and tense, the energy is so compact and controlled, it's just incredible. . . . This drawing is so graceful, so delicate . . . This drawing indicates a grasp of the essential mark that makes the emotion.'

Witkin applauded Siri's balance of positive and negative space, and her placement and orientation of images. Having seen the drawings but knowing nothing about who made them, he guessed correctly that the artist was female and interested in Asian calligraphy. But Witkin did not guess that Siri was 8 feet tall and weighed 4 tons. She was an Asian elephant who drew by holding a pencil in her trunk.

de Kooning's response to being told Siri's identity was, 'That's a damned talented elephant.' Actually, Siri was not extraordinary by elephant standards. Wild elephants often use their trunks to make drawing motions in the dust, while captive elephants often spontaneously scratch marks on the ground with a stick or stone. Hanging in many doctors' and lawyers' offices are paintings by an elephant named Carol, who sold dozens of her works at prices of up to 500 dollars.

Supposedly, art is the noblest distinctively human attribute – one that sets us apart from animals at least as sharply as does spoken language, by differing in basic ways from anything that any animal does. Art ranks as even nobler than language, since language is really 'just' a highly sophisticated advance on animal communication systems, serves an obvious biological function in helping us to survive, and obviously developed from the sounds made by other primates. In contrast, art serves no such transparent function, and its origins are considered a sublime mystery. But it is clear that elephant art could have implications for our own. At the minimum, it is a similar physical activity resulting in products that even experts could not distinguish from human products accepted as constituting art. Of course, there are also huge differences between Siri's art and ours, not least of which is that Siri was not trying to communicate her message to other elephants. Nevertheless, we cannot just dismiss her art as a quirk of one individual beast.

In this chapter I shall go beyond elephants to examine art-like activities of some other animals. I believe that the comparisons will help us understand the original functions of human art. Thus, although we usually consider art to be the antithesis of science, there may really be a science of art.

*

To appreciate that our art must have some animal precursors, recall from Chapter One that it is only about seven million years since we branched off from our closest living relatives, the chimpanzees. Seven million years sound like a lot on the scale of a human lifetime, but they are barely one per cent of the history of complex life on Earth. We still share over ninety-eight per cent of our genes with chimps. Art and those other features that we consider uniquely human must be due to just a tiny fraction of our genes. They must have arisen only a few moments ago on the evolutionary time clock.

Modern studies of animal behaviour have been shrinking the list of features once considered uniquely human, so much so that most differences between us and so-called animals now appear to be only matters of degree. For example, I described in Chapter Eight how vervet monkeys have a rudimentary language. You may not have considered vampire bats allied with us in nobility, but they prove to practise reciprocal altruism regularly (towards other vampire bats, of course). Among our darker qualities, murder has now been documented in innumerable animal species, genocide in wolves and chimps, rape in ducks and orangutans, and organized warfare and slave raids in ants.

As absolute distinctions between us and animals, these discoveries leave us few characteristics besides art, which we managed to dispense with for the first 6,960,000 of the seven million years since we diverged from chimps. Perhaps the earliest art forms were wood carving and body painting, but we would not know because they are not preserved. The first preserved, even questionable, hints of human art consist of some flower remains around Neanderthal skeletons, and some scratches on animal bones at Neanderthal campsites. However, their interpretation as having been arranged or scratched intentionally is in doubt. Not until the Cro-Magnons, beginning around 40,000 years ago, do we have un-equivocal evidence for art surviving in the form of the famous cave paintings, statues, necklaces, and musical instruments.

If we are going to claim that true art is unique to humans, then in what ways do we claim that it differs from superficially similar productions of animals, like bird-songs? Three supposed distinctions are often put forward: that human art is non-utilitarian, that it is only for aesthetic pleasure, and that it is transmitted by learning rather than through our genes. Let's scrutinize these claims more closely.

Firstly, as Oscar Wilde said, 'All art is quite useless.' The implicit meaning a biologist sees behind this quip is that art is non-utilitarian in a narrow sense employed within the fields of animal behaviour and evolutionary biology. That is, human art does not help us to survive or to pass on our genes, which are the readily discernible functions of most animal behaviour. Of course, most human art is utilitarian in the broader sense that the artist thereby communicates something to fellow humans, but transmitting one's thoughts to the next generation is not the same thing as transmitting one's genes. In contrast, bird-song serves the obvious functions of wooing a mate, defending a territory, and thereby transmitting genes.

Regarding the second claim that human art is instead motivated by aesthetic pleasure, Webster's dictionary defines art as 'the making or doing of things that have form or beauty'. While we cannot ask mockingbirds and nightingales if they similarly enjoy the form or beauty of their songs, it is suspicious that they sing mainly during the breeding season. Hence they are probably not singing just for aesthetic pleasure.

As for human art's third claimed distinction, each human group has a distinctive art style, and the knowledge of how to make and enjoy that particular style is learned, not inherited. For example, it is easy to distinguish typical songs being sung today in Tokyo and in Paris. But those stylistic differences are not hard-wired in our genes, as are the differences, say, in the eyes of Parisians and Japanese. Parisians and Japanese can and often do visit each other's cities and learn each other's songs. In contrast, many species of birds (the so-called nonpasserine birds) inherit the knowledge of how to produce and respond to the particular song of their species. Each of those birds would produce the right song even if it had never heard it, and even if it had heard only the songs of other species. It is as if a French baby adopted by Japanese parents, flown in infancy to Tokyo, and educated there began spontaneously to sing the 'Marseillaise'.

At this point, we may seem to be light-years removed from elephant art. Elephants are not even closely related to us evolutionarily. Much more relevant to us are the artworks that were produced by two captive chimpanzees named Congo and Betsy, a gorilla called Sophie, an orangutan named Alexander, and a monkey named Pablo. These primates variously mastered the media of brush or finger-painting and pencil, chalk, or crayon drawing. Congo did up to thirty-three paintings in one day, apparently for his own satisfaction, as he did not show his work to other chimps and threw a tantrum when his pencil was taken away. For human artists, the ultimate proof of success is a one-man show, but Congo and Betsy were honoured by a two-chimp show of their paintings in 1957 at London's Institute of Contemporary Art. The following year, Congo had a one-chimp show at London's Royal Festival Hall. What is more, almost all of the paintings on exhibit at those chimp shows sold (to human buyers); plenty of human artists cannot make that boast. Still other ape paintings were surreptitiously entered into exhibits by human artists and were enthusiastically acclaimed by unsuspecting art critics for their dynamism, rhythm, and sense of balance.

Equally unsuspecting were child psychologists who were given paintings by chimps from the Baltimore Zoo and were asked to diagnose the painters' problems. The psychologists guessed that a painting by a three-year-old male chimp was instead by an aggressive seven- or eight-year-old boy with paranoid tendencies. Two paintings by the same one-year-old female chimp were attributed to different ten-year-old girls, one painting indicating a belligerent girl of the schizoid type, the other a paranoid girl with strong father identification. It is a tribute to the psychologists' insight that they intuited the artist's sex correctly in each case; they were only wrong about the artist's species.

These paintings by our closest relatives do start to blur the distinction between human art and animal activities. Like human paintings, the ape paintings served no narrow utilitarian function of transmitting genes, and were instead just produced for satisfaction. One could object that the ape artists, like the elephant Siri, made their pictures just for their own satisfaction, while most human artists aim to communicate to other humans. The apes did not even keep their paintings to enjoy themselves but simply discarded them. Yet that objection does not strike me as fatal, since the simplest human art (doodling) is also regularly discarded, and since one of the best pieces of art I own is a wood statue carved by a New

Guinea villager who discarded it under his house after carving it. Even some human art that later became famous was created by artists for their private satisfaction: the composer Charles Ives published little of his music, and Franz Kafka not only did not publish his three great novels but even forbade his executor to do so. (Fortunately, the executor disobeyed, thereby forcing Kafka's novels to take on a communicative function posthumously.)

However, there is a more serious objection against claiming a parallel between ape art and human art. Ape painting is just an unnatural activity of captive animals. One might insist that, since it is not natural behaviour, it could not illuminate the animal origins of art. Let us therefore turn now to some undeniably natural and illuminating behaviour: bowerbirds' building of bowers, the most elaborate structures built and decorated by any animal species other than humans.

*

If I had not already heard of bowers, I would have mistaken the first one I saw for something man-made, as did nineteenth-century explorers in New Guinea. I had set out that morning from a New Guinea village, with its circular huts, neat rows of flowers, people wearing decorative beads, and little bows and arrows carried by children in imitation of their fathers' larger ones. Suddenly, in the jungle, I came across a beautifully woven, circular hut 8 feet in diameter and 4 feet high, with a doorway large enough for a child to enter and sit inside. In front of the hut was a lawn of green moss, clean of debris except for hundreds of natural objects of various colours that had obviously been placed there intentionally as decorations. They mainly consisted of flowers and fruits and leaves, but also some butterfly wings and fungi. Objects of similar colour were grouped together, such as red fruits next to a group of red leaves. The largest decorations were a tall pile of black fungi facing the door, with another pile of orange fungi a few yards further from the door. All blue objects were grouped inside the hut, red ones outside, and yellow, purple, black, and a few green ones in other locations.

That hut was not a child's playground. It had instead been built and decorated by an otherwise unimpressive jay-sized bird called a bower-bird, a member of a family of eighteen species confined to New Guinea and Australia. Bowers are erected by males for the sole purpose of seducing females, who then bear the sole responsibility for building the nest and rearing the young. Males are polygamous, try to mate with as many females as possible, and provide the female with nothing except sperm. Females, often in groups, cruise around the bowers and inspect all the ones in the vicinity before selecting one at which to mate. Human equivalents of such scenes are played out every night on Sunset Strip, a few miles from my home in Los Angeles.

Female bowerbirds select their bedmate by the quality of his bower, its number of decorations, and its conformity to local rules, which vary among species and populations of bowerbirds. Some populations prefer blue decorations, others red or green or grey, while some replace the hut with one or two towers, or a two-walled avenue, or a four-walled box. There are populations that paint their bowers with crushed leaves or else with oils that they excrete. These local differences in rules appear not to be hard-wired into the birds' genes. Instead, they are learned through younger birds observing older birds during the many years that it takes a bowerbird to reach adulthood. Males learn the locally correct way to decorate, while females learn those same rules for the purpose of choosing a male.

I tested the males' finickiness by moving decorations, whereupon the bower owner restored them to their original places. When I put out poker chips of various colours, the hated white chips were heaved off into the jungle, the beloved blue ones stacked inside the hut, and the red ones stacked on the lawn next to red leaves and fruits.

At first, this system strikes us as absurd. After all, what a female bowerbird is trying to do is to pick a good mate. The evolutionary winner in such a mate-selection contest is that female bowerbird who picks that male bowerbird who makes it possible for her to leave the largest number of surviving offspring. What good does it do her to pick the guy with the blue fruits?

All animals face similar problems of mate selection. I have already discussed our own problems and solutions in Chapter Five. Consider those species (such as most European and North American songbirds) whose males carve out mutually exclusive territories that each male will share with his mate. The territory contains the nest site and food resources for the female to use in rearing her young. Hence a part of the female's task is to assess the quality of each male's territory. Alternatively, suppose that the male himself will assist in feeding and protecting the young, and in hunting cooperatively with the female. Then the female and the male must assess each other's parenting and hunting skills and the quality of their relationship. All these things are hard enough to assess, but it is even harder for the female to assess a male when he provides nothing but sperm and genes, as is the case with male bowerbirds. How on earth is an animal to assess a prospective mate's genes, and what have blue fruits to do with good genes?

Animals do not have the time to produce ten offspring with each of many prospective mates, and to compare the outcomes (the eventual number of surviving offspring). Instead, they have to resort to shortcuts by relying on mating signals such as songs or ritualized displays. As I shall discuss at more length in Chapter Eleven, it is now a hotly contested problem in animal behaviour to understand why, or even whether, those mating signals serve as veiled indicators of good genes. We have only to reflect on our own difficulties in selecting mates and in assessing the true wealth, parenting skills, and genetic quality of our various prospective partners.

In this light, reflect what it means when a female bowerbird finds a male with a good bower. She knows at once that that male is strong, since the bower he assembled weighs hundreds of times his own weight, and since he had to drag some individual decorations half his weight from dozens of yards distant. She knows that the male has the mechanical dexterity needed to weave hundreds of sticks into a hut, tower, or walls. He must have a good brain, to carry out the complex design correctly. He must have good vision and memory to search out the required hundreds of decorations in the jungle. He must be good at coping with life, to have survived to the age of perfecting all those skills. He must also be dominant over other males – since males spend much of their leisure time trying to wreck and steal from each other's bowers, only the best males end up with intact bowers and many decorations.

Thus, bower-building provides a comprehensive test of male genes. It is as if women put each of their suitors in sequence through a weight-lifting contest, sewing contest, chess tournament, eye test, and boxing tournament, and finally went to bed with the winner. By comparison with bowerbirds, our efforts to identify mates with good genes are pathetic. We grasp at external bagatelles like facial features and ear lobe lengths (Chapter Five), or like sex appeal and Porsche ownership, which tell nothing about intrinsic genetic worth. Think of all the human suffering caused by the sad truth that beautiful sexy women or handsome Porsche-owning men often prove to have miserable genes for other traits. It is no wonder that so many marriages end in divorce, as we belatedly realize how badly we chose and how flimsy our criteria were.

How did bowerbirds evolve to use art so cleverly for such important purposes? Most male birds woo females by advertising their colourful bodies, songs, displays, or offerings of food, as dim indicators of good genes. Males of two groups of birds of paradise in New Guinea go one step further by clearing areas on the jungle floor, as bowerbirds do, to enhance their displays and show off their fancy plumage. Males of one of those birds of paradise have gone still further by decorating their cleared areas with objects useful to a nesting female: pieces of snakeskin to line her nest, pieces of chalk or mammal faeces to eat for their minerals, and fruits to eat for their calories. Finally, bowerbirds have learned that decorative objects useless in themselves may nevertheless be useful indicators of good genes, if the objects are ones that were difficult to acquire and keep.

We can easily relate to that concept. Just think of all those advertisements showing a handsome man presenting a diamond ring to a seemingly fertile young woman. You cannot eat a diamond ring, but a woman knows that the gift of such a ring tells far more about the resources that her suitor commands (and might devote to her offspring and herself) than a gift of a box of chocolates would tell. Yes, chocolates provide a few useful calories, but they are quickly gone and any idiot can afford to buy them. In contrast, the man who can afford that inedible diamond ring has money to support the woman and her kids, and also has whatever genes (for intelligence, persistence, energy, etc.) that it took to acquire or hold on to the money.

Comparisons of different bowerbird species and their bowers show that male bowerbirds achieve through bowers what other birds achieve through bright plumage. Bowerbird species differ in the conspicuousness of adult male plumage. For example, males of the five species that build towers or huts sport brilliant yellow-orange crests, whose lengths vary among the species from 4 inches to nothing at all. The shorter the crest, the bigger the bower, and the more numerous and diverse its decorations. It makes sense that a male whose manly ornament is reduced to a runty 2 inches should go to great lengths to compensate in other ways.

Thus, in the course of bowerbird evolution the less resplendent males have lured the female's attention from ornaments that are permanent parts of the male's body to ornaments that the male gathers. Whereas sexual selection in most species has produced differences between males and females in their bodily ornaments (Chapter Six), in bowerbirds it has shifted towards causing males to emphasize collected ornaments separate from their bodies. From this perspective, bowerbirds are rather human. We, too, rarely court (or at least rarely initiate courtship) by displaying the beauties of our unadorned naked bodies. Instead, we swathe ourselves in coloured cloths, spray or daub ourselves with perfumes and paints and powders, and augment our beauty with decorations ranging from jewels to sports cars. The parallel between bowerbirds and humans may be even closer if, as friends of mine who are into sports cars assure me, duller young men tend to decorate themselves with fancier sports cars.

*

Now let's re-examine, in the light of bowerbirds, those three criteria supposedly separating human art from any animal production. Both bower styles and our art styles are learned rather than inherited, so that there is no difference by the third criterion. As for the second criterion (doing it for aesthetic pleasure), it is unanswerable. We cannot ask bowerbirds whether they get pleasure out of their art, and I suspect that many humans who claim to do so are just putting on cultural affectations. That leaves only the first criterion: Oscar Wilde's assertion that art is useless, in a narrow biological sense. His statement is definitely untrue of bower art, which serves a sexual function. But it is absurd to pretend any longer that our own art lacks biological functions. Instead, there are several ways in which art helps us to survive and to pass on our genes.

Firstly, art often brings direct sexual benefits to its owner. It is not just a joke that men bent on seduction invite a woman to view their etchings. In real life, dance and music and poetry are common preludes to sex.

Secondly, and much more importantly, art brings indirect benefits to its owner. Art is a quick indicator of status, which – in human as in animal societies – is a key to acquiring food, land, and sexual partners. Yes, bowerbirds get the credit for discovering the principle that ornaments separate from one's body are more flexible status symbols than ornaments that one has to grow, but we still get credit for running away with that principle. Cro-Magnons decorated their bodies with bracelets, pendants, and ochre; New Guinea villagers today decorate theirs with shells, fur, and bird-of-paradise plumes. In addition to these art forms for bodily adornment, both Cro-Magnons and New Guinea villagers produced larger art (carvings and paintings) of world quality. We know that New Guinea art signals superiority and wealth, because birds of paradise are hard to hunt, beautiful statues take talent to make, and both are very expensive to buy. These badges of distinction are essential for marital sex in New Guinea: brides are bought, and part of the price consists of luxury art. Elsewhere as well, art is often viewed as a signal of talent, money, or both.

In a world where art is a coin of sex, it is only a small further step for some artists to be able to convert art into food. There are whole societies that support themselves by making art for trade to food-producing groups. For example, the Siassi islanders, who lived on tiny islets with little room for gardens, survived by carving beautiful bowls that other tribes coveted for bride payments and paid for in food.

The same principles hold even more strongly in the modern world. Where we once signalled our status with bird feathers on our bodies and a giant clam shell in our hut, we now do it with diamonds on our bodies and a Picasso on our wall. Where Siassi islanders sold a carved bowl for the equivalent of twenty dollars, Richard Strauss built himself a villa with the proceeds from his opera *Salome* and earned a fortune from *Der Rosenkavalier*. Nowadays we read increasingly often of art sold at auction for tens of millions of dollars, and of art theft. In short, precisely because it serves as a signal of good genes and ample resources, art can be cashed in for still more genes and resources.

So far, I have considered only the benefits that art brings to individuals, but art also helps define human groups. Humans have always formed competing groups whose survival is essential if the individuals in that group are to pass on their genes. Human history largely consists of the details of groups killing, enslaving, or expelling other groups. The winner takes the loser's land, sometimes also the loser's women, and thus the loser's opportunity to perpetuate genes. Group cohesion depends on the group's distinctive culture – especially its language, religion, and art (including stories and dances), hence art is a significant force behind group survival. Even if you have better genes than most of your fellow tribesmen, it will do you no good should your whole tribe (including you) get annihilated by some other tribe.

*

By now, you're probably protesting that I have gone completely overboard in ascribing utility to art. What about all of us who just enjoy art, without converting it to status or sex? What about all the artists who remain celibate? Are there not easier ways to seduce a sex partner than to take piano lessons for ten years? Is private satisfaction not a (the?) main reason for our art, just as for Siri and Congo?

Of course. Such expansion of behavioural patterns far beyond their original role is typical of animal species whose foraging efficiency gives them much leisure time, and who have brought their survival problems under control. Bowerbirds and birds of paradise have much leisure time, because they are big and feed on wild fruit trees out of which they can kick smaller birds. We have much leisure time because we use tools to obtain food. Animals with leisure time can channel it into more lavish signals to outdo each other. Those types of behaviour may then come to serve other purposes, such as representing information (a suggested function of Cro-Magnon cave paintings of hunted animals), relieving boredom (a real problem for captive apes and elephants), channelling neurotic energy (a problem for us as well as for them), and just providing pleasure. To maintain that art is useful is not to deny that art provides pleasure. Indeed, if we were not programmed to enjoy art, it could not serve most of its useful functions for us.

Perhaps we can now answer the question why art as we know it characterizes us, but no other animal. Since chimps paint in captivity, why do they not do so in the wild? As a solution, I suggest that wild chimps still have their day filled with problems of finding food, surviving, and fending off rival chimp groups. If wild chimps had more leisure time plus the means to manufacture paints, they would be painting. The proof of my theory is that it actually happened: we are still ninety-eight per cent chimps in our genes.

Thus, human art has come far beyond its original functions. But let us not forget that even the greatest art may still serve those primal functions. As evidence, may I quote excerpts from a letter that an English lady named Rebecca Schröter wrote to the famous musician who was her lover:

> My Dear
> I cannot close my eyes to sleep till I have returned you ten thousand thanks for the inexpressible delight I have received from your ever enchanting compositions and your incomparably charming performance of them. Be assured, my dear, that no one can have such high veneration for your most brilliant talents as I have. Indeed, my dear love, no tongue can express the gratitude I feel for the infinite pleasure your Music has given me. Let me assure you also, with heartfelt affection, that I shall ever consider the happiness of your acquaintance as one of the chief blessings of my life. I shall be happy to see you for dinner, and if you can come at three o'clock, it would give me great pleasure, as I should be particularly glad to see you, my dear, before the rest of our friends come.
>
> Most sincerely, faithfully, and affectionately yours,
> REBECCA SCHRÖTER.

This letter of surrender was addressed to the composer Franz Josef Haydn, who, at the same time as he was enjoying this doting English lover, also boasted of an Italian mistress and an Austrian wife. Haydn knew how to use great art for its original purposes.

TEN

Agriculture's Two-Edged Sword

Agriculture is conventionally regarded as the human hallmark whose adoption made the biggest material contribution to the improvement in our lifestyle over that of apes. In fact, recent archaeological studies have made it clear that agriculture brought many of the curses as well as the blessings of modern civilization.

*

To science, we owe dramatic changes in our smug self-image. Astronomy taught us that our Earth is not the centre of the universe but merely one of nine planets circling one of billions of stars. From biology, we learned that humans were not specially created by God but evolved along with tens of millions of other species. Now, archaeology is demolishing another sacred belief: that human history over the last million years has been a long tale of progress.

In particular, recent discoveries suggest that the adoption of agriculture (plus animal husbandry), supposedly our most decisive step towards a better life, was actually a milestone for the worse as well as for the better. With agriculture came not only greatly increased food production and food storage, but also the gross social and sexual inequality, the disease and despotism, that curse modern human existence. Thus, among the human cultural hallmarks being discussed in Part Three of this book, agriculture represents in its mixed blessings a halfway station between our noble traits discussed in Chapters Eight and Nine (art and language) and our unmitigated vices, discussed in many of the remaining chapters (drug abuse, genocide, and environmental destructiveness).

At first, the evidence for progress and against this revisionist interpretation will strike twentieth-century Americans and Europeans as irrefutable. We are better off in almost every respect than people of the Middle Ages, who in turn had it easier than Ice-Age cavemen, who were still better off than apes. If you are inclined to be cynical, just count our advantages. We enjoy the most abundant and varied food, the best tools and material goods, the longest and healthiest lives in human history. Most of us are safe from starvation and predators. We obtain most of our energy from oil and machines, not just from our sweat. What neo-Luddite among us would really trade the life of today for that of a medieval peasant, caveman, or ape?

For most of our history, all humans had to practise a primitive lifestyle termed 'hunting and gathering': they hunted wild animals and gathered wild plant food. That hunter-gatherer lifestyle is often characterized by anthropologists as 'nasty, brutish, and short'. Since no food is grown and little is stored, there is (according to this view) no respite from the time-consuming struggle that starts anew each day to find wild foods and avoid starving. Our escape from this misery was launched only after the end of the last Ice Age, when people began independently in different parts of the world to domesticate plants and animals (see Chapter Fourteen). The agricultural revolution gradually spread until today it is nearly universal and few tribes of hunter-gatherers survive.

From the progressivist perspective on which I was brought up, the question 'Why did almost all our hunter-gatherer ancestors adopt agriculture?' is silly. Of course they adopted it because agriculture is an efficient way to get more food for less work. Our planted crops yield far more tons per acre than do wild roots and berries. Just imagine savage hunters, exhausted from searching for nuts and chasing wild animals, suddenly gazing for the first time at a fruit-laden orchard or a pasture full of sheep. How many milliseconds do you think it took those hunters to appreciate the advantages of agriculture?

The progressivist party line goes further and credits agriculture with giving rise to art, the noblest flowering of the human spirit. Since crops can be stored, and since it takes less time to grow food in gardens than to find it in the jungle, agriculture gave us free time that hunter-gatherers never had. But free time is essential for creating art and enjoying it. Ultimately it was agriculture that, as its greatest gift, enabled us to build the Parthenon and compose the B Minor Mass.

*

Among our major cultural hallmarks, agriculture is especially recent, having begun to emerge only 10,000 years ago. None of our primate relatives practises anything remotely resembling agriculture. For the most similar animal precedents, we must turn to ants, which invented not only plant domestication but also animal domestication.

Plant domestication is practised by a group of several dozen related species of New World ants. All those ants cultivate specialized species of yeasts or fungi in gardens within the ants' nest. Rather then relying on natural soil, each gardener ant species gathers its own particular type of compost: some ants grow their crop on caterpillar faeces, others on insect corpses or dead plant material, and still others (the so-called leaf-cutter ants) on fresh leaves, stems, and flowers. For example, leaf-cutter ants clip off leaves, slice them into pieces, scrape off foreign fungi and bacteria, and take the pieces into underground nests. There the leaf fragments are crushed into moist pellets of a paste-like consistency, manured with ant saliva and faeces, and seeded with the ants' preferred species of fungus, which serves as the ants' main or sole food. In an operation the equivalent of weeding a garden, the ants continually remove any spores or threads of other fungus species that they may find growing on their leaf paste. When a queen ant goes off to found a new colony, she carries with her a starting culture of the precious fungus, just as human pioneers bring along seeds to plant.

As for animal domestication, ants obtain a concentrated sugary secretion termed honeydew from diverse insects, ranging from aphids, caterpillars, and mealybugs to scale insects, treehoppers, and spittle insects. In return for the honeydew, the ants protect their 'cows' from predators and parasites. Some aphids have evolved into virtually the insect equivalent of domestic cattle: they lack offensive structures of their own, excrete honeydew from their anus, and have a specialized anal anatomy designed to hold the droplet in place while an ant drinks it. To milk their cow and stimulate honeydew flow, ants stroke the aphid with their antennae. Some ants care for their aphids in the ants' nest during the cold winter, then in the spring carry the aphids at the correct stage of development to the correct part of the correct food plant. When aphids eventually develop wings and disperse in search of a new habitat, some lucky ones are discovered by ants and 'adopted'.

Obviously, we did not inherit plant and animal domestication directly from ants but reinvented it. Actually, 're-evolved' is a better term than 'reinvented', since our early steps towards agriculture did not consist of conscious experimentation towards an articulated goal. Instead, agriculture grew from human behaviours, and from responses or changes in plants and animals, leading unforeseen towards domestication. For example, animal domestication arose partly from people keeping captive wild animals as pets, partly from wild animals learning to profit from the proximity of people (such as wolves following human hunters to catch crippled prey). Similarly, early stages of plant domestication included people harvesting wild plants and discarding seeds, which were thereby accidentally 'planted'. The inevitable result was unconscious selection of those plant and animal species and individuals most useful to humans. Eventually, conscious selection and care followed.

*

Now let's return to the progressivist view of this agricultural revolution of ours. As I explained at the outset of this chapter, we are accustomed to assuming that the transition from the hunter-gatherer lifestyle to agriculture brought us health, longevity, security, leisure, and great art. While the case for this view *seems* overwhelming, it is hard to prove. How do you actually show that lives of people 10,000 years ago got better when they abandoned hunting for farming? Until recently, archaeologists could not test this question directly. Instead, they had to resort to indirect tests, whose results (surprisingly) failed to support the view of agriculture as an unmixed blessing.

Here is one example of such an indirect test. If agriculture had been visibly such a great idea, you would expect it to have spread quickly, once it arose in some source area. In fact, the archaeological record shows that agriculture advanced across Europe at literally a snail's pace: barely 1,000 yards per year! From its origins in the Near East around 8000 BC, agriculture crept north-westwards to reach Greece around 6000 BC and Britain and Scandinavia only 2,500 years later. That is hardly what you can call a wave of enthusiasm. As recently as the Nineteenth Century, all the Indians of California, now the fruit-basket of America, remained hunter-gatherers, even though they knew of agriculture through trade with farming Indians in Arizona. Were California Indians really blind to their self-interest? Or, could it instead be that they were smart enough to see, hidden beyond agriculture's glittering facade, the drawbacks that ensnared the rest of us?

Another indirect test of the progressivist view is to study whether surviving twentieth-century hunter-gatherers really are worse off than farmers. Scattered throughout the world, mainly in areas unsuitable for agriculture, several dozen groups of so-called 'primitive people', like the Kalahari Desert Bushmen, continued to live as hunter-gatherers in recent years. Astonishingly, it turns out that these hunters generally have leisure time, sleep a lot, and work no harder than their farming neighbours. For instance, the average time devoted each week to obtaining food has been reported to be only twelve to nineteen hours for Bushmen; how many readers of this book can boast of such a short working week? As one Bushman replied when asked why he had not emulated neighbouring tribes by adopting agriculture, 'Why should we plant, when there are so many mongongo nuts in the world?'

Of course, one's belly is not filled only by finding food; the food also has to be processed for eating, and that can take a lot of time for things like mongongo nuts. It would be a mistake to swing to the opposite extreme from the progressivist view and to regard hunter-gatherers as living a life of leisure, as some anthropologists have done. However, it would also be a mistake to view them as working much harder than farmers. Compared to my physician and lawyer friends today, and to my shopkeeper grandparents in the early Twentieth Century, hunter-gatherers really do have more free time.

While farmers concentrate on high-carbohydrate crops like rice and potatoes, the mixture of wild plants and animals in the diets of surviving hunters provides more protein and a better balance of other nutrients. The Bushmen's average daily food intake is 2,140 calories and 93 grams of protein, considerably greater than the US RDA (Recommended Daily Allowance) for people of their small size but vigorous activity. Hunters are healthy, suffer from little disease, enjoy a very diverse diet, and do not experience the periodic famines that befall farmers dependent on few crops. It is almost inconceivable for Bushmen, who utilize eighty-five edible wild plants, to die of starvation, as did about a million Irish farmers and their families during the 1840s when a blight attacked potatoes, their staple crop.

Thus, the lives of at least the surviving modern hunter-gatherers are not 'nasty, brutish, and short', even though farmers have pushed them into the world's worst real-estate. Hunters of the past, who still occupied fertile lands, could hardly have been worse off than modern hunters. However, all those modern hunter societies have been affected by farming societies for thousands of years and do not tell us about the condition of hunters before the agricultural revolution. The progressivist view is really making a claim about the distant past: that the lives of people in each part of the world got better when they switched from hunting to farming. Archaeologists can date that switch by distinguishing remains of wild plants and animals from remains of domestic ones in prehistoric rubbish dumps. How can one deduce the health of the prehistoric rubbish makers, and thereby test directly for agriculture's supposed blessings?

*

That question has become answerable only in recent years, through the newly emerging science of 'paleopathology': looking for signs of disease (the science of pathology) in remains of ancient peoples (from the Greek word *paleo* meaning 'ancient', as in paleontology). In some lucky situations, the paleopathologist has almost as much material to study as does a pathologist. For example, archaeologists in the deserts of Chile found well-preserved mummies whose medical condition at time of death could be determined by an autopsy, just as one would do on a fresh corpse in a hospital today. Faeces of long-dead Indians who lived in dry caves in Nevada remained sufficiently well-preserved to examine for hookworm and other parasites.

Usually, though, the only human remains available for paleopathologists to study are skeletons, but they still permit a surprising number of deductions about health. To begin with, a skeleton identifies its owner's sex, and his/her weight and approximate age at time of death. Thus, with enough skeletons, one can construct mortality tables like those used by life insurance companies to calculate expected lifespan and risk of death at any given age. Paleopathologists can also calculate growth rates by measuring bones of people of different ages, can examine teeth for cavities (signs of a high-carbohydrate diet) or enamel defects (signs of a poor diet in childhood), and can recognize scars that many diseases such as anaemia, tuberculosis, leprosy, and osteoarthritis leave on bones.

One straightforward example of what paleopathologists have learned from skeletons concerns historical changes in height. Many modern cases illustrate how improved childhood nutrition leads to taller adults: for instance, we stoop to pass through doorways of medieval castles built for a shorter, malnourished population. Paleopathologists studying ancient skeletons from Greece and Turkey found a striking parallel. The average height of hunter-gatherers in that region towards the end of the Ice Age was a generous 5 foot 10 inches for men, 5 foot 6 inches for women. With the adoption of agriculture, height crashed, reaching by 4000 BC a low value of only 5 foot 3 inches for men, 5 foot 1 inch for women. By classical times, heights were very slowly on the rise again, but modern Greeks and Turks have still not regained the heights of their healthy hunter-gatherer ancestors.

Another example of paleopathologists at work is the study of thousands of American Indian skeletons excavated from burial mounds in the Illinois and Ohio River valleys. Corn, first domesticated in Central America thousands of years ago, became the basis of intensive farming in those valleys around 1000 AD. Until then, Indian hunter-gatherers had skeletons 'so healthy it is somewhat discouraging to work with them', as one paleopathologist complained. With the arrival of corn, Indian skeletons suddenly became interesting to study. The number of cavities in an average adult's mouth jumped from less than one to nearly seven, and tooth loss and abscesses became rampant. Enamel defects in children's milk teeth imply that pregnant and nursing mothers were severely undernourished. Anaemia quadrupled in frequency; tuberculosis became established as an epidemic disease; half the population suffered from yaws or syphilis; and two-thirds suffered from osteoarthritis and other degenerative diseases. Mortality rates at every age increased, with the result that only one per cent of the population survived past the age of fifty, as compared to five per cent in the golden days before corn. Almost one-fifth of the whole population died between the ages of one and four, probably because weaned toddlers succumbed to malnutrition and infectious diseases. Thus, corn, usually considered among the New World's blessings, actually proved to be a public health disaster. Similar conclusions about the transition from hunting to farming emerge from studies of skeletons elsewhere in the world.

There are at least three sets of reasons to explain these findings that agriculture was bad for health. Firstly, hunter-gatherers enjoyed a varied diet with adequate amounts of protein, vitamins, and minerals, while farmers obtained most of their food from starchy crops. In effect, the farmers gained cheap calories at the cost of poor nutrition. Today just three high-carbohydrate plants – wheat, rice, and corn – provide more than fifty per cent of the calories consumed by the human species.

Secondly, because of that dependence on one or a few crops, farmers ran a greater risk of starvation if one food crop failed than did hunters. The Irish potato famine is merely one of many examples.

Finally, most of today's leading infectious diseases and parasites of mankind could not become established until after the transition to agriculture. These killers persist only in societies of crowded, mal-nourished, sedentary people constantly reinfected by each other and by their own sewage. The cholera bacterium, for example, does not survive for long outside the human body. It spreads from one victim to the next through contamination of drinking water with faeces of cholera patients. Measles dies out in small populations once it has either killed or immunized most potential hosts; only in populations numbering at least a few hundred thousand people can it maintain itself indefinitely. Such crowd epidemics could not persist in small, scattered bands of hunters who often shifted camp. Tuberculosis, leprosy, and cholera had to await the rise of farming, while smallpox, bubonic plague, and measles appeared only in the past few thousand years with the rise of cities.

*

Besides malnutrition, starvation, and epidemic diseases, farming brought another curse to humanity – class divisions. Hunter-gatherers have little or no stored food, and no concentrated food sources such as an orchard or herd of cows. Instead, they live off the wild plants and animals that they obtain each day. Everybody except for infants, the sick, and the old joins in the search for food. Thus, there can be no kings, no full-time professionals, no class of social parasites who grow fat on food seized from others.

Only in a farming population could contrasts between the disease-

ridden masses and a healthy, non-producing, elite develop. Skeletons
from Greek tombs at Mycenae around 1500 BC suggest that royals
enjoyed a better diet than commoners, since the royal skeletons were two
or three inches taller and had better teeth (on the average, one instead of
six cavities or missing teeth). Among mummies from Chilean cemeteries
around 1000 AD, the elite were distinguished not only by ornaments and
gold hairclips, but also by a four-fold lower rate of bone lesions
stemming from infectious diseases.

These signs of health differentials within local communities of farmers
in the past appear on a global scale in the modern world. To most
American and European readers, the argument that humanity could on
the average be better off as hunter-gatherers than we are today sounds
ridiculous, because most people in industrial societies today enjoy better
health than most hunter-gatherers. However, Americans and Europeans
are an elite in today's world, dependent on oil and other materials
imported from countries with large peasant populations and much lower
health standards. If you could choose between being a middle-class
American, a Bushman hunter, and a peasant farmer in Ethiopia, the first
choice would undoubtedly be the healthiest one, but the third choice
might be the least healthy.

While giving rise to class divisions for the first time, farming may also
have exacerbated sexual inequality already in existence. With the advent
of agriculture, women often became beasts of burden, were drained by
more frequent pregnancies (see below), and thus suffered poorer health.
For example, among the Chilean mummies from 1000 AD, women
exceeded men in osteoarthritis and in bone lesions from infectious
diseases. In New Guinea farming communities today I often see women
staggering under a load of vegetables and firewood while the men walk
empty-handed. In one case I offered to pay some villagers to carry
supplies from an airstrip to my mountain camp, and a group of men,
women, and children volunteered. The heaviest item was a 110-pound
bag of rice, which I lashed to a pole and assigned to a team of four men to
shoulder the pole together. When I eventually caught up with the
villagers, the men were carrying light loads, while one small woman
weighing less than the bag of rice was bent under it, supporting its weight
by a cord across her temples.

As for the claim that agriculture laid the foundations of art by
providing us with leisure time, modern hunter-gatherers have on the
average at least as much free time as do farmers. I grant that some people
in industrial and farming societies enjoy more leisure than hunter-
gatherers, at the expense of many others who support them and have far
less leisure. Farming undoubtedly made it possible to sustain full-time
craftsmen and artists, without whom we would not have such large-scale

art projects as the Sistine Chapel and Cologne Cathedral. However, the whole emphasis on leisure time as a critical factor in explaining artistic differences among human societies seems to me misguided. It is not lack of time that prevents us today from surpassing the beauty of the Parthenon. While post-agricultural technological advances did make new art forms possible and art preservation easier, great paintings and sculptures on a smaller scale than that of Cologne Cathedral were already being produced by Cro-Magnon hunter-gatherers 15,000 years ago. Great art was still being produced in modern times by hunter-gatherers such as Eskimos and Pacific Northwest Indians. In addition, when we count up the specialists whom society became able to support after the advent of agriculture, we should recall not only Michelangelo and Shakespeare but also standing armies of professional killers.

*

Thus, with the advent of agriculture an elite became healthier, but many people became worse off. Instead of the progressivist party line that we chose agriculture because it was good for us, a cynic might ask how we got trapped by agriculture despite its being such a mixed blessing.

The answer boils down to the adage, 'Might makes right.' Farming could support far more people than hunting, whether or not it also brought on the average more food per mouth. (Population densities of hunter-gatherers are typically one person or less per square mile, while densities of farmers average at least ten times higher.) Partly, this is because an acre of field planted entirely in edible crops produces far more tons of food, and allows one to feed far more mouths, than an acre of forest with scattered edible wild plants. Partly, too, it is because nomadic hunter-gatherers have to keep their children spaced at four-year intervals by infanticide and other means, since a mother must carry her toddler until it is old enough to keep up with the adults. Because sedentary farmers do not have that problem, they can and do have a child every two years. Perhaps the main reason we find it so hard to shake off the traditional view that farming was unequivocally good for us is that there is no doubt that it meant more tons of food per acre. We forget that it also meant more mouths to feed, and that health and quality of life depend on the amount of food per mouth.

As population densities of hunter-gatherers slowly rose at the end of the Ice Age, bands had to 'choose', whether consciously or unconsciously, between feeding more mouths by taking the first steps towards agriculture, or else finding ways to limit growth. Some bands adopted the former solution, unable to anticipate the evils of farming, and seduced by the transient abundance they enjoyed until population growth caught up with increased food production. Such bands outbred and then drove off or killed the bands that chose to remain hunter-gatherers, because ten malnourished farmers can still outfight one healthy hunter. It is not that hunter-gatherers abandoned their lifestyle, but that those sensible enough not to abandon it were forced out of all areas except ones that farmers did not want. Modern hunter-gatherers persisted only in scattered areas useless for agriculture, such as the Arctic, deserts, and some rainforests.

At this point it is ironic to recall the common complaint that archaeology is an expensive luxury, concerned with the remote past, and offering no lessons of present relevance. Archaeologists studying the rise of farming have reconstructed for us a stage where we made one of the most crucial decisions in human history. Forced to choose between limiting population growth or trying to increase food production, we opted for the latter and ended up with starvation, warfare, and tyranny. The same choice faces us today, with the difference that we now can learn from the past.

*

Hunter-gatherers practised the most successful and long-persistent lifestyle in the career of our species. In contrast, we are still struggling with the problems into which we descended with agriculture, and it is unclear whether we can solve them. Suppose that an archaeologist who had visited us from outer space were trying to explain human history to his fellow spacelings. The visitor might illustrate the results of his digs by a twenty-four-hour clock on which one hour of clock-time represents 100,000 years of real past time. If the history of the human race began at midnight, then we would now be almost at the end of our first day. We lived as hunter-gatherers for nearly the whole of that day, from midnight through dawn, noon, and sunset. Finally, at 11:54 pm we adopted agriculture. In retrospect, the decision was inevitable, and there is now no question of turning back. But as our second midnight approaches, will the present plight of African peasants gradually spread to engulf all of us? Or, will we somehow achieve those seductive blessings that we imagine behind agriculture's glittering facade, and that have so far eluded us except in mixed form?

ELEVEN

WHY DO WE SMOKE, DRINK, AND USE DANGEROUS DRUGS?

Self-destructive chemical abuse by humans has precedents in animal displays that are costly or dangerous to the displaying animal. Such behaviour may have originated from the dilemma that signals available to any individual lend themselves to cheating. But costly or dangerous signals carry a built-in guarantee of honesty and are thus useful – as long as their benefits outweigh their costs. Unfortunately, this old evolutionary framework has gone awry in us.

*

Chernobyl – formaldehyde in drywalls – asbestos – lead poisoning – smog – the Valdez oil spill – Love Canal – Agent Orange . . . Hardly a month goes by without our learning of yet another way in which we and our children have been exposed to toxic chemicals through the negligence of others. The public's outrage, sense of helplessness, and demand for change are growing. Why, then, do we do to ourselves that which we cannot stand for others to do to us? How do we explain the paradox that many people intentionally consume, inject, or inhale toxic chemicals, such as alcohol, cocaine, and the chemicals in tobacco smoke? Why are various forms of this wilful self-damage native to many contemporary societies, from primitive tribes to high-tech urbanites, and extending back into the past as far as we have written records?

Like the subjects of the preceding three chapters, drug abuse is also a hallmark virtually unique to the human species, albeit an evil one rather than a noble one (like language and art) or a mixed blessing (agriculture). It is not the worst of our evil hallmarks; it does not threaten the survival of civilization, as do our genocidal tendencies and our environmental destructiveness. But it is still damaging and widespread enough to beg the question of its origins.

The problem is not so much in understanding why we continue to take toxic chemicals once we have started. In part, that is because our drugs of abuse are addictive. Instead, the greater mystery is what impels us to begin at all. Evidence for the damaging or lethal effects of alcohol, cocaine, and tobacco is by now overwhelming and familiar. Only the existence of some strong countervailing motives could explain why people consume these poisons voluntarily, even eagerly. It is as if unconscious programmes were driving us to do something we know to be dangerous. What could those programmes be?

Naturally, there is no single explanation: different motives carry different weight with different people or in different societies. For instance, some people drink to overcome their inhibitions, others to deaden their feelings or drown their sorrows, still others because they like the taste of alcoholic beverages. Naturally, too, differences among human populations and social classes in their options for achieving satisfying lives largely account for geographic and class differences in chemical abuse. It is not surprising that self-destructive alcoholism is a bigger problem in high-unemployment areas of Ireland than in Southeast England, or that cocaine and heroin addiction is commoner in Harlem than in affluent suburbs. Hence it is tempting to dismiss drug abuse as a human hallmark with obvious social and cultural causes, and in no need of a search for animal precedents.

However, none of the motives that I have just mentioned goes to the heart of the paradox of our actively seeking what we know to be harmful. In this chapter I shall propose one other contributing motive which does address that paradox. It relates our chemical self-assaults to a wide range of seemingly self-destructive traits in animals, and to a general theory of animal signalling. It unifies a wide range of phenomena in our culture, from smoking and alcoholism to drug abuse. It has potential cross-cultural validity, for it may explain not just phenomena of the Western world but also some otherwise mystifying customs elsewhere, such as kerosene drinking by Indonesian kung-fu experts. I will also reach into the past and apply the theory to the seemingly bizarre practice of ceremonial enemas in ancient Mayan civilization.

*

Let me begin by relating how I arrived at this idea. One day, I was abruptly struck by the puzzle that companies manufacturing toxic chemicals for human use advertise their use explicitly. This business practice would seem a sure route to bankruptcy. Yet, while we do not tolerate advertisements for cocaine, advertisements for tobacco and alcohol are so widespread that we cease to regard their existence as puzzling. It hit me only after I had been living with New Guinea hunters in the jungle for many months, far from any advertising.

Day after day, my New Guinea friends had been asking me about Western customs, and I had come to realize through their astonished responses how senseless many of our customs are. Then the months of fieldwork ended with one of those sudden transitions that modern transportation has made possible. On 25 June I was still in the jungle, watching a brilliantly coloured male bird of paradise flap awkwardly across a clearing, dragging its 3-foot-long tail behind it. On 26 June I was sitting in a Boeing 747 jet, reading the magazines and catching up on the wonders of Western civilization.

I leafed through the first magazine. It fell open to a page with a photograph of a tough-looking man on horseback chasing cows, and the name of a brand of cigarette in large letters below. The American in me knew what the photograph was about, but part of me was still in the jungle, looking at that photo naively. Perhaps my reaction will not seem so strange to you if you try to imagine yourself completely unfamiliar with Western society, seeing the advertisement for the first time, and trying to fathom the connection between chasing cows and smoking (or not smoking) cigarettes.

The naive part of me, fresh out of the jungle, thought: such a brilliant anti-smoking ad! It is well known that smoking impairs athletic ability and causes cancer and early death. Cowboys are widely regarded as athletic and admirable. This advertisement must be a devastating new appeal by the anti-smoking forces, telling us that if we smoke that particular brand of cigarette, we will not be fit to be cowboys. What an effective message to our youth!

But then it became obvious that the advertisement had been put there by the cigarette company itself, which somehow hoped that readers would draw exactly the opposite message from the advertisement. How on earth did the company let its public relations department talk it into such a disastrous miscalculation? Surely, that advertisement would dissuade any person concerned about his/her strength and self-image from starting to smoke.

Still half immersed in the jungle, I turned to another page. There I saw a photo of a whisky bottle on a table, a man sipping presumably the bottle's contents from a glass, and an obviously fertile young woman gazing at him admiringly as if she were on the verge of sexual surrender. How can that be, I asked myself? Everyone knows that alcohol interferes with sexual function, tends to make men impotent, makes one likely to stumble, impairs judgement, and predisposes to cirrhosis of the liver and other debilitating conditions. In the immortal words of the porter in Shakespeare's *Macbeth*, 'It [drink] provokes the desire, but it takes away the performance.' A man with such handicaps should conceal them at all costs from a woman he aims to seduce. Why is the man in the photograph intentionally displaying those handicaps? Do whisky manufacturers think that pictures of this impaired individual will help sell their product? One could expect that Mothers Against Drunk Driving would be the ones producing such advertisements, and that the whisky companies would be suing to prevent publication.

Page after page of advertisements flaunted the use of cigarettes or strong alcohol, and hinted at their benefits. There were even pictures of young people smoking in the presence of attractive members of the opposite sex, as if to imply that smoking too brought sexual opportunities. Yet any non-smoker who has ever been kissed by (or tried to kiss) a smoker knows how severely the smoker's bad breath compromises his or her sex appeal. The advertisement paradoxically implied not just sexual benefits but also platonic friendships, business opportunities, vigour, health, and happiness, when the direct conclusion to be drawn from the advertisements was actually the reverse.

As the days passed and I reimmersed myself in Western civilization, I gradually stopped noticing its apparently self-defeating advertisements. I retreated into analysing my field data and wondering instead about an entirely different paradox, involving bird evolution. That paradox was what led me finally to understand one rationale behind cigarette and whisky advertisements.

*

The new paradox concerned the reason that male bird of paradise I had been watching on 25 June had evolved the impediment of a tail 3 feet long. Males of other bird of paradise species evolved other bizarre impediments, such as long plumes growing out of their eyebrows, the habit of hanging upside-down, and brilliant colours and loud calls likely to attract hawks. All those features must impair male survival, yet they also serve as the advertisements by which male birds of paradise woo female birds of paradise. Like many other biologists, I found myself wondering why male birds of paradise use such handicaps as advertisements, and why females find the handicaps attractive.

At that point I came across a remarkable paper by an Israeli biologist, Amotz Zahavi, who had conceived a novel general theory about the role of costly or self-destructive signals in animal behaviour. For example, Zahavi attempted to explain how deleterious male traits might attract a female precisely because they constitute handicaps. On reflection, I decided that Zahavi's hypothesis might apply to the birds of paradise I studied. Suddenly I realized, with growing excitement, that his theory perhaps could also be extended to explain the paradox of our use of toxic chemicals, and our touting it in advertisements.

Zahavi's theory as he proposed it concerned the broad problem of animal communication. All animals need to devise quick, easily understood signals for conveying messages to their mates, potential mates, offspring, parents, rivals, and would-be predators. For example, consider a gazelle that notices a lion stalking it. It would be in the gazelle's interests to give a signal that the lion would interpret to mean, 'I am a superior, fast gazelle! You'll never succeed in catching me, so don't waste your time and energy on trying.' Even if that gazelle really is able to outrun a lion, giving a signal that dissuades the lion from trying would save time and energy for the gazelle too.

But what signal will unequivocally tell the lion that it is hopeless? The gazelle cannot take the time to run a demonstration 100-yard dash in front of every lion that shows up. Perhaps gazelles could agree on some quick arbitrary signal that lions learn to understand, such as that pawing the ground with the left hind foot means 'I claim that I'm fast!' However, such a purely arbitrary signal opens the door to cheating; any gazelle can easily give the signal regardless of its speed. Lions will then catch on that many slow gazelles giving the signal are lying, and lions will learn to ignore the signal. It is in the interests both of lions and of fast gazelles that the signal be believable. What type of signal could convince a lion of the gazelle's honesty?

The same dilemma arises in the problem of sexual selection and mate choice that I discussed in Chapters Five, Six, and Nine. This is especially a problem of how females pick males, since females invest more in reproduction, have more to lose, and have to be choosier. Ideally, a female should pick a male for his good genes to pass on to her offspring. Since genes themselves are hard to assess, a female should look for quick indicators of good genes in a male, and a superior male should provide such indicators. In practice, male traits such as plumage, songs, and displays usually serve as indicators. Why do males 'choose' to advertise with those particular indicators, why should females trust a male's honesty and find those indicators attractive, and why do they imply good genes?

I have described the problem as if a gazelle or courting male voluntarily picks out some indicator from among many possible ones, and as if a lion or a female decides on reflection whether it is really a valid indicator of speed or good genes. In practice, of course, those 'choices' are the result of evolution and become specified by genes. Those females who select males on the basis of indicators that really denote good male genes, and those males that use unambiguous indicators of good genes for self-advertisement, tend to leave the most offspring, as do those gazelles and lions that spare themselves unnecessary chases.

As it turns out, many of the advertising signals evolved by animals pose a paradox similar to that posed by cigarette advertisements. The indicators often seem to be ones that do not suggest speed or good genes but instead constitute handicaps, expenses, or sources of risk. For example, a gazelle's signal to a lion that it sees approaching consists of a peculiar form of behaviour termed 'stotting'. Instead of running away as fast as possible, the gazelle runs slowly while repeatedly jumping high into the air with stiff-legged leaps. Why on earth should the gazelle indulge in this seemingly self-destructive display, which wastes time and energy and gives the lion a chance to catch up? Or think of the males of many animal species which sport large structures, such as a peacock's tail or a bird of paradise's plumes, that make movement difficult. Males of many more species have bright colours, loud songs, or conspicuous displays that attract predators. Why should a male advertise such an impediment, and why should a female like it? These paradoxes remain an important unsolved problem in animal behaviour today.

Zahavi's theory, which remains controversial among biologists, goes to the heart of this paradox. According to his theory, those deleterious structures and forms of behaviour constitute valid indicators that the signalling animal is being honest in its claim of superiority, precisely *because* those traits themselves impose handicaps. A signal that entails no cost lends itself to cheating, since even a slow or inferior animal can afford to give the signal. Only costly or deleterious signals are guarantees of honesty. For example, a slow gazelle that stotted at an approaching lion would seal its fate, whereas a fast gazelle could still outrun the lion after stotting. By stotting, the gazelle boasts to the lion, 'I'm so fast that I can escape you even after giving you this head start.' The lion thereby has grounds for believing in the gazelle's honesty, and both the lion and the gazelle profit by not wasting time and energy on a chase whose outcome is certain.

Similarly, as applied to males displaying towards females, Zahavi's theory reasons that any male that has managed to survive despite the handicap of a big tail or conspicuous song must have terrific genes in other respects. He has proved that he must be *especially* good at escaping predators, finding food, and resisting disease. The bigger the handicap, the more rigorous the test that he has passed. The female who selects such a male is like the medieval damsel testing her knight suitors by watching them slay dragons. When she sees a one-armed knight who can still slay a dragon, she knows that she has finally found a knight with great genes. That knight, by flaunting his handicap, is actually flaunting his superiority.

It seems to me that Zahavi's theory applies to much costly or dangerous human behaviour aimed at achieving status in general or at sexual benefits in particular. For instance, men who woo women with costly gifts and other displays of wealth are in effect saying, 'I have plenty of money to support you and children, and you can believe my boast because you see how much money I'm spending now without blanching.' People who show off expensive jewels, sports cars, or works of art gain status because the signal cannot be faked; everyone else knows what those ostentatious objects cost. American Indians of the Pacific Northwest used to seek status by competing to give away as much wealth as possible in ceremonies known as potlatch rituals. In the days before modern medicine, tattooing was not only painful but dangerous because of the risk of infection; hence tattooed people in effect were advertising two facets of their strength, resistance to disease plus tolerance of pain. Men on the Pacific island of Malekula show off by the insanely dangerous practice of building a high tower and jumping off it head first, after tying one end of some stout vines to their ankles and the other end to the top of the tower. The length of the vines is calculated to stop the braggart's plunge while his head is still a few feet above the ground. Survival guarantees that the jumper is courageous, carefully calculating, and a good builder.

Zahavi's theory can also be extended to human abuse of chemicals. Especially in adolescence and early adulthood, the age when drug abuse is most likely to begin, we are devoting much energy to asserting our status. I suggest that we share the same unconscious instinct that leads birds to indulge in dangerous displays. Ten thousand years ago, we 'displayed' by challenging a lion or a tribal enemy. Today, we do it in other ways, such as by fast driving or by consuming dangerous drugs.

The messages of our old and new displays nevertheless remain the same: I'm strong and superior. Even to take drugs only once or twice, I must be strong enough to get past the burning, choking sensation of my first puff on a cigarette, or to get past the misery of my first hangover. To do it chronically and remain alive and healthy, I must be superior (so I imagine). It is a message to our rivals, our peers, our prospective mates – and to ourselves. The smoker's kiss may taste awful, and the drinker may be impotent in bed, but he or she still hopes to impress peers or attract mates by the implicit message of superiority.

Alas, the message may be valid for birds, but for us it is a false one. Like so many animal instincts in us, this one has become maladaptive in modern human society. If you can still walk after drinking a bottle of whisky, it may prove that you have high levels of liver alcohol dehydrogenase, but it implies no superiority in other respects. If you have not developed lung cancer after chronically smoking several packs of cigarettes daily, you may have a gene for resistance to lung cancer, but that gene does not convey intelligence, business acumen, or the ability to create happiness for your spouse and children.

It is true that animals with only brief lives and courtships have no alternative except to develop quick indicators, since prospective mates don't have enough time to measure each other's real quality. But we, with our long lives and courtships and business associations, have ample time to scrutinize each other's worth. We need not rely on superficial, misleading indicators. Drug abuse is a classic instance of a once-useful instinct – the reliance on handicap signals – that has turned foul in us. It is that old instinct to which the tobacco and whisky companies are directing their clever, obscene advertisements. If we legalized cocaine, the drug lords too would soon have advertisements appealing to the same instinct. You can easily picture it: the photo of the cowboy on his horse, or the suave man and the attractive maiden, above the tastefully displayed packet of white powder.

*

Now, let's test my theory by jumping from Western Industrialized Society to the other side of the world. Drug abuse did not begin with the Industrial Revolution. Tobacco was a native American Indian crop, native alcoholic beverages are widespread in the world, and cocaine and opium came to us from other societies. The oldest preserved code of laws, that of the Babylonian king Hammurabi (1792–50 BC), already contained a section regulating drinking houses. Hence my theory, if it is valid, should apply to other societies as well. As an instance of its cross-cultural explanatory power, I shall cite a practice you may not have heard of – kung-fu kerosene drinking.

I learned of this practice when I was working in Indonesia with a wonderful young biologist named Ardy Irwanto. Ardy and I had come to like and admire each other, and to look out for each other's well-being. At one point, when we reached a troubled area and I expressed concern about dangerous people we might encounter, Ardy assured me, 'No problem, Jared. I have kung-fu grade eight.' He explained that he practised the Oriental martial art of kung-fu and had reached a high level of proficiency, such that he could single-handedly fight off a group of eight attackers. To illustrate, Ardy showed me a scar in his back stemming from an attack by eight ruffians. One had knifed him, whereupon Ardy broke the arms of two and the skull of a third and the remainder fled. I had nothing to fear in Ardy's company, he told me.

One evening at our campsite, Ardy walked with his drinking cup up to our jerrycans. As usual, we had two cans: a blue one for water, and a red one for kerosene for our pressure lamp. To my horror, I watched Ardy pour from the red jerrycan and raise the cup to his lips. Remembering an awful moment during a mountaineering expedition when I had taken a sip of kerosene by mistake and spent all the next day coughing it back up, I screamed to Ardy to stop. But he raised his hand and said calmly, 'No problem, Jared. I have kung-fu grade eight.'

Ardy explained that kung-fu gave him strength, which he and his fellow kung-fu masters tested each month by drinking a cup of kerosene. Without kung-fu, of course, kerosene would make a weaker person sick. Heaven forbid that I, Jared, for instance, should try it. But it did him, Ardy, no harm, because he had kung-fu. He calmly retired to his tent to sip his kerosene and emerged the next morning, happy and healthy as usual.

I cannot believe that kerosene did Ardy no harm. I wish that he could have found a less damaging way to make periodic tests of his preparedness. But for him and his kung-fu associates, it served as an indicator of their strength and their advanced level of kung-fu. Only a really robust person could get through that test. Kerosene drinking illustrates the handicap theory of toxic chemical use, in a form as startlingly repellent to us as our cigarettes and alcohol horrified Ardy.

*

In my last example, I shall generalize my theory further by extending its application to the past – in this case, to the civilization of Mayan Indians that flourished in Central America one or two thousand years ago. Archaeologists have been fascinated by Mayan success at creating an advanced society in the middle of tropical rainforest. Many Mayan achievements, such as their calendar, writing, astronomical knowledge, and agricultural practices, are now understood to varying degrees. However, archaeologists were long puzzled by slender tubes of unknown purpose that they kept finding in Mayan excavations.

The tubes' function finally became clear with the discovery of painted vases showing scenes of the tubes' use: to administer intoxicating enemas. The vases depict a high-status figure, evidently a priest or a prince, receiving a ceremonial enema in the presence of other people. The enema tube is shown as connected to a bag of a frothy beer-like beverage – probably containing either alcohol or hallucinogens or both, as suggested by practices of other Indian groups. Many Central and South American Indian tribes formerly practised similar ritual enemas when first encountered by European explorers, and some still do so today. The substances known to be administered range from alcohol (made by fermenting agave sap or a tree bark) to tobacco, peyote, LSD derivatives, and mushroom-derived hallucinogens. Thus, the ritual enema is similar to our consumption of intoxicants by mouth, but there are four reasons why an enema constitutes a more effective and valid indicator of strength than does drinking.

Firstly, it is possible to relapse into solitary drinking and thus to lose all opportunity for signalling one's high status to others. However, it is more difficult for a solitary person to administer the same beverage to himself or herself unassisted as an enema. An enema encourages one to enlist associates and thereby automatically creates an occasion for self-advertisement. Secondly, more strength is required to handle alcohol as an enema than as a drink, since the alcohol goes directly into the intestine and thence to the bloodstream, and it is not first diluted with food in the stomach. Thirdly, drugs absorbed from the small intestine after ingestion by mouth pass first to the liver, where many drugs are detoxified before they can reach the brain and other sensitive organs. But drugs absorbed from the rectum after an enema bypass the liver. Finally, nausea may limit one's intake of drinks but not of enemas. Hence an enema seems to me a more convincing advertisement of superiority than are our whisky advertisements. I recommend this concept to an ambitious public relations firm competing for the account of one of the large distilleries.

*

Let's now step back and summarize the perspective on chemical abuse that I have suggested. Although frequent self-destruction by chemicals may be unique to humans, I see it as fitting into a broad pattern of animal behaviour and thus as having innumerable animal precedents. All animals have had to evolve signals for quickly communicating messages to other animals. If the signals were ones that any individual animal could master or acquire, they would lend themselves to rampant cheating and hence to disbelief. To be valid and believable, a signal must be one that guarantees the honesty of the signaller, by entailing a cost, risk, or burden that only superior individuals can afford. Many animal signals that would otherwise strike us as counterproductive – such as stotting by gazelles, or costly structures and risky displays with which males court females – can be understood in this light.

It seems to me that this perspective has contributed to the evolution not only of human art, already discussed in Chapter Nine, but also of human chemical abuse as discussed in this chapter. Both art and chemical abuse are widespread human hallmarks characteristic of most known human societies. Both beg explanation, since it is not immediately obvious why they promote our survival through natural selection, or why they help us acquire mates through sexual selection. I argued in Chapter Nine that art often serves as a valid indicator of an individual's superiority or status, since art requires skill to create and requires status or wealth to acquire. But those individuals perceived by their fellows as enjoying status thereby acquire enhanced access to resources and mates. I have argued in this chapter that humans seek status through many other costly displays besides art, and that some of those displays (like jumping from towers, fast driving, and chemical abuse) are surprisingly dangerous. The former costly displays advertise status or wealth; the latter, dangerous ones advertise that the displaying individual can master even such risks and hence must be superior.

I do not claim, though, that this perspective affords a total understanding of art or chemical abuse. As I mentioned in Chapter Nine in connection with art, complex patterns of behaviour acquire a life of their own, go far beyond their original purpose (if there ever was just a single purpose), and may even originally have served multiple functions. Just as art is now motivated far more by pleasure than by need for advertisement, chemical abuse too is now clearly much more than an advertisement. It is also a way to get past inhibitions, drown sorrows, or just enjoy a good-tasting drink.

I also do not deny that, even from an evolutionary perspective, there remains a basic difference between human abuse of chemicals and its animal precedents. Stotting, long tails, and all the animal precedents that I described involve costs, but those forms of behaviour persist because the costs are outweighed by the benefits. A stotting gazelle loses a possible head start in a chase, but gains by decreasing the likelihood that a lion will embark on a serious chase at all. A long-tailed male bird is encumbered in finding food or escaping predators, but those survival disadvantages imposed by natural selection are more than compensated by mating advantages gained through sexual selection. The net balance is more rather than fewer offspring to pass on the male's genes. These animal traits only appear to be self-destructive; they are actually self-promoting.

In the case of our chemical abuse, though, the costs outweigh the benefits. Drug addicts and drunkards not only lead shorter lives, but they lose rather than gain attractiveness in the eyes of potential mates and lose the ability to care for children. These traits do not persist because of hidden advantages outweighing costs; they persist mainly because they are chemically addicting. Thus, on balance, they are self-destructive, not self-promoting, patterns of behaviour. While gazelles may occasionally miscalculate in stotting, they do not commit suicide through addiction to the excitement of stotting. In that respect, our self-destructive abuse of chemicals diverged from its animal precursors to become truly a human hallmark.

TWELVE

ALONE IN A CROWDED UNIVERSE

While humans are unique among Earth's species, the enormous number of stars suggests that intelligent creatures like us must have evolved elsewhere in the universe. If so, why have we not been visited by their flying saucers? The insights that woodpeckers provide into the supposed inevitability of convergent evolution help us reassess the possibility that we are unique in the accessible universe as well as on Earth.

*

The next time you are outdoors on a clear night away from city lights, look up at the sky and get a sense of the myriads of stars. Next, find a pair of binoculars, train them on the Milky Way, and appreciate how many more stars escaped your naked eye. Then look at a photo of the Andromeda Nebula as seen through a powerful telescope to realize how enormous is the number of stars that escaped your binoculars as well.

Once all those numbers have sunk in, you will finally be ready to ask how humans could possibly be unique in the universe. How many civilizations of intelligent beings like ourselves must be out there, looking back at us? How long before we are in communication with them, before we visit them, or before we are visited?

On Earth, we certainly are unique. No other species possesses language, art, or agriculture of a complexity remotely approaching ours. No other species abuses drugs. But we have seen in the last four chapters that, for each of those human hallmarks, there are many animal precedents or even precursors. Accept for the moment the assumption that the universe contains innumerable other planets on which life evolved. Do not those considerations suggest that some other species on some other planets have also extended such widespread precursors as far as the level of our own intelligence, technical ability, and communication skills? While no other species on Earth is now wondering where else in the universe there exists intelligent life, such species must exist elsewhere.

Alas, most human hallmarks lack effects detectable at a distance of many light-years. If there were creatures enjoying art or addicted to drugs on planets orbiting even the nearest stars, we would never know it. But fortunately there are two signs of intelligent beings elsewhere that might be detectable on Earth – space probes and radio signals. We ourselves are already becoming effective at sending out both, so surely other intelligent creatures have mastered the necessary skills. Where, then, are the expected flying saucers?

This seems to me one of the greatest puzzles in all of science. Given the billions of stars, and given the abilities that we know did develop in our own species, we ought to be detecting flying saucers or at least radio signals. There is no question about there being billions of stars. What is there about the human species, then, that could explain the missing saucers? Could we really be unique not only on Earth, but also in the accessible universe? In this chapter I shall argue that we can obtain a fresh perspective on our uniqueness by looking carefully at some other well-known creatures here on Earth – woodpeckers!

*

For a long time, people have asked themselves such questions. Already around 400 BC the philosopher Metrodorus wrote, 'To consider the Earth the only populated world in an infinite space is as absurd as to assert that in an entire field sown with millet, only one grain will grow.' Not until 1960, however, did scientists make a serious first attempt to find the answer, by listening (unsuccessfully) for radio transmissions from two nearby stars. In 1974 astronomers at the giant Arecibo radio telescope tried to establish an interstellar dialogue, by beaming a powerful radio signal to the star cluster M13 in the constellation Hercules. The signal described to Hercules' denizens what we earthlings look like, how many of us there are, and where the Earth is located in our solar system. Two years later the search for extraterrestrial life provided the main motivation behind the Viking missions to Mars, whose cost of about a billion dollars dwarfed all the US National Science Foundation's expenditures (since its inception) for classifying the life known to exist on Earth. More recently the US government has decided to spend a further hundred million dollars to detect radio signals from any intelligent beings who might exist outside our solar system. Several spacecraft that we launched are now heading out of our solar system, carrying sound tapes and photographic records of our civilization to inform spacelings who might be encountered.

It is easy to understand why lay people as well as biologists would consider the detection of extraterrestrial life as possibly the most exciting scientific discovery ever made. Just imagine how it would affect our self-image to find that the universe holds other intelligent creatures, with complex societies, languages, and learned cultural traditions, and capable of communicating with us. Among those of us who believe in an afterlife and an ethically concerned deity, most would agree that an afterlife awaits humans but not beetles (or even chimpanzees). Creationists believe that our species had a separate origin through divine creation. Suppose, though, that we should detect on another planet a society of seven-legged creatures more intelligent and ethical than we are, and able to converse with us, but having a radio receiver and transmitter in place of eyes and a mouth. Shall we believe that those creatures (but still not chimps) share the afterlife with us, and that they too were divinely created?

Many scientists have tried to calculate the odds of there being intelligent creatures out there, somewhere. Those calculations have spawned a whole new field of science termed exobiology – the sole scientific field whose subject matter has not yet been shown to exist. Let's now consider the numbers that encourage exobiologists to believe in their subject matter.

Exobiologists calculate the number of advanced technical civilizations in the universe by an equation known as the Green Bank formula, which multiplies a string of estimated numbers. Some of those numbers can be estimated with considerable confidence. There are billions of galaxies, each with billions of stars. Astronomers conclude that many stars probably have one or more planets each, and that many of those planets probably have an environment suitable for life. Biologists conclude that, where conditions suitable for life exist, life will probably evolve eventually. Multiplying all of those probabilities or numbers together, we conclude that there are likely to be billions of billions of planets supporting living creatures.

Now let's estimate what fraction of those planetary biotas include intelligent beings with an advanced technical civilization, which we will define operationally as a civilization capable of interstellar radio communication. (This is a less demanding definition than flying saucer capability, since our own development suggests that interstellar radio communication will precede interstellar probes.) Two arguments suggest that that fraction may be considerable. Firstly, the sole planet where we are certain that life evolved – our own – did evolve an advanced technical civilization. We have already launched interplanetary probes. We have made progress with techniques for freezing and thawing life and for making life from DNA – techniques relevant to preserving life as we know it for the long duration of an interstellar trip. Technical progress in recent decades has been so rapid that manned interstellar probes surely will be feasible within a few centuries at the very most, since some of our unmanned interplanetary probes are already on their way out of our solar system.

However, this first argument suggesting that many planetary biotas have evolved advanced technical civilizations is not a compelling one. To use the language of statisticians, it suffers from the obvious flaw of very small sample size (how can you generalize from one case?) and very high ascertainment bias (we picked out that one case precisely because it evolved our own advanced technical civilization).

A second, stronger argument is that life on Earth is characterized by what biologists term convergent evolution. That is, seemingly whatever ecological niche or physiological adaptation you consider, many groups of creatures have converged by evolving independently to exploit that niche, or to acquire that adaptation. An obvious example is the independent evolution of flight by birds, bats, pterodactyls, and insects. Other spectacular cases are the independent evolution of eyes, and even of devices for electrocuting prey, by many animals. Within the past two decades, biochemists have recognized convergent evolution at the molecular level, such as the repeated evolution of similar protein-splitting enzymes or membrane-spanning proteins. It is now difficult to pick up any issue of any journal in any field of biology without encountering further examples. So common is convergent evolution of anatomy, physiology, biochemistry, and behaviour that whenever biologists observe two species to be similar in some respect, one of the first questions they now ask is: did that similarity result from common ancestry or from convergence?

There is nothing surprising about the seeming ubiquity of convergent evolution. If you expose millions of species for millions of years to similar selective forces, of course you can expect similar solutions to emerge time and time again. We know that there has been much convergence among species on Earth, but by the same reasoning there should also be much convergence between Earth's species and those elsewhere. Hence although radio communication is one of those things that happens to have evolved here only once so far, considerations of convergent evolution lead us to expect its evolution on some other planets as well. As the *Encyclopaedia Britannica* puts it, 'It is difficult to imagine life evolving on another planet without progressing towards intelligence.'

That conclusion brings us back to the puzzle I mentioned earlier. If many or most stars have a planetary system, and if many of those systems include at least one planet with conditions suitable for life, and if life is likely eventually to evolve where suitable conditions exist, and if about one per cent of planets with life include an advanced technical civilization – then one estimates that our own galaxy alone contains about a million planets supporting advanced civilizations. But within only a few dozen light-years of us are several hundred stars, some (most?) of which surely have planets like ours, supporting life. Then where are all the flying saucers that we would expect? Where are the intelligent beings that should be visiting us, or at least directing radio signals at us?

If intelligent beings from elsewhere had visited Earth after literate civilizations began to develop here several thousand years ago, those beings would probably have searched out the most interesting civilizations on Earth, and we would now have written records of the visit. If the visitors had arrived in the pre-literate or prehuman past, they might have colonized Earth, and we would know of it as an abrupt arrival of drastically different life forms in our fossil record. We are bombarded by Hollywood films depicting such visits, and by tabloids actually claiming them. You will see the headlines at any US supermarket checkout counter: 'Woman kidnapped by UFO', 'Flying saucer terrorizes family', and so on. But compare that pseudo-bombardment, or our expectations, with reality. The silence is deafening.

Something must be wrong with the astronomers' calculations. They know what they are talking about when they estimate the number of planetary systems, and the fraction of those likely to be supporting life. I find these estimates plausible. Instead, the problem is likely to lie in the argument, based on convergent evolution, that a significant fraction of biotas will evolve advanced technical civilizations. Let's scrutinize more closely the inevitability of convergent evolution.

*

This brings me at last to woodpeckers. The 'woodpecker niche' is based on digging holes in live wood and on prying off pieces of bark. It is a terrific niche that offers much more food than do flying saucers or radios. Thus, we might expect convergence among many species that evolved independently to exploit the woodpecker niche. The niche provides dependable food sources in the form of insects living under bark, insects burrowing into wood, and sap. Since wood contains insects and sap all year round, occupants of the woodpecker niche would not have to migrate.

The other advantage of the woodpecker niche is that it provides a terrific place for a nest. A hole in a tree is a stable environment with relatively constant temperature and humidity, protected from wind and rain and desiccation and temperature fluctuations, and concealed and protected from predators. Other bird species can pull off the easier feat of digging nest holes in dead wood, but there are many fewer dead trees than live trees available. Many other species nest in natural holes, but such holes too are few in number, quickly become known to predators, get reused year after year, and breed infections. Hence it is a big advantage to be able to excavate a clean new nest hole in a live tree, instead of having to use a dead tree or natural hole. Other birds often pay tribute (unsought by woodpeckers) to that advantage, by usurping woodpeckers' holes.

All these considerations mean that if we are counting on convergent evolution of radio communication, we can surely count on convergent evolution of woodpecking. Not surprisingly, woodpeckers are very successful birds. There are nearly 200 species, many of them common. They come in all sizes, from tiny birds the size of kinglets up to crow-sized species. They are widespread over most of the world, with a few exceptions that I shall mention later. They do not have to migrate in winter. Some species have even exploited their woodpecking skills to live in treeless places, excavate nest holes in the ground, and feed on ants. While the earliest known fossil woodpeckers date only from the Pliocene (about seven million years ago), molecular evidence indicates that woodpeckers evolved about fifty million years ago.

How hard is it to evolve to become a woodpecker? Two considerations seem to suggest, 'Not very hard'. Woodpeckers are not an extremely distinctive old group without close relatives, like egg-laying mammals. Instead ornithologists have agreed for a long time that their closest relatives are the honey-guides of Africa, the toucans and barbets of tropical America, and the barbets of the tropical Old World, to which woodpeckers are fairly similar except in their special adaptations for woodpecking. Woodpeckers have numerous such adaptations, but none is remotely as extraordinary as building radios, and all are readily seen as extensions of adaptations possessed by other birds. The adaptations fall into four groups.

First and most obvious are the adaptations for drilling in live wood. These include a strong, straight, chisel-like bill with a hard, horny covering at the tip; nostrils protected with feathers to keep out sawdust; a thick skull; strong head and neck muscles; a broad base of the bill, and a hinge between that base and the front of the skull, to help spread the shock of pounding; and possibly a brain/skull design like a bicycle helmet, to protect the brain from shock. These features for drilling in live wood can be traced to features of other birds much more easily than our radios can be traced to any primitive radios of chimpanzees. Many other birds, such as parrots, peck or bite holes in dead wood. Some barbets can actually excavate in live wood, but they are much slower, clumsier, and less neat than woodpeckers and peck from the side rather than straight. Within the woodpecker family there is a gradation of drilling ability – from wrynecks, which cannot excavate at all, to the many woodpeckers that drill in softer wood, to hardwood specialists like sapsuckers and the pileated woodpecker.

Another set of adaptations are those for perching vertically on bark, such as a stiff tail to press against bark as a brace, strong muscles for manipulating the tail, short legs, long curved toes, and a pattern of moulting the tail feathers that saves the central pair of tail feathers (crucial in bracing) as the last to be moulted. The evolution of these adaptations can be traced even more easily than can the adaptations for woodpecking. Even within the woodpecker family, wrynecks and piculets do not have stiff tails for use as braces. Many birds outside the woodpecker family, including creepers and pygmy parrots, do have stiff tails that they evolved to prop themselves on bark.

The third adaptation is an extremely long and extensible tongue, fully as long as our own tongue in some woodpeckers. Once a woodpecker has broken into the tunnel system of wood-dwelling insects at one point, the bird uses its tongue to lick out many branches of the system without having to drill a new hole for each branch. Some woodpeckers have barbs at the tip of the tongue to spear insects, while others have big salivary glands to catch insects by making the tongue sticky. Woodpeckers' tongues have many animal precedents, including the similarly long insect-catching tongues of frogs, anteaters, and aardvarks and the brush-like tongues of nectar-drinking lories.

Finally, woodpeckers have tough skins to withstand insect bites plus the stresses from pounding and from strong muscles. Anyone who has skinned and stuffed birds knows that some birds have much tougher skins than others. Taxidermists groan when given a pigeon, whose paper-thin skin tears almost as soon as you look at it, but smile when given a woodpecker, hawk, or parrot.

*

While woodpeckers have many adaptations for woodpecking, most of those adaptations have also evolved convergently in other birds or animals, and the unique skull adaptations can at least be traced to precursors. You might therefore expect the whole package of wood-pecking to have evolved repeatedly, with the result that there would now be many groups of large animals capable of excavating into live wood for food or nest sites. Some animal groups defined initially by distinctive ways of feeding have proved to be polyphyletic, meaning that the group is actually an unnatural one, consisting of several groups that evolved similar adaptations from different ancestors. For instance, vultures are now known, and bats and seals are suspected, to be polyphyletic. But all the classical evidence, and now the newer molecular evidence, have uncovered no hint of polyphyly for woodpeckers. Modern woodpeckers are all more closely related to each other than to any non-woodpecker. Woodpecking thus appears to have evolved only once.

Picologists, the scientists who specialize in studying woodpeckers, take that conclusion for granted. On reflection, though, it is startling to the rest of us non-picologists who had convinced ourselves that woodpecking would evolve repeatedly. Could it be that other pseudo-woodpeckers did evolve, but that our surviving woodpeckers were so superior that they exterminated their unrelated competitors? For example, separate groups of mammalian carnivores evolved in South America, Australia, and the Old World. But the Old World carnivores (our cats and dogs and weasels) proved so superior that they exterminated South America's carnivorous mammals millions of years ago and are now in the process of exterminating Australia's carnivorous marsupials. Was there a similar shootout in the woodpecker niche?

Fortunately, we can test that theory. True woodpeckers do not fly far over water, with the result that they never colonized remote oceanic land masses like Australia/New Guinea (formerly joined in a single land mass), New Zealand, and Madagascar. Similarly, placental terrestrial mammals other than bats and rodents were never able to reach Australia/New Guinea, where instead marsupials evolved good functional equivalents of moles, mice, cats, wolves, and anteaters. Evidently it was not so hard to fill those mammalian niches by convergent evolution. Let's see what happened to the woodpecker niche in Australia/New Guinea.

We find there a diverse array of birds that evolved convergently to feed on or under bark, including pygmy parrots, birds of paradise, honey-eaters, Australian creepers, Australian nuthatches, ploughbills, ifritas, and flycatchers. Some of those birds have powerful bills used to dig into dead wood. Some of them have evolved elements of the woodpecker anatomical syndrome, such as stiff tails and tough skins. The species that has come the closest to filling the woodpecker niche is not a bird at all but a mammal, the striped possum, which taps on dead wood to detect insect tunnels, rips open the wood with its incisor teeth, then inserts its long tongue or very long fourth finger to pull out the insects.

However, none of these would-be woodpeckers has actually made it into the woodpecking niche. None can excavate live wood. Many are visibly inefficient; I recall seeing a black-throated honeyeater trying to hop up a tree trunk and repeatedly falling off. The ploughbill and striped possum seem to be the would-be's most effective at digging in dead wood, but both are quite uncommon and evidently cannot make a good living by their efforts. New Zealand's and Madagascar's pseudo-woodpeckers are no better. In a stunning instance of convergent evolution, Madagascar's best would-be is also a mammal, a primate called the aye-aye, that operates like a striped possum except for having a very long third instead of a fourth finger. But just as in Australia/New Guinea, none of the would-be's in New Zealand or Madagascar can excavate in live wood.

Thus, in the absence of woodpeckers, many try, and none succeeds. The woodpecker niche is flagrantly vacant on those masses not reached by woodpeckers. If woodpeckers had not evolved that one time in the Americas or Old World, a terrific niche would be flagrantly vacant over the whole Earth, just as it has remained vacant in Australia/New Guinea, New Zealand, and Madagascar.

*

I have dwelt on woodpeckers at length to illustrate that convergence is not universal, and that not all opportunities are seized. I could have illustrated the same point with other, equally flagrant examples. The most ubiquitous opportunity available to animals is to consume plants, much of whose mass consists of cellulose. Yet no higher animal has managed to evolve a cellulose-digesting enzyme. Those animal herbivores that digest cellulose instead have to rely on microbes housed within their intestines. Among such herbivores, none comes close to achieving the efficiency of ruminants, the cud-chewing mammals exemplified by cows. To take another example that I discussed in Chapter Ten, growing your own food would seem to offer obvious advantages for animals, but the only animals to master the trick before the dawn of human agriculture 10,000 years ago were leaf-cutter ants and their relatives plus a few other insects, which cultivate fungi or domesticate aphid 'cows'.

Thus, it has proved extraordinarily difficult to evolve even such obviously valuable adaptations as woodpecking, digesting cellulose efficiently, or growing one's own food. Radios do much less for one's food needs and would seem far less likely to evolve. Are our radios a fluke, unlikely to have been duplicated on any other planet?

Consider what biology might have taught us about the inevitability of radio evolution on Earth. If radio-building were like woodpecking, some species might have evolved certain elements of the package or evolved them in inefficient form, although only one species managed to evolve the complete package. For instance, we might have found today that turkeys build radio transmitters but no receivers, while kangaroos build receivers but no transmitters. The fossil record might have shown dozens of now-extinct animals experimenting over the last half-billion years with metallurgy and increasingly complex electronic circuits, leading to electric toasters in the Triassic, battery-operated rat traps in the Oligocene, and finally radios in the Holocene. Fossils might have revealed 5-watt transmitters built by trilobites, 200-watt transmitters amidst bones of the last dinosaurs, and 500-watt transmitters in use by sabertooths, until humans finally upped the power output enough to be the first to broadcast into space.

But none of that happened. Neither fossils nor living animals – not even our closest living relatives, the common and pygmy chimpanzees – had even the most remote precursors of radios. It is instructive to consider the experience of the human line itself. Neither australopithecines nor early *Homo sapiens* developed radios. As recently as 150 years ago, modern *Homo sapiens* did not even have the concepts that would lead to radios. The first practical experiments did not begin until around 1888; it is still less than 100 years since Marconi built the first transmitter capable of broadcasting one mere mile; and we still are not sending signals targeted at other stars, though the 1974 Arecibo experiment was our first attempt.

I mentioned early in this chapter that the existence of radios on the one planet known to us seemed at first to suggest a high probability of radios evolving on other planets. In fact, closer scrutiny of Earth's history supports exactly the opposite conclusion: radios had a vanishingly low probability of evolving here. Only one of the billions of species that have existed on Earth showed any proclivities towards radios, and even it failed to do so for the first 69,999/70,000ths of its seven-million-year history. A visitor from outer space who had come to Earth as recently as the year 1800 AD would have written off any prospects of radios being built here.

You might object that I am being too stringent in looking for early precursors of radios themselves, when I should instead look just for the two qualities necessary to make radios, intelligence and mechanical dexterity. The situation there is little more encouraging. Based on the very recent evolutionary experience of our own species, we arrogantly assume intelligence and dexterity to be the best way of taking over the world, and to have evolved inevitably. Think again about that quote from the *Encyclopaedia Britannica*: 'It is difficult to imagine life evolving on another planet without progressing towards intelligence.' Earth history again supports exactly the opposite conclusion. In reality, vanishingly few animals on Earth have bothered with much of either intelligence or dexterity. No animal has acquired remotely as much of either as have we; those that have acquired a little of one (smart dolphins, dexterous spiders) have acquired none of the other; and the only other species to acquire a little of both (common and pygmy chimpanzees) have been rather unsuccessful. Earth's really successful species have instead been dumb and clumsy rats and beetles, who found better routes to their current dominance.

*

We have only still to consider the last missing variable in the Green Bank formula for calculating the likely number of civilizations capable of interstellar radio communication. That variable is the lifetime of such a civilization. The intelligence and dexterity required to build radios are useful for other purposes that have been our species' hallmark for much longer than have radios and that will be the subject of the remaining chapters in this book: purposes such as mass-killing devices and means of environmental destruction. We have become so potent at doing both that we are gradually stewing in our civilization's juices. We may not enjoy the luxury of an end by slow stewing. Half-a-dozen countries now possess the means for bringing us all to a quick end, and still other countries are eagerly seeking to acquire those means. The wisdom of some past leaders of bomb-possessing nations, or of some present leaders of bomb-seeking nations, does not encourage us to believe that there will be radios on Earth for much longer.

It was an extremely unlikely fluke that we developed radios at all, and more of a fluke that we developed them before we developed the technology that will end us in a slow stew or fast bang. While Earth's history thus offers little hope that radio civilizations exist elsewhere, it also suggests that any that might exist are short-lived.

We are very lucky that that is so. I find it mind-boggling that the astronomers now eager to spend a hundred million dollars on the search for extraterrestrial life have never thought seriously about the most obvious question: what would happen if we found it, or if it found us. The astronomers tacitly assume that we and the little green monsters would welcome each other and settle down to fascinating conversations. Here again, our own experience on Earth offers useful guidance. We have already discovered two species that are very intelligent but technically less advanced than us – the common chimpanzee and pygmy chimpanzee. Has our response been to sit down and try to communicate with them? Of course not. Instead we shoot them, stuff them, dissect them, cut off their hands for trophies, put them on exhibit in cages, inject them with AIDS virus as a medical experiment, and destroy or take over their habitat. That response was predictable, because human explorers who discovered technically less advanced humans also regularly responded by shooting them, decimating their populations with new diseases, and destroying or taking over their habitat.

Any advanced extraterrestrials who discovered us would surely treat us in the same way. Think again of those astronomers who beamed radio signals into space from Arecibo, describing Earth's location and its inhabitants. In its suicidal folly that act rivalled the folly of the last Inca emperor, Atahuallpa, who described to his gold-crazy Spanish captors the wealth of his capital and provided them with guides for the journey. If there really are any radio civilizations within listening distance of us, then for heaven's sake let's turn off our own transmitters and try to escape detection, or we are doomed.

Fortunately for us, the silence from outer space is deafening. Yes, out there are billions of galaxies with billions of stars. Out there must be some transmitters as well, but not many, and they do not last long. Probably there are no others in our galaxy, and surely none within hundreds of light-years of us. What woodpeckers teach us about flying saucers is that we are unlikely ever to see one. For practical purposes, we are unique and alone in a crowded universe. Thank God!

PART FOUR

WORLD CONQUERORS

PART THREE DISCUSSED SOME OF OUR CULTURAL HALLMARKS AND THEIR animal precedents or precursors. Several of those hallmarks are ones that we are proud of, though one (agriculture) has proved to be a mixed blessing and another (chemical abuse) an unmitigated evil. Those cultural hallmarks – especially language, agriculture, and advanced technology – have been the causes of our rise. They are what permitted us to expand over the globe and become world conquerors.

That expansion, though, consisted of more than our conquering areas previously unoccupied by the human species. It also involved the expansion of particular human populations that conquered, expelled, or killed other populations. We became conquerors of each other, as well as of the world. Thus, our expansion has been marked by yet another human hallmark that has animal precursors and that we have taken far beyond its animal limits – namely, our propensity to kill other members of our species *en masse*. Along with our environmental destructiveness, it now poses one of the two potential causes for our fall.

To appreciate our rise to the status of world conquerors, recall that most animal species are distributed over only a small part of the Earth's surface. For example, Hamilton's frog is confined to one forest patch of thirty-seven acres plus one rock-pile covering 720 square yards in New Zealand. The most widespread wild land mammal other than humans used to be the lion, which as of 10,000 years ago occupied most of Africa, much of Eurasia, North America and northern South America. Even at the time of its greatest extent, though, the lion did not reach Southeast Asia, Australia, southern South America, the polar regions, or islands. There are even more widespread bird species that occur on all continents except Antarctica, such as the barn owl or peregrine falcon, but they too are absent from many islands, high elevations, cold climates, and all the oceans.

Humans used to have a typically mammalian restricted distribution, in warm, non-forested areas of Africa. As recently as 50,000 years ago, we were still confined to tropical and mild-temperature areas of Africa and Eurasia. Then we expanded in turn to Australia and New Guinea (around 50,000 years ago), cold parts of Europe (by 30,000 years ago), Siberia (by

20,000 years ago), North and South America (around 11,000 years ago), and Polynesia (between 3,600 and 1,000 years ago). One dramatic stage in this expansion of ours into a large realm formerly without people – the New World – will be the subject of a later chapter, Chapter Eighteen. Today we occupy or at least visit not only all lands but also the surface of all the oceans, and we are starting to probe into space and the oceans' abysses.

In the process of this world conquest, our species underwent a basic change in the relations between its populations. Most animal species with sufficiently wide geographic ranges fall into populations that have contact with neighbouring populations but have little or no contact with distant ones. In this respect, too, humans used to be just another species of big mammal. Until relatively recently, most people spent their entire lives within a few dozen miles of their birthplace, and had no way of learning even of the existence of people living at much greater distances. Relations between neighbouring tribes were marked by an uneasy shifting balance between trade and xenophobic hostility.

This fragmentation promoted, and was reinforced by, the tendency for each human population to develop its own language and culture. Initially, the massive expansion of our species' geographic range involved a massive increase in our linguistic and cultural diversity. Among the 'new' parts of our range occupied only within the last 50,000 years, New Guinea and North and South America alone came to account for about half of the modern world's languages. Much of that long heritage of cultural diversity has been erased in the last 5,000 years by the expansion of centralized political states. Freedom of travel – a modern invention – is now accelerating that homogenization of our language and culture. However, in a few areas of the world, notably New Guinea, stone-age technology and our traditional xenophobic outlook persisted into the Twentieth Century, giving us a last glimpse of what the rest of the world used to be like. Chapter Thirteen will try to convey some feeling for our pre-homogenized condition, and for what we have lost as well as gained through our new-found mobility.

The outcome of conflicts between expanding human groups has been heavily influenced by group differences in our cultural hallmarks. Especially decisive have been differences in military and maritime technology, in political organization, and in agriculture. Groups with more advanced agriculture thereby acquired the military advantage of larger population numbers, ability to support a permanent military caste, and resistance to infectious diseases against which sparser populations had evolved no defence.

All those cultural differences used to be ascribed to genetic superiority of conquering 'advanced' peoples over conquered 'primitive' ones.

However, no evidence of such genetic superiority has been forthcoming. The likelihood of genetics playing such a role is refuted by the ease with which the most dissimilar human groups have mastered each other's cultural techniques, given adequate learning opportunities. New Guineans born of stone-age parents now pilot aeroplanes, while Amundsen and his Norwegian crew mastered Eskimo dog-sledding methods to reach the South Pole.

Instead, one has to ask why some people acquired the cultural advantages that let them conquer other people, despite lack of any evident genetic advantages. For example, was it purely by chance that Bantu peoples originating from equatorial Africa displaced Khoisan people over most of southern Africa, rather than vice versa? While we cannot expect to identify ultimate environmental factors behind small-scale conquests, chance should play less of a role and ultimate factors should be more compelling if we focus on large-scale population shifts over long times. Hence Chapters Fourteen and Fifteen will examine two of the largest-scale shifts in recent history: the modern expansion of Europeans over the New World and Australia; and the perennial puzzle of how Indo-European languages managed earlier to overrun so much of Eurasia from an initially restricted homeland. We shall see clearly in the former case, and more speculatively in the latter, how each human society's culture and competitive position have been shaped by its biological and geographical heritage, especially by the plant and animal species available to it for domestication.

Competition among members of the same species is not unique to humans. Among all animal species as well, the closest competitors are inevitably members of the same species, because they share the closest ecological similarity. What varies greatly among species is the form that competitive strife takes. In the most inconspicuous form, rival animals compete merely by consuming food potentially available to each other and exhibit no overt aggression. Mild escalation involves ritualized displays, or chasing. As a last resort, now documented in many species, rival animals kill each other.

The competing units also vary greatly among animal species. In most songbirds, such as American or European robins, individual males or else male/female pairs face off. Among lions and common chimpanzees, small groups of males who may be brothers fight, sometimes to the death. Packs of wolves or hyenas do battle, while ant colonies engage in large-scale wars with other colonies. Although for some species these contests may end in deaths, there is no animal species whose survival as a species is even remotely threatened by such deaths.

Humans compete with each other for territory as do members of most animal species. Because we live in groups, much of our competition has taken the form of wars between adjacent groups, on the model of the wars between ant colonies rather than the small-scale contests between robins. As with adjacent groups of wolves and common chimps, relations of adjacent human tribes were traditionally marked by xeno-phobic hostility, intermittently relaxed to permit exchanges of mates (and, in our species, of goods as well). Xenophobia comes especially naturally to our species, because so much of our behaviour is culturally rather than genetically specified, and because cultural differences among human populations are so marked. Those features make it easy for us, unlike wolves and chimps, to recognize members of other groups at a glance by their clothes or hair style.

What makes human xenophobia much more lethal than chimp xenophobia is of course our recent development of weapons for mass killing at a distance. While Jane Goodall described males of one group of common chimps gradually killing off individuals of the neighbouring group and usurping their territory, those chimps had no means to kill chimps of a more remote group, nor to exterminate all chimps (including themselves). Thus, xenophobic murder has innumerable animal pre-cursors, but only we have developed it to the point of threatening to bring about our fall as a species. Threatening our own existence has now joined art and language as a human hallmark. Hence Chapter Sixteen will survey the history of human genocide, to make clear the ugly tradition from which Dachau's ovens and modern nuclear warfare spring.

THIRTEEN

THE LAST FIRST CONTACTS

For most of human history, human populations lived in a state of xenophobic isolation from each other, tempered by the need for trade and for exchanging spouses, but reinforced by differences in language and culture. In the modern world, 'first contacts' of isolated populations by outsiders have been accelerating, to the point where the last first contact is expected within the present decade. The end of our mutual isolation is bringing a tragic loss in our cultural diversity. Yet it also brings the hope that we may not continue destroying each other with increasingly powerful weapons.

*

On 4 August 1938, an exploratory biological expedition from the American Museum of Natural History made a discovery that hastened towards its end a long phase of human history. That was the date on which the advance patrol of the Third Archbold Expedition (named after its leader, Richard Archbold) became the first outsiders to enter the Grand Valley of the Balim River, in the supposedly uninhabited interior of western New Guinea. To everyone's astonishment, the Grand Valley proved to be densely populated – by 50,000 Papuans, living in the Stone Age, previously unknown to the rest of humanity and themselves unaware of others' existence. In search of undiscovered birds and mammals, Archbold had found an undiscovered human society.

To appreciate the significance of Archbold's finding, we need to understand the phenomenon of 'first contact'. As I mentioned on page 198, most animal species occupy a geographic range confined to a small fraction of the Earth's surface. Of those species occurring on several continents (such as lions and grizzly bears), it is not the case that individuals from one continent visit one another. Instead, each continent, and usually each small part of a continent, has its own distinctive population, in contact with close neighbours but not with distant members of the same species. (Migratory songbirds constitute an apparently glaring exception. But while they do commute seasonally between continents, it is only along a traditional path, and both the summer breeding range and the winter non-breeding range of a given population tend to be quite circumscribed.)

This geographic fidelity of animals is reflected in the geographic variability that I discussed in Chapter Six. Populations of the same species in different geographic areas tend to evolve into different-looking subspecies, because most breeding remains within the same population. For example, no gorilla of the East African lowland sub-species has ever turned up in West Africa or vice versa, though the eastern and western subspecies look sufficiently different that biologists could recognize a wanderer if there were any.

In these respects, we humans have been typical animals throughout most of our evolutionary history. Like other animals, each human population is genetically moulded to its area's climate and diseases, but human populations are also impeded from freely mixing by linguistic and cultural barriers far stronger than in other animals. As mentioned in Chapter Six, an anthropologist can identify roughly where a person originates from the person's naked appearance, and a linguist or student of dress styles can pinpoint origins much more closely. That is testimony to how sedentary human populations have been.

While we think of ourselves as travellers, we were quite the opposite throughout several million years of human evolution. Every human group was ignorant of the world beyond its own lands and those of its immediate neighbours. Only in recent millenia did changes in political organization and technology permit some people routinely to travel afar, to encounter distant peoples, and to learn first-hand about places and peoples that they had not personally visited. This process accelerated with Columbus's voyage of 1492, until today there remain only a few tribes in New Guinea and South America still awaiting first contact with remote outsiders. The Archbold Expedition's entry into the Grand Valley will be remembered as one of the last first contacts of a large human population. It was thus a landmark in the process by which humanity became transformed from thousands of tiny societies, collect-ively occupying only a fraction of the globe, to world conquerors with world knowledge.

How could such a numerous people as the Grand Valley's 50,000 Papuans remain completely unknown to outsiders until 1938? How could those Papuans in turn remain completely ignorant of the outside world? How did first contact change human societies? I shall argue that this world before first contact – a world that is finally ending within our own generation – holds a key to the origins of human cultural diversity. As world conquerors, our species now numbers over five billion, compared to the mere ten million people who existed before the advent of agriculture. Ironically, though, our cultural diversity has plunged even as our numbers have soared.

*

To anyone who has not been to New Guinea, the long concealment of 50,000 people there seems incomprehensible. After all, the Grand Valley lies only 115 miles from both New Guinea's north coast and its south coast. Europeans discovered New Guinea in 1526, Dutch missionaries took up residence in 1852, and European colonial governments were established in 1884. Why did it take another fifty-four years to find the Grand Valley?

The answers – terrain, food, and porters – become obvious as soon as one sets foot in New Guinea and tries to walk away from an established trail. Swamps in the lowlands, endless series of knife-edge ridges in the mountains, and jungle that covers everything reduce one's progress to a few miles per day under the best conditions. On my 1983 expedition into the Kumawa Mountains, it took me and a team of twelve New Guineans two weeks to penetrate seven miles inland. Yet we had it easy compared to the British Ornithologists' Union Jubilee Expedition. On 4 January 1910 they landed on New Guinea's coast and set off for the snow-capped mountains that they could see only a hundred miles inland. On 12 February 1911 they finally gave up and turned back, having covered less than half the distance (forty-five miles) in those thirteen months.

Compounding those terrain problems is the impossibility of living off the land, because of New Guinea's lack of big game animals. In lowland jungle the staple of New Guineans is a tree called the sago palm, whose pith yields a substance with the consistency of rubber and the flavour of vomit. However, not even New Guineans can find enough wild foods to survive in the mountains. This problem was illustrated by a horrible sight on which the British explorer Alexander Wollaston stumbled while descending a New Guinea jungle trail: the bodies of thirty recently dead New Guineans and two dying children, who had starved while trying to return from the lowlands to their mountain gardens without carrying enough provisions.

The paucity of wild foods in the jungle compels explorers going through uninhabited areas, or unable to count on obtaining food from native gardens, to bring their own rations. A porter can carry forty pounds, the weight of the food necessary to feed himself for about fourteen days. Thus, until the advent of planes made airdrops possible, all New Guinea expeditions that penetrated more than seven days' walk from the coast (fourteen days' round trip) did so by having teams of porters going back and forth, building up food depots inland. Here is a typical plan: fifty porters start from the coast with 700 man-days of food, deposit 200 man-days' food five days inland, and return in another five days to the coast, having consumed the remaining 500 man-days' food (fifty men times ten days) in the process. Then fifteen porters march to that first depot, pick up the cached 200 man-days' food, deposit fifty man-days' food a further five days' march inland, and return to the first depot (reprovisioned in the meantime), having consumed the remaining 150 man-days' food in the process. Then. . . .

The expedition that came closest to discovering the Grand Valley before Archbold, the 1921–22 Kremer Expedition, used 800 porters, 200 tons of food, and ten months of relaying to get four explorers inland to just beyond the Grand Valley. Unfortunately for Kremer, his route happened to pass a few miles west of the valley, whose existence he did not suspect because of intervening ridges and jungle.

Apart from these physical difficulties, the interior of New Guinea seemed to hold no attractions for missionaries or colonial governments, because it was believed to be virtually uninhabited. European explorers landing on the coast or rivers discovered many tribes in the lowlands living off sago and fish, but few people eking out an existence in the steep foothills. From either the north or south coast, the snow-capped Central Cordillera that forms New Guinea's backbone presents steep faces. It was assumed that the northern and southern faces meet in a ridge. What remained invisible from the coasts was the existence of broad inter-montane valleys, hidden behind those faces and suitable for agriculture.

For eastern New Guinea, the myth of an empty interior was shattered on 26 May 1930, when two Australian miners, Michael Leahy and Michael Dwyer, scaled the crest of the Bismarck Mountains in search of gold, looked down at night on the valley beyond, and were alarmed to see countless dots of light: the cooking fires of thousands of people. For western New Guinea, the myth ended with Archbold's second survey flight on 23 June 1938. After hours of flying over jungle with few signs of humans, Archbold was astonished to spot the Grand Valley, looking like Holland: a cleared landscape devoid of jungle, neatly divided into small fields outlined by irrigation ditches, and with scattered hamlets. It took six more weeks before Archbold could establish camps at the nearest lake and river where his seaplane could land, and before patrols from those camps could reach the Grand Valley to make first contact with its inhabitants.

*

That is why the outside world did not know of the Grand Valley till 1938.

Why did the valley's inhabitants, now referred to as the Dani people, not know of the outside world?

Part of the reason, of course, is the same logistic problems that faced the Kremer Expedition on its march inland, but in reverse. Yet those problems would be minor in areas of the world with gentler terrain and more wild foods than New Guinea, and they do not explain why all other human societies in the world also used to live in relative isolation. Instead, at this point we have to remind ourselves of a modern perspective that we take for granted. Our perspective did not apply to New Guinea until very recently, and it did not apply anywhere in the world 10,000 years ago.

Recall that the whole globe is now divided into political states, whose citizens enjoy more or less freedom to travel within the boundaries of their state and to visit other states. Anyone with the time, money, and desire can visit almost any country except for a few xenophobic exceptions, such as Albania and North Korea. As a result, people and goods have diffused around the globe, and many items such as Coca-cola are now available on every continent. I recall with embarrassment my visit in 1976 to a Pacific island called Rennell, whose isolated location, vertical sea cliffs without beaches, and fissured coral landscape had preserved its Polynesian culture unchanged until recently. Setting out at dawn from the coast, I plodded through jungle with not a trace of humans. When in the late afternoon I finally heard a woman's voice ahead and glimpsed a small hut, my head whirled with fantasies of the beautiful, unspoilt, grass-skirted, bare-breasted Polynesian maiden who awaited me at this remote site on this remote island. It was bad enough that the lady proved to be fat and with her husband. What humiliated my self-image as intrepid explorer was the 'University of Wisconsin' sweatshirt that she wore.

In contrast, for all but the last 10,000 years of human history, unfettered travel was impossible, and diffusion of sweatshirts was very limited. Each village or band constituted a political unit, living in a perpetually shifting state of wars, truces, alliances, and trade with neighbouring groups. New Guinea highlanders spent their entire lives within twenty miles of their birthplace. They might occasionally enter lands bordering their village lands by stealth during a war raid, or by permission during a truce, but they had no social framework for travel beyond immediately neighbouring lands. The notion of tolerating unrelated strangers was as unthinkable as the notion that any such stranger would dare appear.

Even today, the legacy of this no-trespassing mentality persists in many parts of the world. Whenever I go bird-watching in New Guinea, I take pains to stop at the nearest village to request permission to bird-watch on that village's land or rivers. On two occasions when I neglected that precaution (or asked permission at the wrong village) and proceeded to boat up the river, I found the river barred on my return by canoes of stone-throwing villagers, furious that I had violated their territory. When I was living among Elopi tribespeople in western New Guinea and wanted to cross the territory of the neighbouring Fayu tribe to reach a nearby mountain, the Elopis explained to me matter-of-factly that the Fayus would kill me if I tried. From a New Guinean perspective, it seemed perfectly natural and self-explanatory. Of course the Fayus will kill any trespasser; you surely do not think they are so stupid that they would admit strangers to their territory? Strangers would just hunt their game animals, molest their women, introduce diseases, and reconnoitre the terrain in order to stage a raid later.

While most pre-contact peoples had trade relations with their neighbours, many thought they were the only humans in existence. Perhaps the smoke of fires on the horizon, or an empty canoe floating past down a river, did prove the existence of other people. But to venture out of one's territory to meet those humans, even if they lived only a few miles away, was equivalent to suicide. As one New Guinea highlander recalled his life before first arrival of whites in 1930, 'We had not seen far places. We knew only this side of the mountains. And we thought that we were the only living people.'

Such isolation bred great genetic diversity. Each valley in New Guinea has not only its own language and culture, but also its own genetic abnormalities and local diseases. The first valley where I worked was the home of the Foré people, famous to science for their unique affliction with a fatal viral disease called *kuru* or laughing sickness, which accounted for over half of all deaths (especially among women) and left men outnumbering women three-to-one in some Foré villages. At Karimui, sixty miles to the west of the Foré area, *kuru* is completely unknown, and the people are instead affected with the world's highest incidence of leprosy. Still other tribes are unique in their high frequency of deaf mutes, or of male pseudo-hermaphrodites lacking a penis, or of premature aging, or of delayed puberty.

Today we can picture areas of the globe that we have not visited, from films and television. We can read about them in books. English dictionaries exist for all the world's major languages, and most villages speaking minor languages contain individuals who have learned one of the world's major languages. For example, missionary linguists have studied literally hundreds of New Guinea and South American Indian languages in recent decades, and I have found some inhabitant speaking either Indonesian or Neo-Melanesian in every New Guinea village that I have visited, no matter how remote. Linguistic barriers no longer impede the worldwide flow of information. Almost every village in the world today has thereby obtained fairly direct accounts of the outside world and has yielded fairly direct accounts of itself.

In contrast, pre-contact peoples had no way to picture the outside world, or to learn about it directly. Information instead arrived via long chains of languages, with accuracy lost at each step – as in the children's game called 'telephone' or 'Chinese whispers', where one child in a circle whispers a message to the next child, who in turn whispers it to her neighbour, until by the time the message is whispered back to the first child its meaning has become changed beyond recognition. As a result, New Guinea highlanders had no concept of the ocean a hundred miles distant, and knew nothing about the white men who had been prowling their coasts for several centuries. When highlanders tried to figure out why the first arriving white men wore trousers and belts, one theory was that the clothes served to conceal an enormously long penis coiled around the waist. Some Dani believed that a neighbouring group of New Guineans munched grass and had their hands joined behind their back.

Thus, first-contact patrols had a traumatic effect that is difficult for all of us living in the modern world to imagine. Highlanders 'discovered' by Michael Leahy in the 1930s, and interviewed fifty years later, still recalled perfectly where they were and what they were doing at that moment of first contact. Perhaps the closest parallel, to modern Americans and Europeans, is our recollection of one or two of the most important political events of our lives. Most Americans of my age recall that moment on 7 December 1941 when they heard of the Japanese attack on Pearl Harbor. We knew at once that our lives would be very different for years to come, as a result of the news. Yet even the impact of Pearl Harbor and of the resulting war on American society was minor, compared to the impact of a first-contact patrol on New Guinea highlanders. On that day, their world changed forever.

The patrols revolutionized the highlanders' material culture by bringing steel axes and matches, whose superiority over stone axes and fire drills was immediately obvious. The missionaries and government administrators who followed the patrols suppressed ingrained cultural practices like cannibalism, polygyny, homosexuality, and war. Other practices were discarded spontaneously by tribespeople themselves, in favour of new practices that they saw. But there was also a more profoundly unsettling revolution, in the highlanders' view of what comprised the universe. They and their neighbours were no longer the sole humans, with the sole way of life.

A book by Bob Connolly and Robin Anderson, entitled *First Contact*, poignantly relates that moment in the eastern highlands, as recalled in their old age by New Guineans and whites who met there as young adults

·or children in the 1930s. Terrified highlanders took the whites for returning ghosts, until the New Guineans dug up and scrutinized the whites' buried faeces, sent terrified young girls to have sex with the intruders, and discovered that whites defaecated and were men like themselves. Leahy wrote in his diaries that highlanders smelled bad, while at the same time the highlanders were finding the whites' smell strange and frightening. Leahy's obsession with gold was as bizarre to the highlanders as their obsession with their own form of wealth and currency – cowry shells – was to him. For the survivors of those Grand Valley Dani and Archbold Expedition members who met in 1938, such an account of first contact has yet to be written.

*

I said at the outset that Archbold's entry into the Grand Valley was not only a watershed for the Dani, but also part of a watershed in human history. What difference did it make that all human groups used to live in relative isolation, awaiting first contact, while only a few such groups remain today? We can infer the answer by comparing those areas of the world where isolation ended long ago, with those other areas where it persisted into modern times. We can also study the rapid changes that followed historical first contacts. These comparisons suggest that contact between distant peoples gradually obliterated much of the human cultural diversity that had arisen during millennia of isolation.

Take artistic diversity as one obvious example. Styles of sculpture, music, and dance used to vary greatly from village to village within New Guinea. Some villagers along the Sepik River and in the Asmat swamps produced carvings that are now world-famous because of their quality. But New Guinea villagers have been increasingly coerced or seduced into abandoning their artistic traditions. When I visited an isolated tribelet of 578 people at Bomai in 1965, the missionary controlling the only store had just manipulated the people into burning all their art. Centuries of unique cultural development ('heathen artifacts', as the missionary put it) had been destroyed in one morning. On my first visit to remote New Guinea villages in 1964, I heard log drums and traditional songs; on my visits in the 1980s, I heard guitars, rock music, and battery-operated boom boxes. Anyone who has seen the Asmat carvings at New York's Metropolitan Museum of Art, or heard log drums played in antiphonal duet at breathtaking speed, can appreciate the enormous tragedy of post-contact loss of art.

There has been massive loss of languages as well. For example, as I shall describe in Chapter Fifteen, Europe today has only about fifty languages, most of them belonging to a single language family (Indo-European). In contrast, New Guinea, with less than one-tenth of Europe's area and less than one-hundredth of its population, has about 1,000 languages, many of them unrelated to any other known language in New Guinea or elsewhere! The average New Guinea language is spoken by a few thousand people living within a radius of ten miles. When I travelled sixty miles from Okapa to Karimui in New Guinea's eastern highlands, I passed through six languages, starting with Foré (a language with postpositions, like Finnish) and ending with Tudawhe (a language with alternative tones and nasalized vowels, like Chinese).

New Guinea shows linguists what the world used to be like, with each isolated tribe having its own language, until the rise of agriculture permitted a few groups to expand and spread their tongue over large areas. It was only about 6,000 years ago that the Indo-European expansion began, leading to the extermination of all prior western European languages except Basque. The Bantu expansion within the last few millennia similarly exterminated most other languages of tropical and sub-Saharan Africa, just as the Austronesian expansion did in Indonesia and the Philippines. In the New World alone, hundreds of American Indian languages have become extinct in recent centuries.

Is language loss not a good thing, because fewer languages mean easier communication among the world's people? Perhaps, but it is a bad thing in other respects. Languages differ in structure and vocabulary, in how they express causation and feelings and personal responsibility, and consequently in how they shape our thoughts. There is no single-purpose 'best' language; instead, different languages are better suited for different purposes. For instance, it may not have been an accident that Plato and Aristotle wrote in Greek, while Kant wrote in German. The grammatical particles of those two languages, plus their ease in forming compound words, may have helped make them the pre-eminent languages of Western philosophy. Another example, familiar to all of us who studied Latin, is that highly inflected languages (ones in which word endings suffice to indicate sentence structure) can use variations of word order to convey nuances impossible with English. Our English word order is severely constrained by having to serve as the main clue to sentence structure. If English becomes a world language, that would not be because English was necessarily the best language for diplomacy.

The range of cultural practices in New Guinea also eclipses that within equivalent areas elsewhere in the modern world, because isolated tribes were able to live out social experiments that others would find utterly unacceptable. Forms of self-mutilation and cannibalism varied from tribe to tribe. At the time of first contact, some tribes went naked, others concealed their genitals and practised extreme sexual prudery, and still others (including the Grand Valley Dani) flagrantly advertised the penis and testes with various props. Child-rearing practices ranged from extreme permissiveness (including freedom for Foré babies to grab hot objects and burn themselves), through punishment of misbehaviour by rubbing a Baham child's face with stinging nettles, to extreme repression resulting in Kukukuku child suicide. Barua men pursued institutionalized bisexuality by living in a large, communal, homosexual house with the young boys, while each man had a separate, small, heterosexual house for his wife and daughters and infant sons. Tudawhes instead had two-storey houses in which women, infants, unmarried girls, and pigs lived in the lower storey, while men and unmarried boys lived in the upper storey accessed by a separate ladder from the ground.

We would not mourn the shrinking cultural diversity of the modern world if it only meant the end of self-mutilation and child suicide. But the societies whose cultural practices have now become dominant were selected only for economic and military success. Those qualities are not necessarily the ones that foster happiness or promote long-term human survival. Our consumerism and our environmental exploitation serve us well at present but bode ill for the future. Features of American society that already rate as disasters in anyone's book include our treatment of old people, adolescent turmoil, abuse of psychotropic chemicals, and gross inequality. For each of these problem areas, there are (or were before first contact) many New Guinea societies that found far better solutions to the same issues.

*

Unfortunately, alternative models of human society are rapidly disappearing, and the time has passed when humans could try out new models in isolation. Surely there are no remaining uncontacted populations anywhere as large as the one encountered by Archbold's patrol on that August day of 1938. When I worked on New Guinea's Rouffaer River in 1979, missionaries nearby had just found a tribe of a few hundred nomads, who reported another uncontacted band five days' travel upstream. Small bands have also been turning up in remote parts of Peru and Brazil. However, at some point within this last decade of the Twentieth Century, we can expect the last first contact, and the end of the last separate experiment at designing human society.

While that last first contact will not mean the end of human cultural diversity, much of which is proving capable of surviving television and travel, it certainly does mean a drastic reduction. That loss is to be mourned, for the reasons that I have just been discussing. But our xenophobia was tolerable only as long as our means to kill each other were too limited to bring about our fall as a species. When I try to think of reasons why nuclear weapons will not inexorably combine with our genocidal tendencies to break the records we have already set for genocide in the first half of the Twentieth Century, our accelerating cultural homogenization is one of the chief grounds for hope that I can identify. Loss of cultural diversity may be the price that we have to pay for survival.

FOURTEEN

ACCIDENTAL CONQUERORS

The largest-scale human population shift of the past millenium has been the European conquest of the Americas and of Australia, formerly settled by other peoples. Why did conquest go in that direction rather than in the reverse direction? It was largely an accident of biogeography: Europeans inherited the most useful suite of wild plant and animal species suitable for domestication, on which subsequent technological and political development depended.

*

Some of the most obvious features of our daily lives pose the hardest questions for scientists. If you look around you at most locations in the US or Australia, most of the people you see will be of European ancestry. At the same locations 500 years ago, everyone without exception would have been an American Indian in the US, or a native (aboriginal) Australian in Australia. Why is it that Europeans came to replace most of the native population of North America and Australia, instead of Indians or native Australians coming to replace most of the original population of Europe?

This question can be rephrased to ask: why was the ancient rate of technological and political development fastest in Eurasia, slower in the Americas (and in Africa south of the Sahara), and slowest in Australia? For example, in 1492 much of the population of Eurasia used iron tools, had writing and agriculture, had large centralized states with ocean-going ships, and was on the verge of industrialization. The Americas had agriculture, only a few large centralized states, writing in only one area, no ocean-going ships or iron tools, and were technologically and politically a few thousand years behind Eurasia. Australia lacked agriculture, writing, states, and ships, was still in a pre-first-contact condition, and used stone tools comparable to ones made over ten thousand years earlier in Eurasia. It was those technological and political differences – not the biological differences determining the outcome of competition among animal populations – that permitted Europeans to expand to other continents.

Nineteenth-century Europeans had a simple, racist answer to such questions. They concluded that they acquired their cultural head start through being inherently more intelligent, and that they therefore had a manifest destiny to conquer, displace, or kill 'inferior' peoples. The trouble with this answer is that it was not just loathsome and arrogant, but also wrong. It is obvious that people differ enormously in the knowledge they acquire, depending on their circumstances as they grow up. But no convincing evidence of genetic differences in mental ability among peoples has been found, despite much effort.

Because of this legacy of racist explanations, the whole subject of human differences in level of civilization still reeks of racism. Yet there are obvious reasons why the subject begs to be properly explained. Those technological differences led to great tragedies in the past 500 years, and their legacies of colonialism and conquest still powerfully shape our world today. Until we can come up with a convincing alternative explanation, the suspicion that racist genetic theories might be true will linger.

In this chapter I shall argue that continental differences in level of civilization arose from geography's effect on the development of our cultural hallmarks, not from human genetics. Continents differed in the resources on which civilization depended – especially, in the wild animal and plant species that proved useful for domestication. Continents also differed in the ease with which domesticated species could spread from one area to another. Even today, Americans and Europeans are painfully aware how distant geographical features, like the Persian Gulf or the Isthmus of Panama, affect our lives. But geography and biogeography have been moulding human lives even more profoundly, for hundreds of thousands of years.

Why do I emphasize plant and animal species? As the biologist J. B. S. Haldane remarked, 'Civilization is based, not only on men, but on plants and animals.' Agriculture and herding, though they also brought the disadvantages discussed in Chapter Ten, still made it possible to feed far more people per square mile of land than could live on the wild foods available in that same area. Storable food surpluses grown by some individuals permitted other individuals to devote themselves to metallurgy, manufacturing, writing – and to serving in full-time professional armies. Domestic animals provided not only meat and milk to feed people, but also wool and hides to clothe people, and power to transport people and goods. Animals also provided power to pull ploughs and carts, and thus to increase agricultural productivity greatly over that previously attainable by human muscle power alone.

As a result, the world's human population rose from about ten million around 10,000 BC, when we were all still hunter-gatherers, to over five billion today. Dense populations were prerequisite to the rise of centralized states. Dense populations also promoted the evolution of infectious diseases, to which exposed populations then evolved some resistance but other populations did not. All these factors determined who colonized and conquered whom. Europeans' conquest of America and Australia was due not to their better genes but to their worse germs (especially smallpox), more advanced technology (including weapons and ships), information storage through writing, and political organization – all stemming ultimately from continental differences in geography.

*

Let's start with the differences in domestic animals. By around 4000 BC western Eurasia already had its 'Big Five' domestic livestock that continue to dominate today: sheep, goats, pigs, cows, and horses. Eastern Asians domesticated four other cattle species that locally replace cows: yaks, water buffalo, gaur, and banteng. As already mentioned, these animals provided food, power, and clothing, while the horse was also of incalculable military value. (It was both the tank, the truck, and the jeep of warfare until the Nineteenth Century.) Why did American Indians not reap similar benefits by domesticating the corresponding native American mammal species, such as mountain sheep, mountain goats, peccaries, bison, and tapirs? Why did Indians mounted on tapirs, and native Australians mounted on kangaroos, not invade and terrorize Eurasia?

The answer is that, even today, it has proved possible to domesticate only a tiny fraction of the world's wild mammal species. This becomes clear when one considers all the attempts that failed. Innumerable species reached the necessary first step of being kept captive as tame pets. In New Guinea villages I routinely find tamed possums and kangaroos, while I saw tamed monkeys and weasels in Amazonian Indian villages. Ancient Egyptians had tamed gazelles, antelopes, cranes, and even hyenas and possibly giraffes. Romans were terrorized by the tamed African elephants with which Hannibal crossed the Alps (*not* Asian elephants, the tame elephant species in circuses today).

But all these incipient efforts at domestication failed. Since the domestication of horses around 4000 BC and reindeer a few thousand years later, no large European mammal has been added to our repertoire of successful domesticates. Thus, our few modern species of domestic mammals were quickly winnowed from hundreds of others that had been tried and abandoned.

Why have efforts at domesticating most animal species failed? It turns out that a wild animal must possess a whole suite of unusual characteristics for domestication to succeed. Firstly, in most cases it must be a social species living in herds. A herd's subordinate individuals have instinctive submissive behaviours that they display towards dominant individuals, and that they can transfer towards humans. Asian mouflon sheep (the ancestors of domestic sheep) have such behaviour but North American bighorn sheep do not – a crucial difference that prevented Indians from domesticating the latter. Except for cats and ferrets, solitary territorial species have not been domesticated.

Secondly, species such as gazelles and many deer and antelopes, which instantly take flight at signs of danger instead of standing their ground, prove too nervous to manage. Our failure to domesticate deer is especially striking, since there are few other wild animals with which humans have been so closely associated for tens of thousands of years. Although deer have always been intensively hunted and often tamed, reindeer alone among the world's forty-one deer species were successfully domesticated. Territorial behaviour, flight reflexes, or both eliminated the other forty species as candidates. Only reindeer had the necessary tolerance of intruders and gregarious, non-territorial behaviour.

Finally, domestication requires being able to breed an animal in captivity. As zoos often discover to their dismay, captive animals that are docile and healthy may nevertheless refuse to breed in cages. You yourself would not want to carry out a lengthy courtship and copulate under the watchful eyes of others; many animals do not want to either.

This problem has derailed persistent attempts to domesticate some potentially very valuable animals. For example, the finest wool in the world comes from the vicuña, a small camel species native to the Andes. But neither the Incas nor modern ranchers have ever been able to domesticate it, and wool must still be obtained by capturing wild vicuñas. Many potentates, from ancient Assyrian kings to nineteenth-century Indian maharajahs, have tamed cheetahs, the world's swiftest land mammal, for hunting. However, every prince's cheetah had to be captured from the wild, and not even zoos were able to breed them until 1960.

Collectively, these reasons help explain why Eurasians succeeded in domesticating the Big Five but not other closely related species, and why American Indians did not domesticate bison, peccaries, tapirs, and mountain sheep or goats. The military value of the horse is especially interesting in illustrating what seemingly slight differences make one species uniquely prized, another useless. Horses belong to the group of mammals termed Perissodactyla, which consists of the hoofed mammals with an odd number of toes: horses, tapirs, and rhinoceroses. Of the seventeen living species of Perissodactyla, all four tapirs and all five rhinos, plus five of the eight wild horse species, have never been domesticated. Africans or Indians mounted on rhinos or tapirs would have trampled any European invaders, but it never happened.

A sixth wild horse relative, the wild ass of Africa, gave rise to domestic donkeys, which proved splendid as pack animals but useless as military chargers. The seventh wild horse relative, the onager of western Asia, may have been used to pull wagons for some centuries after 3000 BC. But all accounts of the onager blast its vile disposition with adjectives like 'bad-tempered', 'irascible', 'unapproachable', 'unchangeable', and 'inherently intractable'. The vicious beasts had to be kept muzzled to prevent them from biting their attendants. When domesticated horses reached the Middle East around 2300 BC, onagers were finally kicked onto the scrapheap of failed domesticates.

Horses revolutionized warfare in a way that no other animal, not even elephants or camels, ever rivalled. Soon after their domestication, they may have enabled herdsmen speaking the first Indo-European languages to begin the expansion that would eventually stamp their languages on much of the world (Chapter Fifteen). A few millenia later, hitched to battle chariots, horses became the unstoppable Sherman tanks of ancient war. After the invention of saddles and stirrups, they enabled Attila the Hun to devastate the Roman Empire, Genghis Khan to conquer an empire from Russia to China, and military kingdoms to arise in West Africa. A few dozen horses helped Cortes and Pizarro, leading only a few hundred Spaniards each, to overthrow the two most populous and advanced New World states, the Aztec and Inca empires. With futile Polish cavalry charges against Hitler's invading armies in September 1939, the military importance of this most universally prized of all domestic animals finally came to an end after 6,000 years.

Ironically, relatives of the horses that Cortes and Pizarro rode had formerly been native to the New World. Had those horses survived, Montezuma and Atahuallpa might have shattered the conquistadores with cavalry charges of their own. But, in a cruel twist of fate, America's horses had become extinct long before that, along with eighty or ninety per cent of the other large animal species of the Americas and Australia. It happened around the time that the first human settlers – ancestors of modern Indians and native Australians – reached those continents. The Americas lost not only their horses but also other potentially domestic-atable species like large camels, ground sloths, and elephants. Australia lost all its giant kangaroos, giant wombats, and rhinoceros-like diprotodonts. Australia and North America ended up with no domesti-catable mammal species at all, unless Indian dogs were derived from

North American wolves. South America was left with only the guinea-pig (used for food), alpaca (used for wool), and llama (used as a pack animal, but too small to carry a rider).

As a result, domestic mammals made no contribution to the protein needs of native Australians and Americans except in the Andes, where their contribution was still much slighter than in the Old World. No native American or Australian mammal ever pulled a plough, cart, or war chariot, gave milk, or bore a rider. The civilizations of the New World limped forward on human muscle power alone, while those of the Old World ran on the power of animal muscle, wind, and water.

Scientists still debate whether the prehistoric extinctions of most large American and Australian mammals were due to climatic factors or were caused by the first human settlers themselves (Chapters Seventeen to Nineteen). Whichever was the case, the extinctions may have virtually ensured that the descendants of those first settlers would be conquered over 10,000 years later by people from Eurasia and Africa, the continents that retained most of their large mammal species.

*

Do similar arguments apply to plants? Some parallels jump out immediately. As true of animals, only a tiny fraction of all wild plant species have proved suitable for domestication. For example, plant species in which a single hermaphroditic individual can pollinate itself (like wheat) were domesticated earlier and more easily than cross-pollinated species (like rye). The reason is that self-pollinating varieties are easier to select and then maintain as true strains, since they are not continually mixing with their wild relatives. As another example, although acorns of many oak species were a major food source in prehistoric Europe and North America, no oak has ever been domesticated, perhaps because squirrels remained much better than humans at selecting and planting acorns. For every domesticated plant that we still use today, many others were tried in the past and discarded. (What living American has eaten sumpweed, which Indians in the eastern US domesticated for its seeds by around 2000 BC?)

Such considerations help explain the slow rate of human technological development in Australia. That continent's relative poverty in wild plants appropriate for domestication, as in appropriate wild animals, undoubtedly contributed to the failure of aboriginal Australians to develop agriculture. But it is not so obvious why agriculture in the Americas lagged behind that in the Old World. After all, many food plants now of worldwide importance were domesticated in the New

World: corn, potatoes, tomatoes, and squash, to name just a few. The answer to this puzzle requires closer scrutiny of corn, the New World's most important crop.

Corn is a cereal – that is, a grass with edible starchy seeds, like barley kernels or wheat grains. Cereals still provide most of the calories consumed by the human race. While all civilizations have depended on cereals, different native cereals have been domesticated by different civilizations: for instance, wheat, barley, oats, and rye in the Near East and Europe; rice, foxtail millet, and broomcorn millet in China and Southeast Asia; sorghum, pearl millet, and finger millet in sub-Saharan Africa; but only corn in the New World. Soon after Columbus discovered America, corn was brought back to Europe by early explorers and spread around the globe, and it now exceeds all other crops except wheat in world acreage planted. Why, then, did corn not enable American Indian civilizations to develop as fast as the Old World civilizations fed by wheat and other cereals?

It turns out that corn was a much bigger pain in the neck to domesticate and grow, and gave an inferior product. Those will be fighting words to all of you who, like me, love hot, buttered corn-on-the-cob. Throughout my childhood, I looked forward to late summer as the season to stop at roadside stands and pick out the best-looking fresh ears. Corn is the most important crop in the US today, worth twenty-two billion dollars to us and fifty billion dollars to the world. But before you charge me with slander, please hear me out on the differences between corn and other cereals.

The Old World had over a dozen wild grasses that were easy to domesticate and grow. Their large seeds, favoured by the Near East's highly seasonal climate, made their value obvious to incipient farmers. They were easy to harvest *en masse* with a sickle, easy to grind, easy to prepare for cooking, and easy to sow. Another subtle advantage was first recognized by University of Wisconsin botanist Hugh Iltis: we did not have to figure out for ourselves that they could be stored, since wild rodents in the Near East already made caches of up to sixty pounds of those wild grass seeds.

The Old World grains were already productive in the wild, and one can still harvest up to 700 pounds of grain per acre from wild wheat growing naturally on hillsides in the Near East. In a few weeks a family could harvest enough to feed itself for a year. Even before wheat and barley were domesticated, there were sedentary villages in Palestine that had already invented sickles, mortars and pestles, and storage pits, and that were supporting themselves on wild grains.

Domestication of wheat and barley was not a conscious act. It was not the case that several hunter–gatherers sat down one day, mourned the extinction of big game animals, discussed which particular wheat plants were best, planted the seeds of those plants, and thereby became farmers the next year. Instead, as I mentioned in Chapter Ten, the process we call domestication – the changes in wild plants under cultivation – was an unintended by-product of people preferring some wild plants over others, and hence accidentally spreading seeds of the preferred plants. In the case of wild cereals, people naturally preferred to harvest ones with big seeds, ones whose seeds were easy to remove from the seed-coverings, and ones with firm non-shattering stalks that held all the seeds together. It took only a few mutations, favoured by this unconscious human selection, to produce the large-seeded, non-shattering cereal varieties that we refer to as domesticated rather than wild.

By around 8000 BC, wheat and barley remains from archaeological digs at ancient Near Eastern village sites are beginning to show these changes. The development of bread wheats, other domestic varieties, and intentional sowing soon followed. Gradually, fewer remains of wild foods are found at the sites. By 6000 BC, crop cultivation had been integrated with animal herding into a complete food production system in the Near East. For better or worse (in some major respects worse, as I argued in Chapter Ten), people were no longer hunter–gatherers but farmers and herders, en route to being civilized.

Now contrast these relatively straightforward Old World developments with what happened in the New World. The parts of the Americas where farming began lacked the Near East's highly seasonal climate, and so lacked large-seeded grasses that were already productive in the wild. North American and Mexican Indians did start to domesticate three small-seeded wild grasses called maygrass, little barley, and a wild millet, but these were displaced by the arrival of corn and then of European cereals. Instead, the ancestor of corn was a Mexican wild grass that did have the advantage of big seeds but in other respects hardly seemed like a promising food plant: annual teosinte.

Teosinte ears look so different from corn ears that scientists argued about teosinte's precise role in corn's ancestry till recently, and even now some scientists remain unconvinced. No other crop underwent such drastic changes on domestication as did teosinte. It has only six to twelve kernels per ear, and they are inedible, because they are enclosed in stone-hard cases. One can chew teosinte stalks like sugar cane, as Mexican farmers still do. But no one uses its seeds today, and there is no indication that anyone did prehistorically either.

Hugh Iltis identified the key step in teosinte's becoming useful: a permanent sex change! In teosinte the side branches end in a tassel composed of male flowers; in corn they end in a female structure, the ear. Although that sounds like a drastic difference, it is really a simple hormonally-controlled change that could have been started by a fungus, virus, or change in climate. Once some flowers on the tassel had changed sex to female, they would have produced edible naked grains likely to catch the attention of hungry hunter-gatherers. The tassel's central branch would then have been the beginning of a corn cob. Early Mexican archaeological sites have yielded remains of tiny ears, barely an inch-and-a-half long and much like the tiny ears of our 'Tom Thumb' corn variety.

With that abrupt sex change, teosinte (alias corn) was now finally on the road to domestication. However, in contrast to the case with Near Eastern cereals, thousands of years of development still lay ahead before high-yield corns capable of sustaining villages or cities resulted. The final product was still much more difficult for Indian farmers to manage than were the cereals of Old World farmers. Corn ears had to be harvested individually by hand, rather than *en masse* with a sickle; the cobs had to be shucked; the kernels did not fall off but had to be scraped or bitten off; and sowing the seeds involved planting them individually, rather than scattering them *en masse*. The result was still poorer nutritionally than Old World cereals: lower protein content, deficiencies of nutritionally important amino acids, deficiency of the vitamin niacin (tending to cause the disease pellagra), and need for alkali treatment of the grain to partially overcome these deficiencies.

In short, characteristics of the New World's staple food crop made its potential value much harder to discern in the wild plant, harder to develop by domestication, and harder to extract even after domestication. Much of the lag between New World and Old World civilization may have been due to those peculiarities of one plant.

*

So far, I have discussed geography's biogeographic role, in providing the local wild animal and plant species suitable for domestication. But there is another major role of geography that deserves mention. Each civilization has depended not only on its own food plants domesticated locally, but also on other food plants that arrived after having been first domesticated elsewhere. The predominantly north/south axis of the New World made such diffusion of food plants difficult; the predominantly east/west axis of the Old World made it easy (see map overleaf).

Today, we take plant diffusion so much for granted that we seldom stop to think where our foods originated. A typical American or European meal might consist of chicken (of Southeast Asian origin) with corn (from Mexico) or potatoes (from the southern Andes), seasoned

AXES OF THE OLD AND NEW WORLDS

with pepper (from India), accompanied by a piece of bread (from Near Eastern wheat) and butter (from Near Eastern cattle), and washed down by a cup of coffee (from Ethiopia). But this diffusion of valued plants and animals did not begin just in modern times: it has been happening for thousands of years.

Plants and animals spread quickly and easily within a climate zone to which they are already adapted. To spread out of this zone, they have to develop new varieties with different climate tolerances. A glance at the map of the Old World on this page shows how species could shift long distances without encountering a change of climate. Many of these shifts proved enormously important in launching farming or herding in new areas, or enriching it in old areas. Species moved between China, India, the Near East, and Europe without ever leaving temperate latitudes of the northern hemisphere. Ironically, the US patriotic song 'America the Beautiful' invokes America's own spacious skies, its amber waves of grain. In reality, the most spacious skies of the northern hemisphere were in the Old World, where amber waves of related grains came to stretch for 7,000 miles from the English Channel to the China Sea.

The Romans were already growing wheat and barley from the Near East, peaches and citrus fruits from China, cucumbers and sesame from India, and hemp and onions from central Asia, along with oats and poppies originating locally in Europe. Horses that spread from the Near East to West Africa revolutionized military tactics there, while sheep and cattle spread down the highlands of East Africa to launch herding in southern Africa among the Hottentots, who lacked locally domesticated animals of their own. African sorghum and cotton reached India by around 2000 BC, while bananas and yams from tropical Southeast Asia crossed the Indian Ocean to enrich agriculture in tropical Africa.

In the New World, however, the temperate zone of North America is isolated from the temperate zone of the Andes and southern South America by thousands of miles of tropics, in which temperate-zone species cannot survive. As a result, the llama, alpaca, and guinea-pig of the Andes never spread in prehistoric times to North America or even to Mexico, which consequently remained without any domestic mammals to carry packs or to produce wool or meat (except for corn-fed edible dogs). Potatoes also failed to spread from the Andes to Mexico or North America, while sunflowers never spread from North America to the Andes. Many crops that were apparently shared prehistorically between North and South America actually occurred as different varieties or even species in the two continents, suggesting that they were domesticated independently in both areas. This seems true, for instance, of cotton, beans, lima beans, chili peppers, and tobacco. Corn did spread from Mexico to both North and South America, but it evidently was not easy, perhaps because of the time it took to develop varieties suited to other latitudes. Not until around 900 AD – thousands of years after corn had emerged in Mexico – did corn become a staple food in the Mississippi Valley, thereby triggering the belated rise of the mysterious mound-building civilization of the American Midwest.

Thus, if the Old and New Worlds had each been rotated ninety degrees about their axes, the spread of crops and domestic animals would have been slower in the Old World, faster in the New World. The rates of rise of civilization would have been correspondingly different. Who knows whether that difference would have sufficed to let Montezuma or Atahuallpa invade Europe, despite their lack of horses?

*

I have argued, then, that continental differences in the rates of rise of civilization were not an accident caused by a few individual geniuses. They were not produced by the biological differences determining the outcome of competition among animal populations – for example, some populations being able to run faster or digest food more efficiently than others. They also were not the result of average differences among whole peoples in inventiveness; there is no evidence for such differences anyway. Instead, they were determined by biogeography's effect on cultural development. If Europe and Australia had exchanged their human populations twelve thousand years ago, it would have been the former native Australians, transplanted to Europe, who eventually invaded America and Australia from Europe.

Geography sets ground rules for the evolution, both biological and cultural, of all species, including our own. Geography's role in determining our modern political history is even more obvious than the role I have discussed in determining the rate at which we domesticate plants and animals. From this perspective, it is almost funny to read that half of all American schoolchildren do not know where Panama is, but not at all funny when politicians display comparable ignorance. Among the many notorious examples of disasters brought on by politicians ignorant of geography, two must suffice: the unnatural boundaries drawn on the map of Africa by nineteenth-century European colonial powers, thereby undermining the stability of some modern African states that inherited those borders; and the borders of Eastern Europe drawn at the Treaty of Versailles in 1919 by politicians who knew little of that region, thereby helping to fuel the Second World War.

Geography used to be a required subject in US schools and colleges until a few decades ago, when it began to be dropped from many curricula. The mistaken belief arose then that geography consisted of little more than memorizing the names of capital cities. But twenty weeks of geography in the seventh grade is not enough to teach our future politicians about the effects that maps really have on us. The fax machines and satellite communications that span the globe cannot erase the differences among us bred by differences in location. In the long run, and on a broad scale, where we live has contributed heavily to making us who we are.

FIFTEEN

HORSES, HITTITES, AND HISTORY

More than 4,000 years before the recent expansion of Europeans over all other continents, there was an earlier expansion within Europe and western Asia that sired most of the languages spoken in that region today. Although those earlier conquerors were illiterate, much of their language and culture can be reconstructed from shared word roots preserved in modern Indo-European languages. Their conquest of much of Eurasia, like the subsequent overseas expansion of their descendants, appears to have been an accident of biogeography.

*

'*Yksi, kaksi, kolme, neljä, viisi.*'

I watched the little girl counting out five marbles, one by one. Her act was familiar, but her words were strange. Almost anywhere else in Europe, I would have heard words like our English 'one, two, three', '*uno, due, tre*' in Italy, '*ein, zwei, drei*' in Germany, '*odin, dva, tri*' in Russia. But I was vacationing in Finland, and Finnish is one of Europe's few non-Indo-European languages.

Today, most European languages and many Asian languages as far east as India are very similar to each other (*see* table of vocabulary overleaf). No matter how we complain while memorizing French word lists in school, these so-called 'Indo-European' languages resemble English and each other, and differ from all the world's other languages, in vocabulary and grammar. Only 140 of the modern world's 5,000 tongues belong to this language family, but their importance is far out of proportion to their numbers. Thanks to the global expansion of Europeans since 1492 – especially of people from England, Spain, Portugal, France, and Russia – nearly half the world's present population of five billion now speaks an Indo-European language as its native tongue.

To us it may seem perfectly natural, and in no need of further explanation, that most European languages resemble each other. Not until we go to parts of the world with great linguistic diversity do we realize how weird is Europe's homogeneity, and how it cries out for explanation. For example, in areas of the New Guinea highlands where I work and where first contact with the outside world began only in the Twentieth Century, languages as different as Chinese is from English replace each other over short distances (Chapter Thirteen). Eurasia must also have been diverse in its pre-first-contact condition, and gradually become less so until finally some people speaking the mother tongue of the Indo-European language family steamrollered almost all other European languages out of existence.

INDO-EUROPEAN VERSUS NON-INDO-EUROPEAN VOCABULARY

INDO-EUROPEAN LANGUAGES

English	one	two	three	mother	brother	sister
German	ein	zwei	drei	Mutter	Bruder	Schwester
French	un	deux	trois	mère	frère	soeur
Latin	unus	duo	tres	mater	frater	soror
Russian	odin	dva	tri	mat'	brat	sestra
Old Irish	oen	do	tri	mathir	brathir	siur
Tocharian	sas	wu	trey	macer	procer	ser
Lithuanian	vienas	du	trys	motina	brolis	seser
Sanskrit	eka	duva	trayas	matar	bhratar	svasar
PIE	oynos	dwo	treyes	mater	bhrater	suesor

NON-INDO-EUROPEAN LANGUAGES

Finnish	yksi	kaksi	kolme	äiti	veli	sisar
Foré	ka	tara	kakaga	nano	naganto	nanona

PIE stands for proto-Indo-European, the reconstructed mother tongue of the first Indo-Europeans. Foré is a language of the New Guinea Highlands. Note that most words are very similar among the Indo-European languages and totally different among the non-Indo-European languages.

Of all the processes by which the modern world lost its earlier linguistic diversity, the Indo-European expansion has been the most important. Its first stage, which long ago carried Indo-European languages over Europe and much of Asia, was followed by a second stage that began in 1492 and carried them to all other continents (Chapter Fourteen). When and where did the steamroller start, and what gave it its power? Why was Europe not overrun instead by speakers of a language related to, say, Finnish or Assyrian?

While the Indo-European problem is the most famous problem of historical linguistics, it is a problem of archaeology and history as well. In the case of those Europeans who carried out the second stage of the Indo-European expansion beginning in 1492, we know not only their vocabulary and grammar but also the ports where they set out, the dates of their sailings, the names of their leaders, and why they succeeded in conquering (Chapter Fourteen). But the quest to understand the first stage is a search for an elusive people whose language and society lie veiled in the pre-literate past, even though they became world conquerors and founded today's dominant societies. That quest is also a great detective story, whose solution depends on a language discovered behind a secret wall in a Buddhist monastery, and on an Italian language inexplicably preserved on the linen wrappings of an Egyptian mummy.

When you first think about it, you might be excused for dismissing the Indo-European problem as obviously insoluble. Since the Indo-European mother tongue arose before the origin of writing, is it not almost by definition impossible to study? Even if we found the skeletons or pottery of the first Indo-Europeans, how would we recognize them? The skeletons and pottery of modern Hungarians, living in the centre of Europe, are as typically European as goulash is typically Hungarian. A future archaeologist excavating a Hungarian city would never guess that Hungarians speak a non-Indo-European language, if no examples of writing itself were recovered. Even if we could somehow locate the place and time of the first Indo-Europeans, how could we hope to deduce what advantage let their language triumph?

Remarkably, it turns out that linguists have been able to extract answers to these questions from the languages themselves. I shall explain why we are so confident that language distributions today reflect a steamroller in the past, then try to assess when and where the mother tongue was spoken, and how it managed to take over so much of the world.

*

How can we infer that the modern Indo-European languages replaced other, now-vanished languages? I am not talking about the observed second-stage replacements of the past 500 years, which saw English and Spanish dislodge most native tongues of the Americas and Australia. Those modern expansions were obviously due to the advantages Europeans gained from guns, germs, iron, and political organization (Chapter Fourteen). Instead, I am talking about the inferred first-stage replacement that saw Indo-European dislodge older languages of Europe and western Asia, and that must have happened before writing reached those areas.

The map on the following page shows the distribution of Indo-European language branches surviving in 1492, just before Spanish began to leap across the Atlantic with Columbus. Three branches are especially

LANGUAGE MAP OF EUROPE AND WESTERN ASIA

Indo-European	Non-Indo-European
A Albanian	**1** Basque
Ar Armenian	**2** Finno-Ugric
B Baltic	**3** Turkic and Mongolian
C Celtic	**4** Semitic
Ge Germanic	**5** Caucasian
Gr Greek	**6** Dravidian
I Italic	
II Indo-Iranian	
S Slavic	

An Anatolian ⎫ extinct
Toc Tocharian ⎬ before
　　　　　　　⎭ 1492

This map shows language distribution, *circa* 1492, just before the European discovery of the New World. There must have been other Indo-European language branches that had become extinct before then. However, lengthy written texts exist only in languages of the Anatolian branch (including Hittite) and the Tocharian branch, whose homelands became occupied by speakers of Turkic and Mongolian languages before 1492.

familiar to most Europeans and Americans: Germanic (including English and German), Italic (including French and Spanish), and Slavic (including Russian), each branch with twelve to sixteen surviving languages and 300 to 500 million speakers. The largest branch, however, is Indo–Iranian, with ninety languages and nearly 700 million speakers from Iran to India (including Romany, the language of gypsies). Relatively tiny surviving branches are Greek, Albanian, Armenian, Baltic (consisting of Lithuanian and Latvian), and Celtic (including Welsh and Gaelic), each with only two to ten million speakers. In addition, at least two Indo-European branches, Anatolian and Tocharian, vanished long ago but are known from extensive preserved writings, while others disappeared with less trace.

What proves that all these tongues are related to each other and distinct from other language stocks? One obvious clue is shared vocabulary, as illustrated by the table of vocabulary on page 226 and thousands of other examples. A second clue is similar word endings (so-called inflectional endings) used to form verb conjugations and noun declensions. These endings are illustrated by part of the conjugation of 'to be' below. It becomes easier to recognize such similarities when you realize that word roots and endings shared between related languages are generally not shared identically. Instead, a particular sound in one language is often replaced by another sound in the other language. Familiar examples are the frequent equivalence of English 'th' and German 'd' (English 'thing' equals German '*ding*', 'thank' equals '*danke*'), or of English 's' and Spanish 'es' (English 'school' equals '*escuela*', 'stupid' equals '*estupido*').

Those resemblances among the Indo–European languages are detailed, but much grosser features of sounds and word formation set Indo-

INDO-EUROPEAN VERSUS NON-INDO-EUROPEAN VERB ENDINGS:
TO BE OR NOT TO BE

INDO-EUROPEAN LANGUAGES

English	(I) am	(he) is
Gothic	im	ist
Latin	sum	est
Greek	eimi	esti
Sanskrit	asmi	asti
Old Church Slavonic	jesmi	jesti

NON-INDO-EUROPEAN LANGUAGES

Finnish	olen	on
Foré	miyuwe	miye

Not only vocabulary, but also verb and noun endings, connect Indo-European languages and set them apart from other languages.

European languages apart from other language families. For example, my atrocious French accent embarrasses me as soon as I open my mouth to ask, 'Où est le métro?' But my difficulties with French are nothing compared with my total inability to produce the click sounds of some southern African languages, or to produce the eight gradations of vowel pitch in the Lakes Plain languages of the New Guinea lowlands. Naturally, my Lakes Plain friends loved teaching me bird names that differed only in pitch from words for excrement, then watching me ask the next villager I met for more information about that 'bird'.

As distinctive to Indo-European as its sounds is its word formation. Indo-European nouns and verbs have various endings that we memorize assiduously when we learn a new language. (How many of you ex-scholars of Latin can still chant 'amo, amas, amat, amamus, amatis, amant'?) Each such ending conveys several types of information. For example, the 'o' of 'amo' specifies first person singular present active: the lover is I, not my rival; one of me, not two of me; I am giving, not receiving, love; and I am giving it now, not yesterday. Heaven help the serenading lover who gets even a single one of those details wrong! But other languages, like Turkish, use a separate syllable or phoneme for each such type of information, while still other languages, like Vietnamese, virtually dispense with such variations of word form.

Given all these resemblances among Indo-European languages, how could the differences among them have arisen? A clue is that any language whose written documents span many centuries can be seen to change with time. For example, modern English-speakers find eighteenth-century English quaint but completely understandable; we can read Shakespeare (1564–1616), though we need notes to explain many of his words; but Old English texts, such as the poem *Beowulf* (*circa* 700–750 AD), are virtually a foreign language to us. A good example of how English has changed over the last 1,000 years is provided by the Twenty-Third Psalm:

MODERN (1989)

The Lord is my shepherd, I lack nothing.
He lets me lie down in green pastures.
He leads me to still waters.

KING JAMES BIBLE (1611)

The Lord is my shepherd, I shall not want.
He maketh me to lie down in green pastures.
He leadeth me beside the still waters.

MIDDLE ENGLISH (1100–1500)

Our Lord gouerneth me, and nothyng shal defailen to me.
In the sted of pasture he sett me ther.
He norissed me upon water of fyllyng.

OLD ENGLISH (800–1066)

Drihten me raet, ne byth me nanes godes wan.
And he me geset on swythe good feohland.
And fedde me be waetera stathum.

As speakers of one original language spread into different areas with limited contact, the independent changes of words and pronunciation in each area inevitably lead to different dialects, such as those that have arisen in different parts of the US in the few centuries since permanent English settlement began in 1607. With the passing of more centuries, dialects diverge to the point where their speakers can no longer understand each other and they now rank as distinct languages. One of the best documented examples of this process is the development of the Romance languages after the break-up of Latin around 500 AD. Surviving written texts from the Eighth Century onwards show us how the languages of France, Italy, Spain, Portugal, and Rumania gradually diverged from Latin – and from each other.

The derivation of the modern Romance languages from Latin thus illustrates how groups of related languages develop from a shared ancestral tongue. Even if we had no surviving Latin texts, we could still reconstruct much of the Latin mother tongue by comparing traits in its daughter languages today. In the same way, one can reconstruct a family tree of all the Indo-European language branches, based partly on ancient texts and partly on inferences. Hence language evolution proceeds by descent and divergence, just as Darwin demonstrated for biological evolution. In their languages as well as their skeletons, modern Englishmen and Australians, who began to diverge with the colonization of Australia in 1788, are much more similar to each other than cither are to the Chinese, from whom they diverged tens of thousands of years ago.

Given time, the languages within any part of the world will keep on diverging, held back only by contacts between adjacent peoples. An example of the result is New Guinea, which had never been unified politically before European colonization, and where nearly one thousand mutually unintelligible languages – including dozens with no known relation to each other or to any other language in the world – are now spoken in an area the size of Texas. Thus, wherever you find the same language or related languages spoken over a wide area, you know that the clock of language evolution must have been restarted recently. That is, one language must have recently spread, eliminated other languages, and then started to differentiate all over again. Such a process accounts for the close similarities among southern Africa's Bantu languages, and among Austronesian languages of Southeast Asia and the Pacific.

The Romance languages again provide our best documented example. As of 500 BC, Latin was confined to a small area around Rome and was only one of many languages spoken in Italy. The expansion of Latin-speaking Romans eradicated all those other languages of Italy, then eradicated entire branches of the Indo-European family elsewhere in Europe, like the continental Celtic languages. These sister branches were so thoroughly replaced by Latin that we know each of them only by scattered words, names, and inscriptions. With the subsequent overseas expansion of Spanish and Portuguese after 1492, the language spoken initially by a few hundred thousand Romans trampled hundreds of other languages out of existence, as it gave rise to the Romance languages spoken by half a billion people today.

If the Indo-European language family as a whole constituted a similar steamroller, we might expect to find its trampled debris in the form of older non-Indo-European languages surviving here and there. The sole such vestige surviving in Western Europe today is the Basque language of Spain, without known relations to any other language in the world. (The remaining non-Indo-European languages of modern Europe – Hungarian, Finnish, Estonian, and possibly Lapp – are relatively recent invaders of Europe from the east.) However, there were other languages that were spoken in Europe until Roman times, and of which enough words or inscriptions have been preserved to identify them as non-Indo-European. The most extensively preserved of these vanished tongues is the mysterious Etruscan language of northwest Italy, for which we have a 281-line text written on a roll of linen that somehow ended up in Egypt as wrapping for a mummy. All such vanished non-Indo-European languages were part of the debris left from the Indo-European expansion.

Still more linguistic debris was swept up into the surviving Indo-European languages themselves. To understand how linguists can recognize such debris, imagine that you, as a freshly arrived visitor from outer space, were given one book each, written in English by an Englishman, an American, and an Australian, about his or her country.

The language and most of the words in all three books would be the same. But if you compared the American book with the one about England, the American book would contain many place names that were obviously foreign to the basic language of the books – names like Massachusetts, Winnepesaukee, and Mississippi. The Australian book would contain more place names equally foreign to the language but unlike the American names – such as Woonarra, Goondiwindi, and Murrumbidgee. You might guess that English immigrants coming to America and Australia encountered natives who spoke different languages, and from whom the immigrants picked up names for local places and things. You would even be able to infer something about the words and sounds of those unknown native languages. We actually know the native American and Australian languages from which those borrowings took place, and we can confirm that your indirect inferences from the borrowed words alone would have been correct.

Linguists studying several Indo-European languages have similarly detected words borrowed from vanished, apparently non-Indo-European languages. For example, about one-sixth of Greek words whose derivations can be traced appear to be non-Indo-European. These words are just the sort that one might expect to have been borrowed by invading Greeks from the natives they encountered: place names like Corinth and Olympus, words for Greek crops like olive and vine, and names of gods or heroes like Athene and Odysseus. These words may be the linguistic legacy of Greece's pre-Indo-European population to the Greek speakers who overran them.

Thus, at least four types of evidence indicate that Indo-European languages are the products of an ancient steamroller. The evidence includes the family-tree relationship of surviving Indo-European languages; the much greater linguistic diversity of areas like New Guinea, that have not been recently overrun; the non-Indo-European languages that survived in Europe into Roman times or later; and the non-Indo-European legacy in several Indo-European languages.

*

Given this evidence for an Indo-European mother tongue in the distant past, can one reconstruct something of this tongue? At first, the notion of learning how to write a vanished unwritten language seems absurd. In fact, linguists have been able to reconstruct much of the mother tongue by examining word roots shared among its daughter languages.

To take an example, if the word meaning 'sheep' were totally different in each modern Indo-European language branch, we could conclude nothing about the word for 'sheep' in the mother tongue. But if the word were similar in several branches, especially in ones as geographically distant as Indo-Iranian and Celtic, we might infer that the various branches had inherited the same root from the mother tongue. By knowing what sound shifts have taken place among the various daughter tongues, we could even reconstruct the form of the word root in the mother tongue.

A SHEEP IS A SHEEP IS A SHEEP

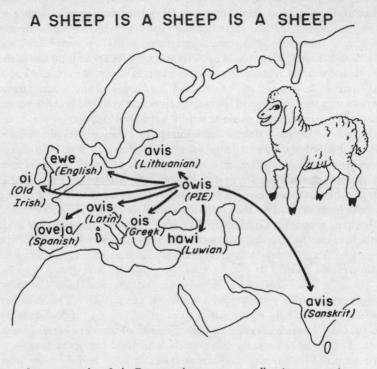

In many modern Indo-European languages, as well as in some ancient ones that we know from preserved writings, the words meaning 'sheep' are quite similar. These words must have been derived from an ancestral form that is inferred to have been *owis* and that was used in proto-Indo-European (PIE), the unwritten mother tongue.

As the figure above shows, the words for 'sheep' in many Indo-European languages from India to Ireland really are very similar: *avis*, *hawis*, *ovis*, *ois*, *oi*, etc. The modern English 'sheep' is obviously from a different root, but English retains the original root in the word 'ewe'. Consideration of the sound shifts that the various Indo-European languages have undergone suggests that the original form was *owis*.

Naturally, the same word root shared among several daughter languages does not automatically prove shared inheritance from the mother tongue. The word could also have spread later from one daughter language to another. Archaeologists sceptical of linguists' attempts to reconstruct mother tongues love to cite words like 'Coca-Cola', shared among many modern European languages. The archaeologists claim that linguists would absurdly attribute 'Coca-Cola' to the mother tongue of

HONOURABLE ROOT, DISHONOURABLE WORD

Just as in the case of words for 'sheep', the words that mean 'to fart loudly' are similar among many written Indo-European languages. This suggests an ancestral form *perd*, used in proto-Indo-European (PIE), the unwritten mother tongue.

thousands of years ago. In fact, 'Coca–Cola' illustrates how linguists weed out recent borrowings from old inheritances: the word is obviously foreign ('coca' is actually from a Peruvian Indian word, 'cola' West African), and it does not exhibit the same sound shifts among languages as do old Indo-European roots (in German it is still 'Coca-Cola', not *Köcherköhler*).

By such methods, linguists have been able to reconstruct much of the grammar and nearly 2,000 word roots of the mother tongue, termed proto-Indo-European but usually abbreviated as PIE. That is not to say that all words in modern Indo-European languages are descended from PIE: most are not, because there have been so many new inventions or borrowings (like the root 'sheep' replacing the old PIE root *owis* in English). Our inherited PIE roots tend to be words for human universals that people surely were already naming thousands of years ago: words for the numbers and human relationships (as in the table on page 226); words for body parts and functions; and ubiquitous objects or concepts like 'sky', 'night', 'summer', and 'cold'. Among the human universals thus reconstructed are such homely acts as 'to break wind', with two distinct roots in PIE depending on whether one does it loudly or softly. The root for doing it loudly (PIE *perd*) gave rise to a series of similar words in modern Indo-European languages (*perdet*, *pardate*, etc.) – including English 'fart' (see figure on page 235, and sample text on pages 248–9).

*

So far, we have seen how linguists have been able to extract, from written languages, evidence of a pre-literate mother tongue and steamroller. The obvious next questions are: when was PIE spoken, where was it spoken, and how was it able to overwhelm so many other languages? Let's begin with the matter of 'when', another seemingly impossible question. It is bad enough that we have to infer the words of an unwritten language; how on earth do we determine when it was spoken?

We can at least start to narrow down the possibilities, by examining the oldest written samples of Indo-European languages. For a long time, the oldest samples that scholars could identify were Iranian texts of around 1000–800 BC, and Sanskrit texts probably composed around 1200–1000 BC but written down later. Texts of a Mesopotamian kingdom called Mitanni, written in a non-Indo-European language but containing some words obviously borrowed from a language related to Sanskrit, push the proven existence of Sanskrit-like languages back to nearly 1500 BC.

The next breakthrough was the late-nineteenth-century discovery of a mass of ancient Egyptian diplomatic correspondence. Most of it was written in a Semitic language, but two letters in an unknown language remained a mystery until excavations in Turkey uncovered thousands of tablets in the same tongue. The tablets proved to be the archives of a kingdom that thrived between 1650 and 1200 BC and that we now refer to by the biblical name 'Hittite'.

In 1917 scholars were astonished by the announcement that the Hittite language proved on deciphering to belong to a previously unknown, very distinctive and archaic, now-vanished branch of the Indo-European family, termed Anatolian. Some obviously Hittite-like names mentioned in earlier letters of Assyrian merchants at a trading post near the Hittite capital's future site push the detective trail back to nearly 1900 BC. This remains our first direct evidence for the existence of any Indo-European language.

Thus, as of 1917, two Indo-European branches – Anatolian and Indo-

Iranian – had been shown to exist by around 1900 and 1500 BC, respectively. A third early branch was established in 1952, when the young British cryptographer, Michael Ventris, showed that the so-called Linear B writing of Crete and Greece, which had resisted deciphering since its discovery around 1900, was an early form of the Greek language. Those Linear B tablets date to around 1300 BC. But Hittite, Sanskrit, and early Greek are very different from each other, certainly more so than are modern French and Spanish, which diverged over a thousand years ago. That suggests that the Hittite, Sanskrit, and Greek branches must have split off from PIE by 2500 BC or earlier.

How much earlier do the differences between those branches imply? How can we obtain a calibration factor that converts 'percentage difference between languages' into 'time since the languages began to diverge'? Some linguists use the rate of word change in historically documented, written languages, like the changes from Anglo–Saxon to Chaucer's English to Modern English. These calculations, which belong to a science called glottochronology (or chronology of languages), yield the rule of thumb that languages replace about twenty per cent of their basic vocabulary every thousand years.

Most scholars reject glottochronology calculations, on the grounds that word replacement rates must vary with social circumstances and with the particular words themselves. However, the same scholars are generally still willing to make a seat-of-the-pants estimate. The usual conclusion from either glottochronology or pants' seats is that PIE may have started to break up by 3000 BC, surely by 2500 BC, and not before 5000 BC.

There is still another, completely independent approach to the dating problem – the science termed linguistic paleontology. Just as paleontologists try to discover what the past was like by looking for relics buried in the ground, linguistic paleontologists do it by looking for relics buried in languages.

To understand how this works, recall that linguists have reconstructed nearly 2,000 words of PIE vocabulary. It is not surprising that these include words like 'brother' and 'sky', which must have existed and been named since the dawn of human language. But PIE should not have had a word for 'gun', since guns were not invented until about 1300 AD, long after PIE-speakers had already scattered to speak distinct languages in Turkey and India. In fact, the word for 'gun' uses different roots in different Indo-European languages: 'gun' in English, *fusil* in French, *ruzhyo* in Russian, and so on. The reason is obvious: different languages could not possibly have inherited the same root for 'gun' from PIE, and they each had to invent or borrow their own word when guns were invented.

The gun example suggests that we should take a series of inventions whose dates we know, and see which of those do and which do not have reconstructed names in PIE. Anything – like gun – that was invented after PIE began to break up should not have a reconstructed name. Anything – like brother – that was invented or known before the break-up might have a name. (It does not *have* to have a name, because plenty of PIE words have surely become lost. We know the PIE words for 'eye' and 'eyebrow' but not 'eyelid', although PIE speakers must have had eyelids.)

Perhaps the earliest major developments *without* PIE names are battle chariots, which became widespread between 2000 and 1500 BC, and iron, whose use became important between 1200 and 1000 BC. The lack of PIE terms for these relatively late inventions does not surprise us, since the distinctness of Hittite had already convinced us that PIE broke up long before 2000 BC. Among earlier developments that do have PIE names, there are words for 'sheep' and 'goat', first domesticated by around 8000 BC; cattle (including separate words for cow, steer, and ox), domesticated by 6400 BC; horses, domesticated by around 4000 BC; and ploughs, invented around the time that horses were domesticated. The latest datable invention with a PIE name is the wheel, invented around 3300 BC.

Therefore, linguistic paleontology, even in the absence of any other evidence, would date the break-up of PIE as before 2000 BC but after 3300 BC. This conclusion agrees well with the one reached by extrapolating the differences between Hittite, Greek, and Sanskrit backwards in time. Hence if we wish to find traces of the first Indo-Europeans, we should be safe concentrating on the archaeological record between 2500 and 5000 BC, and perhaps slightly before 3000 BC.

*

Having reached fair agreement about the 'when' question, let's now ask: *where* was PIE spoken? Linguists have disagreed about the PIE homeland ever since they first began to appreciate its significance. Almost every possible answer has been proposed, from the North Pole to India, and from the Atlantic to the Pacific shores of Eurasia. As the archaeologist J.P. Mallory puts it, the question is not, 'Where do scholars locate the Indo-European homeland?', but 'Where do they put it *now*?'

To understand why this problem has proved so difficult, let's first try to solve it quickly by looking at a map (see page 228). As of 1492, most surviving Indo-European branches were virtually confined to Western Europe, and only Indo-Iranian extended east of the Caspian Sea. Western Europe would be the most parsimonious solution to the search for the

PIE homeland, the solution that required the fewest people to move long distances.

Unfortunately for that solution, in 1900 a 'new' but long-extinct Indo-European language was discovered in a triply unlikely location. Firstly, the language (Tocharian, as it is now known) turned up in a secret chamber behind a wall in a Buddhist cave monastery. The chamber contained a library of ancient documents in the strange language, written around 600–800 AD by Buddhist missionaries and traders. Secondly, the monastery lay in Chinese Turkestan, east of all extant Indo-European speakers and about a thousand miles removed from the nearest ones. Finally, Tocharian was not related to Indo-Iranian, the geographically closest branch of Indo-European, but possibly instead to branches used in Europe itself, thousands of miles to the west. It is as if we suddenly discovered that the early medieval inhabitants of Scotland spoke a language related to Chinese.

Obviously, the Tocharians did not reach Chinese Turkestan by helicopter. They surely walked or rode there, and we have to assume that central Asia formerly had many other Indo-European languages that disappeared without the good fortune to be preserved by documents in secret chambers. A modern linguistic map of Eurasia (see page 228) makes obvious what must have happened to Tocharian and all those other lost Indo-European languages of central Asia. That whole area today is occupied by people speaking Turkic or Mongolian languages, descendants of hordes that overran the area from the time of at least the Huns to Genghis Khan. Scholars debate whether Genghis Khan's armies slaughtered 2,400,000 or only 1,600,000 people when they captured Harat, but scholars agree that such activities transformed the linguistic map of Asia. In contrast, most Indo-European languages known to have disappeared in Europe – like the Celtic languages Caesar found spoken in Gaul – were replaced by other Indo-European languages. The apparently European centre of gravity of Indo-European languages as of 1492 was actually an artifact of recent linguistic holocausts in Asia. If the PIE homeland really was centrally located in what became the Indo-European realm by 600 AD, stretching from Ireland to Chinese Turkestan, then that homeland would have been in the Russian steppes north of the Caucasus, rather than in Western Europe.

Just as the languages themselves gave us some clues to the time of PIE's break-up, so too they contain clues to the location of the PIE homeland. One clue is that the language family to which Indo-European has the clearest connections is Finno-Ugric, the family that includes Finnish and other languages native to the forest zone of north Russia (see map on page 228). Now it is true that the links between Finno-Ugric and Indo-European languages are enormously weaker than those between German and English, which stem from the fact that the English language was brought to England from northwest Germany only 1,500 years ago. The links are also much weaker than those between the Germanic and Slavic language branches of Indo-European, which probably diverged a few thousand years ago. Instead, the links suggest a much older propinquity between the speakers of PIE and of proto-Finno-Ugric. But since Finno-Ugric comes from the north Russian forests, that suggests a PIE homeland in the Russian steppe south of the forests. In contrast, if PIE had arisen much further south (say, in Turkey), the closest affinities of Indo-European might have been with the ancient Semitic languages of the Near East.

A second clue to the PIE homeland is the non-Indo-European vocabulary swept up as debris into quite a few Indo-European languages. I mentioned that this debris is especially noticeable in Greek, and it is also conspicuous in Hittite, Irish, and Sanskrit. That suggests that those areas used to be occupied by non-Indo-Europeans and were later invaded by Indo-Europeans. If so, the PIE homeland was not Ireland or India (which almost no one suggests today anyway), but it also was not Greece or Turkey (which some scholars still do suggest).

Conversely, the modern Indo-European language still most similar to PIE is Lithuanian. Our first preserved Lithuanian texts, from around 1500 AD, contain as high a fraction of PIE word roots as did Sanskrit texts of nearly 3,000 years earlier. The conservatism of Lithuanian suggests that it has been subject to few disturbing influences from non-Indo-European languages and may have remained near the PIE homeland. Formerly, Lithuanian and other Baltic languages were more widely distributed in Russia, until Goths and Slavs pushed the Balts back to their current shrunken domain of Lithuania and Latvia. Thus, this reasoning too suggests a PIE homeland in Russia.

A third clue comes from the reconstructed PIE vocabulary. We already saw how its inclusion of words for things familiar in 4000 BC, but not for things unknown until 2000 BC, helps date the time when PIE was spoken. Might it also pinpoint the place where PIE was spoken? PIE includes a word for snow (*snoighwos*), suggesting a temperate rather than tropical location and providing the root of our English word 'snow'. Of the many wild animals and plants with PIE names (like *mus* meaning mouse), most are widespread in the temperate zone of Eurasia and help to pin down the homeland's latitude but not its longitude.

To me, the strongest clue from the PIE vocabulary is what it lacked rather than included – words for many crops. PIE speakers surely did some farming, since they had words for plough and sickle, but only one word for an unspecified grain has survived. In contrast, the reconstructed proto-Bantu language of Africa, and the proto-Austronesian language of

Southeast Asia, have many crop names. Proto-Austronesian was spoken even longer ago than PIE, so that modern Austronesian languages have had more time to lose those old names for crops than have the modern Indo-European languages. Despite that, the modern Austronesian languages still contain far more old names of crops. Hence PIE speakers probably actually had few crops, and their descendants borrowed or invented crop names as they moved to more agricultural areas.

That conclusion presents us with a double puzzle. Firstly, by 3500 BC farming had become the dominant way of life in almost all of Europe and much of Asia. That severely narrows down the possible choices for the PIE homeland; it must have been an unusual area where farming was not so dominant. Secondly, it begs the question why PIE speakers were able to expand. A major cause of the Bantu and Austronesian expansions was that the first speakers of those language families were farmers, spreading into areas occupied by hunter-gatherers whom they could outnumber or dominate. For PIE speakers to have been rudimentary farmers invading a farming Europe turns historical experience on its head. Thus, we cannot solve the 'where' of Indo-European origins until we have come to grips with the hardest question: why?

*

In Europe just before the age of writing, there were not one but two economic revolutions so far-reaching in impact that they could have caused a linguistic steamroller. The first was the arrival of farming and herding, which originated in the Near East around 8000 BC, leapt from Turkey to Greece around 6500 BC, and then spread north and west to reach Britain and Scandinavia. Farming and herding permitted a large increase in human population numbers over those previously sustainable by hunting and gathering alone (Chapter Ten). Colin Renfrew, Professor of Archaeology at the University of Cambridge in England, recently published a thought-provoking book arguing that those farmers from Turkey were the PIE speakers who brought Indo-European languages to Europe.

My first reaction to reading Renfrew's book was, 'Of course, he must be right!' Farming *had* to produce a linguistic upheaval in Europe, just as it did in Africa and Southeast Asia. This is especially likely since, as geneticists have shown, those first farmers made the biggest contribution to the genes of modern Europeans.

But – Renfrew's theory ignores or dismisses all the linguistic evidence. Farmers reached Europe thousands of years before the estimated arrival of PIE. The first farmers lacked, and PIE speakers possessed, innovations like ploughs, wheels, and domesticated horses. PIE is strikingly deficient in words for the crops that defined the first farmers. Hittite, the oldest known Indo-European language of Turkey, is not the Indo-European language closest to pure PIE, as one might expect from Renfrew's Turkey-based theory, but is instead the most deviant language and the one least Indo-European in its vocabulary. Renfrew's theory rests on nothing more than a syllogism: farming probably caused a steamroller, the PIE steamroller requires a cause, so farming is assumed to have been that cause. Everything else suggests that farming instead brought to Europe the older languages that PIE overran, like Etruscan and Basque.

Yet around 5000–3000 BC – at the right time for PIE origins – there was a second economic revolution in Eurasia. This later revolution coincided with the beginnings of metallurgy and involved a greatly expanded use of domestic animals – not just for meat and hides, as humans had been using wild animals for a million years, but for new purposes that included milk, wool, pulling ploughs, pulling wheeled vehicles, and riding. The revolution is richly reflected in the PIE vocabulary, through words for 'yoke' and 'plough', 'milk' and 'butter', 'wool' and 'weave', and a host of words associated with wheeled vehicles ('wheel', 'axle', 'shaft', 'harness', 'hub', and 'lynch-pin').

The economic significance of this revolution was to increase human population and power far beyond the levels made possible by farming and herding alone. For instance, through milk and its products one cow gradually yielded many more calories than did its meat alone. Ploughing allowed a farmer to plant much more acreage than he could with a hoe or digging stick. Animal-drawn vehicles allowed people to exploit far more land and still bring its produce to their village for processing.

For some of these advances it is hard to say where they arose, because they spread so quickly. For example, wheeled vehicles are unknown before 3300 BC, but within a few centuries of that date they are widely recorded throughout Europe and the Middle East. But there is one crucial advance whose origin can be identified: the domestication of horses. Just before their domestication, wild horses were absent from the Mideast and southern Europe, rare in northern Europe, and abundant only in the steppes of Russia eastwards. The first evidence of horse domestication is for the Sredny Stog culture around 4000 BC, in the steppes just north of the Black Sea, where archaeologist David Anthony has identified wear-marks on horses' teeth that indicate use of a bit for riding.

Throughout the world, wherever and whenever domestic horses have been introduced, they have yielded enormous benefits for human societies (Chapter Fourteen). For the first time in human evolution, people could travel overland faster than their own legs could propel them. Speed helped hunters run down their prey and helped herders

manage their sheep and cattle over large areas. Most importantly, speed
helped warriors to launch quick surprise raids on distant enemies and to
withdraw again before the enemies had time to organize a counterattack.
Throughout the world the horse revolutionized warfare and enabled
horse-owning peoples to terrorize their neighbours. The stereotype that
Americans hold of Great Plains Indians as fearsome mounted warriors
was actually created only recently, within a few generations from 1660 to
1770. Since European horses reached the US West in advance of
Europeans themselves and other European goods, we can be sure that the
horse alone was what transformed Plains Indian society.

Archaeological evidence makes clear that domestic horses had similarly transformed human society on the Russian steppe much earlier,
around 4000 BC. The steppe habitat of open grassland was hard for

HOW INDO-EUROPEAN LANGUAGES
MIGHT HAVE SPREAD

The inferred homeland where proto-Indo-European (PIE), the mother
tongue, was spoken lay in the Russian steppes north of the Black Sea and
east of the Dnieper River.

people to exploit until they could use horses to solve the problems of distance and transport. Human occupation of the Russian steppe accelerated with horse domestication and then exploded with the invention of ox-drawn wheeled vehicles around 3300 BC. The steppe economy came to be based on the combination of sheep and cattle for meat, milk, and wool, plus horses and wheeled vehicles for transport, and supplemented by a little farming.

There is no evidence for intensive agriculture and food storage at those early steppe sites, in marked contrast to the abundant evidence at other European and Mideast sites around the same time. Steppe people lacked large permanent settlements and were evidently highly mobile – again in contrast to the villages with rows of hundreds of two-storey houses in southeast Europe at the time. What the horsemen lacked in architecture, they made up for in military zeal, as attested by their lavish tombs (for men only!), filled with enormous numbers of daggers and other weapons, and sometimes even with wagons and horse skeletons.

Thus, Russia's Dnieper River (see map on page 243) marked an abrupt cultural boundary:, to the east, the well-armed horsemen, to the west, the rich farming villages with their granaries. That proximity of wolves and sheep spelt T–R–O–U–B–L–E. Once the invention of the wheel completed the horsemens' economic package, their artifacts indicate a very rapid spread for thousands of miles eastwards through the steppes of central Asia (see map). From that movement, the ancestors of the Tocharians may have arisen. The steppe peoples' spread westwards is marked by the concentration of European farming villages nearest the steppes into huge defensive settlements, then the collapse of those societies, and the appearance of characteristic steppe graves in Europe as far west as Hungary.

Of the innovations that drove the steppe peoples' steamroller, the sole one for which they clearly get full credit is the domestication of the horse. They might also have developed wheeled vehicles, milking, and wool technology independently of the Mideast's civilizations, but they borrowed sheep, cattle, metallurgy, and probably the plough from the Mideast or Europe. Thus, there was no single 'secret weapon' that alone explains the steppe expansion. Instead, with horse domestication the steppe peoples became the first to put together the economic and military package that came to dominate the world for the next 5,000 years – especially after they added intensive agriculture upon invading south-eastern-Europe. Hence their success, like that of the second-stage European expansion that began in 1492, was an accident of bio-geography. They happened to be the peoples whose homeland combined abundant wild horses and open steppe with proximity to Mideastern and European centres of civilization.

*

As archaeologist Marija Gimbutas, from the University of California, Los Angeles, has argued, the Russian steppe peoples who lived west of the Ural Mountains in the fourth millenium BC fit quite well into our postulated picture of proto-Indo-Europeans. They lived at the right time. Their culture included the important economic elements reconstructed for PIE (like wheels and horses), and lacked the elements lacking from PIE (like battle chariots and many crop terms). They lived in the right place for PIE: the temperate zone, south of Finno-Ugric peoples, near the later homeland of Lithuanians and other Balts.

If the fit is so good, why does the steppe theory of Indo-European origins remain so controversial? There would have been no controversy if archaeologists had been able to demonstrate a rapid expansion of steppe culture from southern Russia all the way to Ireland around 3000 BC. But that did not happen; direct evidence of the steppe invaders themselves extends no further west than Hungary. Instead, around and after 3000 BC, one finds a bewildering array of other cultures developing in Europe and named for their artifacts (for instance, the 'Corded Ware and Battle-axe Culture'). Those emerging Western European cultures combine steppe elements like horses and militarism with old Western European elements, especially settled agriculture. Such facts cause many archaeologists to discount the steppe hypothesis altogether, and to see the emerging Western European cultures as local developments.

However, there is an obvious reason why the steppe culture could not spread intact to Ireland. The steppe itself reaches its western limit in the plains of Hungary. That is where all subsequent steppe invaders of Europe, such as the Mongols, stopped. To spread further, steppe society had to adapt to the forested landscape of Western Europe – by adopting intensive agriculture, or by taking over existing European societies and hybridizing with their peoples. Most of the genes of the resulting hybrid societies may have been the genes of Old Europe.

If steppe people imposed PIE, their mother tongue, on southeastern Europe as far as Hungary, then it was the resulting daughter Indo-European culture, not the original steppe culture itself, that spread to derived granddaughter cultures elsewhere in Europe. Archaeological evidence of major cultural change suggests that such granddaughter cultures may have arisen throughout Europe and east to India between 3000 and 1500 BC. Many non-Indo-European languages held out long enough to be preserved in writing (like Etruscan), and Basque still survives today. Thus, the Indo-European steamroller was not a single wave, but a long chain of events that has taken 5,000 years to unfold.

As an analogy, consider how Indo-European languages came to

dominate North and South America today. We have abundant written records to prove that they stem from invasions of Indo-European speakers from Europe. Those European immigrants did not overrun the Americas in one step, and archaeologists do not find remains of unmodified European culture throughout the sixteenth-century New World. That culture was useless on the US frontier. Instead, the colonists' culture was a highly modified or hybrid one that combined Indo-European languages and much of European technology (such as guns and iron) with American Indian crops and (especially in Central and South America) Indian genes. Some areas of the New World have taken many centuries for Indo-European language and economy to master. The takeover did not reach the Arctic until this century. It is reaching much of the Amazon only now, and the Andes of Peru and Bolivia promise to remain Indian for a long time yet.

Suppose that some future archaeologist should dig in Brazil, after written records have been destroyed and Indo-European languages have disappeared from Europe. The archaeologist will find European artifacts suddenly appearing on the coast of Brazil around 1530, but penetrating the Amazon only very slowly thereafter. The people whom the archaeologist finds living in the Brazilian Amazon will be a genetic mishmash of American Indians, blacks, Europeans, and Japanese, speaking Portuguese. The archaeologist will be unlikely to realize that Portuguese was an intrusive language, contributed by invaders, to a hybrid local society.

*

Even after the PIE expansion of the fourth millenium BC, new interactions of horses, steppe peoples, and Indo-European languages continued to shape Eurasian history. PIE horse technology was primitive and probably involved little more than a rope-bit and bareback rider. For thousands of years thereafter, the military value of horses continued to improve with inventions ranging from metal bits and horse-drawn battle chariots around 2000 BC to the horseshoes, stirrups, and saddle of later cavalry. While most of these advances did not originate in the steppes, steppe peoples were still the ones who profited the most, because they always had more pasture and therefore more horses.

As horse technology evolved, Europe was invaded by many more steppe peoples, among whom the Huns, Turks, and Mongols are best known. These peoples carved out a succession of huge, short-lived empires, stretching from the steppes to Eastern Europe. But never again were steppe peoples able to impose their language on Western Europe.

They enjoyed their biggest advantage at the outset, when PIE bareback riders invaded a Europe entirely without domestic horses.

There was another difference between these later recorded invasions and the earlier unrecorded PIE invasion. The later invaders were no longer Indo-European speakers from the western steppes, but speakers of Turkic and Mongol languages from the eastern steppes. Ironically, horses were what enabled Turkish tribes from central Asia in the Eleventh Century AD to invade the land of the first written Indo-European language, Hittite. The most important innovation of the first Indo-Europeans was thus turned against their descendants. Turks are largely European in their genes, but non-Indo-European (Turkish) in their language. Similarly, an invasion from the east in 896 AD left modern Hungary largely European in its genes but Finno-Ugric in its language. By illustrating how a small invading force of steppe horsemen could impose their language on a European society, Turkey and Hungary provide models of how the rest of Europe came to speak Indo-European.

Eventually, steppe peoples in general, regardless of their language, ceased to win in the face of Western Europe's advancing technology. When the end came, it was swift. In 1241 AD the Mongols achieved the largest steppe empire that ever existed, stretching from Hungary to China. But after about 1500 AD the Indo-European-speaking Russians began to encroach on the steppes from the west. It took only a few more centuries of tsarist imperialism to conquer the steppe horsemen who had terrorized Europe and China for over 5,000 years. Today the steppes are divided between Russia and China, and only Mongolia remains as a vestige of steppe independence.

Much racist nonsense has been written about the supposed superiority of Indo-European peoples themselves. Nazi propaganda invoked a pure Aryan race. In fact, Indo-Europeans have never been unified since the PIE expansion of 5,000 years ago, and even PIE speakers themselves may have been divided among related cultures. Some of the most bitter fighting and vilest deeds of recorded history pitted one Indo-European group against another. The Jews, Gypsies, and Slavs whom the Nazis sought to exterminate conversed in languages as Indo-European as that of their persecutors. Speakers of proto-Indo-European merely happened to be in the right place at the right time to put together a useful package of technology. Through that stroke of luck, theirs was the mother tongue whose daughter languages came to be spoken by half the world today.

Appendix

A PROTO-INDO-EUROPEAN FABLE

Owis Ekwoosque

Gwrreei owis, quesyo wlhnaa ne eest, ekwoons espeket, oinom ghe gwrrum woghom weghontm, oinomque megam bhorom, oinomque ghmmenm ooku bherontm.

Owis nu ekwomos ewewquet: 'Keer aghnutoi moi ekwoons agontm nerm widntei.'

Ekwoos tu ewewquont: 'Kludhi, owei, keer ghe aghnutoi nsmei widntmos: neer, potis, owioom r wlhnaam sebhi gwhermom westrom qurnneuti. Neghi owioom wlhnaa esti.'

Tod kekluwoos owis agrom ebhuget.

[The] Sheep and [the] Horses

On [a] hill, [a] sheep that had no wool saw horses, one [of them] pulling [a] heavy wagon, one carrying [a] big load, and one carrying [a] man quickly.

[The] sheep said to [the] horses: 'My heart pains me, seeing [a] man driving horses.'

[The] horses said: 'Listen, sheep, our hearts pain us when we see [this]: [a] man, the master, makes [the] wool of [the] sheep into [a] warm garment for himself. And [the] sheep has no wool.'

Having heard this, [the] sheep fled into [the] plain.

*

In order to provide some sense of how proto-Indo-European (PIE) might have sounded, I have included the fable above, in reconstructed PIE, together with an English translation. The fable was invented over a century ago by the linguist August Schleicher. The revised version given here is based on one published by W. P. Lehmann and L. Zgusta in 1979, which takes account of the deeper understanding of PIE gained since Schleicher's time. This version has been slightly altered from that of

Lehmann and Zgusta to make it more 'user-friendly' for non-linguists, with the advice of Jaan Puhvel.

While PIE initially looks strange, many words will prove familiar on scrutiny because of similar English or Latin roots derived from PIE. For instance, *owis* means 'sheep' (*cf.* 'ewe', 'ovine'); *wlhnaa* means 'wool'; *ekwoos* means 'horses' (*cf.* 'equestrian', Latin *equus*); *ghmmenm* means 'man' (*cf.* 'human', Latin *hominem*); *que* means 'and', as in Latin; *mega* means 'big' (*cf.* 'megabucks'); *keer* means 'heart' (*cf.* 'core', 'cardiology'); *moi* means 'to me'; and *widntei* and *widntmos'* mean 'see' (*cf.* 'video'). The PIE text lacks definite and indefinite articles ('the' and 'a') and places the verb at the end of the clause or sentence.

While this sample text will show what some linguists think PIE was like, it cannot be taken as an exact sample. Remember: PIE was never written; scholars differ on details of how to reconstruct PIE; and the fable itself is imaginary.

SIXTEEN

IN BLACK AND WHITE

*Genocide, often considered a human hallmark confined to rare
perverts, actually has many animal precedents and used to be
considered socially acceptable or admirable. Whether we will succeed
in curbing our modern power to commit it depends on our coming to
recognize its frequency in human history, the potential for it in all of
us, and the ways in which ordinary people try to rationalize becoming
killers.*

*

While the anniversary of any nation's founding is taken as cause for its
inhabitants to celebrate, Australians had special cause in 1988, their
bicentennial year. Few groups of colonists faced such obstacles as those
who landed with the First Fleet at the future site of Sydney in 1788.
Australia was still Terra Incognita: the colonists had no idea what to
expect or how to survive. They were separated from their mother
country by a sea voyage of 15,000 miles, lasting eight months.
Two-and-a-half years of starvation would pass until a further supply fleet
arrived from England. Many of the settlers were convicts who had
already been traumatized by the most brutal aspects of brutal eighteenth-
century life. Despite those beginnings, the settlers survived, prospered,
filled a continent, built a democracy, and established a distinctive national
character. It is no wonder that Australians felt pride as they celebrated
their nation's founding.

Nevertheless, one set of protests marred the celebrations. The white
settlers were not the first Australians. Instead, Australia had been settled
around 50,000 years ago, by the ancestors of people now usually referred
to as Australian Aborigines and also known in Australia as blacks. In the
course of English settlement, most of those original inhabitants were
killed by the settlers or died of other causes, leading some modern
descendants of the survivors to stage bicentenary protests instead of
celebrations. The celebrations focused implicitly on how Australia
became white. I shall begin this chapter by focusing instead on how

Australia ceased to be black, and how courageous English settlers came to commit genocide.

Lest white Australians take offence, I should make clear that I am not accusing their forefathers of having done something uniquely horrendous. Instead, my reason for discussing the extermination of the Aborigines is precisely because it is not unique: it is a well documented example of a phenomenon whose frequency few people appreciate. While our first association with the word 'genocide' is likely to be the killings in Nazi concentration camps, they did not constitute the largest-scale genocide even of this century. The Tasmanians and hundreds of other peoples were modern targets of successful smaller extermination campaigns. Numerous peoples scattered throughout the world are potential targets in the near future. Yet genocide is such a painful subject that either we would rather not think about it at all, or else we would like to believe that nice people do not commit genocide, only Nazis do. But our refusal to think about it has consequences: we have done little to halt the numerous episodes of genocide since the Second World War, and we are not alert to where it may happen next. Together with our destruction of our own environmental resources, our genocidal tendencies coupled to nuclear weapons now constitute the two most likely means by which the human species may reverse all its progress virtually overnight.

Despite increasing interest in genocide on the part of psychologists and biologists as well as some lay people, basic questions about it remain disputed. Do any animals routinely kill members of their own species, or is that a human invention without animal precedents? Throughout human history, has genocide been a rare aberration, or has it been common enough to rank as a human hallmark along with art and language? Is its frequency now increasing, because modern weapons permit push-button genocide and thereby reduce our instinctive inhibitions about killing fellow humans? Why have so many cases attracted so little attention? Are genocidal killers abnormal individuals, or are they normal people placed in unusual situations?

To understand genocide, we cannot proceed narrowly but must draw on biology, ethics, and psychology. Consequently our exploration of genocide will begin by tracing its biological history, from our animal ancestors to the Twentieth Century. After asking how killers have reconciled genocide with their ethical codes, we can examine its psychological effects on the perpetrators, surviving victims, and on-lookers. But before we search for answers to these questions, it is useful to start with the extermination of the Tasmanians, as a case study typical of a broad class of genocide.

*

Tasmania is a mountainous island similar in area to Ireland and lying 200 miles off Australia's southeast coast. When discovered by Europeans in 1642, it supported about 5,000 hunter-gatherers related to the Aborigines of the Australian mainland and with perhaps the simplest technology of any modern peoples. Tasmanians made only a few types of simple stone and wooden tools. Like the mainland Aborigines, they lacked metal tools, agriculture, livestock, pottery, and bows and arrows. Unlike the mainlanders, they also lacked boomerangs, dogs, nets, knowledge of sewing, and ability to start a fire.

Since the Tasmanians' sole boats were rafts capable of only short journeys, they had had no contact with any other humans since the rising sea level cut off Tasmania from Australia 10,000 years ago. Confined to their private universe for hundreds of generations, they had survived the longest isolation in modern human history – an isolation otherwise depicted only in science fiction. When the white colonists of Australia finally ended that isolation, no two peoples on Earth were less equipped to understand each other than were Tasmanians and whites.

The tragic collision of these two peoples led to conflict almost as soon as British sealers and settlers arrived around 1800. Whites kidnapped Tasmanian children as labourers, kidnapped women as consorts, mutilated or killed men, trespassed on hunting grounds, and tried to clear Tasmanians off their land. Thus, the conflict quickly focused on *lebensraum*, which throughout human history has been among the commonest causes of genocide. As a result of the kidnappings, the native population of northeast Tasmania in November 1830 had been reduced to seventy-two adult men, three adult women, and no children. One shepherd shot nineteen Tasmanians with a swivel gun loaded with nails. Four other shepherds ambushed a group of natives, killed thirty, and threw their bodies over a cliff remembered today as Victory Hill.

Naturally, Tasmanians retaliated, and whites counter-retaliated in turn. To end the escalation, Governor Arthur in April 1828 ordered all Tasmanians to leave the part of their island already settled by Europeans. To enforce this order, government-sponsored groups called roving parties, and consisting of convicts led by police, hunted down and killed Tasmanians. With the declaration of martial law in November 1828, solidiers were authorized to kill on sight any Tasmanian in the settled areas. Next, a bounty was declared on the natives: five British pounds for each adult, two pounds for each child, caught alive. 'Black catching', as it was called because of the Tasmanians' dark skins, became big business pursued by private as well as official roving parties. At the same time a commission headed by William Broughton, the Anglican archdeacon of

Australia, was set up to recommend an overall policy towards the natives. After considering proposals to capture them for sale as slaves, poison or trap them, or hunt them with dogs, the commission settled on continued bounties and the use of mounted police.

In 1830 a remarkable missionary, George Augustus Robinson, was hired to round up the remaining Tasmanians and take them to Flinders Island, thirty miles away. Robinson was convinced that he was acting for the good of the Tasmanians. He was paid 300 pounds in advance, 700 pounds on completing the job. Undergoing real dangers and hardship, and aided by a courageous native woman named Truganini, he succeeded in bringing in the remaining natives – initially, by persuading them that a worse fate awaited them if they did not surrender, but later at gunpoint. Many of Robinson's captives died en route to Flinders, but about 200 reached there, the last survivors of the former population of 5,000.

On Flinders Island Robinson was determined to civilize and christianize the survivors. His settlement was run like a jail, at a windy site with little fresh water. Children were separated from parents to facilitate the work of civilizing them. The regimented daily schedule included Bible reading, hymn singing, and inspection of beds and dishes for cleanliness and neatness. However, the jail diet caused malnutrition, which combined with illness to make the natives die. Few infants survived more than a few weeks. The government reduced expenditures in the hope that the natives would die out. By 1869 only Truganini, one other woman, and one man remained alive.

These last three Tasmanians attracted the interest of scientists, who believed them to be a missing link between humans and apes. Hence when the last man, one William Lanner, died in 1869, competing teams of physicians, led by Dr George Stokell from the Royal Society of Tasmania and Dr W.L. Crowther from the Royal College of Surgeons, alternately dug up and reburied Lanner's body, cutting off parts of it and stealing them back and forth from each other. Dr Crowther cut off the head, Dr Stokell the hands and feet, and someone else the ears and nose, as souvenirs. Dr Stokell made a tobacco pouch out of Lanner's skin.

Before Truganini, the last woman, died in 1876, she was terrified of similar post-mortem mutilation and asked in vain to be buried at sea. As she had feared, the Royal Society dug up her skeleton and put it on public display in the Tasmanian Museum, where it remained until 1947. In that year the Museum finally yielded to complaints of poor taste and transferred Truganini's skeleton to a room where only scientists could view it. That too stimulated complaints of poor taste. Finally, in 1976 – the centenary year of Truganini's death – her skeleton was cremated despite the Museum's objections, and her ashes were scattered at sea as she had requested.

While the Tasmanians were few in number, their extermination was disproportionately influential in Australian history, because Tasmania was the first Australian colony to solve its native problem and achieved the most nearly final solution. It had done so by apparently succeeding in getting rid of all its natives. (Actually, some children of Tasmanian women by white sealers survived, and their descendants today constitute an embarrassment to the Tasmanian government, which has not figured out what to do about them.) Many whites on the Australian mainland envied the thoroughness of the Tasmanian solution and wanted to imitate it, but they also learned a lesson from it. The extermination of the Tasmanians had been carried out in settled areas in full view of the urban press, and had attracted some negative comment. The extermination of the much more numerous mainland Aborigines was instead effected at or beyond the frontier, far from urban centres.

The mainland governments' instrument of this policy, modelled on the Tasmanian government's roving parties, was a branch of mounted police termed Native Police, who used search-and-destroy tactics to kill or drive out Aborigines. A typical strategy was to surround a camp at night, and to shoot the inhabitants in an attack at dawn. White settlers also made widespread use of poisoned food to kill Aborigines. Another common practice was round-ups in which captured Aborigines were kept chained together at the neck while being marched to jail and held there. The British novelist Anthony Trollope expressed the prevailing nineteenth-century British attitude towards Aborigines when he wrote, 'Of the Australian black man we may certainly say that he has to go. That he should perish without unnecessary suffering should be the aim of all who are concerned in the matter.'

These tactics continued in Australia long into the Twentieth Century. In an incident at Alice Springs in 1928, police massacred thirty-one Aborigines. The Australian Parliament refused to accept a report on the massacre, and two Aboriginal survivors (rather than the police) were put on trial for murder. Neck chains were still in use and defended as humane in 1958, when the Commissioner of Police for the state of Western Australia explained to the *Melbourne Herald* that Aboriginal prisoners preferred being chained.

The mainland Aborigines were too numerous to exterminate completely in the manner of the Tasmanians. However, from the arrival of British colonists in 1788 until the 1921 census, the Aboriginal population declined from about 300,000 to 60,000.

Today, the attitudes of white Australians towards their murderous history vary widely. While government policy and many whites' private views have become increasingly sympathetic to the Aborigines, other whites deny responsibility for genocide. For instance, in 1982 one of

Australia's leading news magazines, *The Bulletin*, published a letter by a lady named Patricia Cobern, who denied indignantly that white settlers had exterminated the Tasmanians. In fact, wrote Ms Cobern, the settlers were peace-loving and of high moral character, while Tasmanians were treacherous, murderous, war-like, filthy, gluttonous, vermin-infested, and disfigured by syphilis. Moreover, they took poor care of their infants, never bathed, and had repulsive marriage customs. They died out because of all those poor health practices, plus a death wish and lack of religious beliefs. It was just a coincidence that, after thousands of years of existence, they happened to die out during a conflict with the settlers. The only massacres were of settlers by Tasmanians, not vice versa. Besides, the settlers only armed themselves in self-defence, were unfamiliar with guns, and never shot more than forty-one Tasmanians at one time.

*

To place these cases of the Tasmanians and the Australian Aborigines in perspective, consider the three maps on pages 256–8, depicting for three different time periods some mass killings that have been labelled as genocide. These maps beg a question for which there is no simple answer: how to define genocide. Etymologically, it means 'group killing': the Greek root *genos*, meaning race, and the Latin root *-cide*, meaning killing (as in suicide, infanticide). The victims must be selected because they belong to a group, whether or not each victim as an individual has done something to provoke killing. As for the defining group characteristic, it may be racial (white Australians killing black Tasmanians), national (Russians killing fellow white Slavs, the Polish officers at Katyn in 1940), ethnic (the Hutu and Tutsi, two black African groups, killing each other in Rwanda and Burundi in the 1960s and 1970s), religious (Moslems and Christians killing each other in Lebanon in recent decades), or political (the Khmer Rouge killing their fellow Cambodians from 1975 to 1979).

While collective killing is the essence of genocide, one can argue over how narrow a definition to adopt. The word 'genocide' is often used so broadly that it loses meaning and we become tired of hearing it. Even if it is to be restricted to large-scale cases of collective killing, ambiguities remain. A sample of the ambiguities could run as follows.

How many deaths are needed for a killing to count as genocide rather than mere murder? This is a totally arbitrary question. Australians killed all 5,000 Tasmanians, and American settlers killed the last twenty Susquehanna Indians in 1763. Does the small number of available victims disqualify these killings as genocidal, despite the completeness of extermination?

SOME GENOCIDES, 1492-1900

	Deaths	Victims	Killers	Place	Date
1.	xx	Aleuts	Russians	Aleutian Islands	1745–70
2.	x	Beothuk Indians	French, Micmaws	Newfoundland	1497–1829
3.	xxxx	Indians	Americans	US	1620–1890
4.	xxxx	Caribbean Indians	Spaniards	West Indies	1492–1600
5.	xxxx	Indians	Spaniards	Central & South America	1498–1824
6.	xx	Araucanian Indians	Argentinians	Argentina	1870s
7.	xx	Protestants	Catholics	France	1572
8.	xx	Bushmen, Hottentots	Boers	South Africa	1652–1795
9.	xxx	Aborigines	Australians	Australia	1788–1928
10.	x	Tasmanians	Australians	Tasmania	1800–1876
11.	x	Morioris	Maoris	Chatham Islands	1835

x = less than 10,000; xx = 10,000 or more; xxx = 100,000 or more; xxxx = 1,000,000 or more

Must genocide be carried out by governments, or do private acts also count? The sociologist Irving Horowitz distinguished private acts as 'assassination', and defined genocide as 'a structural and systematic destruction of innocent people by a state bureaucratic apparatus'. However, there is a complete continuum from 'purely' governmental killings (Stalin's purges of his opponents) to 'purely' private killings (Brazilian land development companies hiring professional Indian killers). American Indians were killed by private citizens and the US army alike, while the Ibos in Northern Nigeria were killed both by street mobs and by soldiers. In 1835 the Te Ati Awa tribe of New Zealand Maoris succeeded in a bold plan to capture a ship, load it with supplies, invade the Chatham Islands, kill 300 of the occupants (another Polynesian

SOME GENOCIDES, 1900-1950

	Deaths	Victims	Killers	Place	Date
1.	xxxxx	Jews, gypsies, Poles, Russians	Nazis	occupied Europe	1939–45
2.	xxx	Serbs	Croats	Yugoslavia	1941–45
3.	xx	Polish officers	Russians	Katyn	1940
4.	xx	Jews	Ukrainians	Ukraine	1917–20
5.	xxxxx	political opponents	Russians	Russia	1929–39
6	xxx	ethnic minorities	Russians	Russia	1943–46
7.	xxxx	Armenians	Turks	Armenia	1915
8	xx	Hereros	Germans	Southwest Africa	1904
9.	xxx	Hindus, Moslems	Moslems, Hindus	India, Pakistan	1947

xx = 10,000 or more; xxx = 100,000 or more; xxxx = 1,000,000 or more; xxxxx = 10,000,000 or more

group called the Morioris), enslave the remainder, and thereby take over the islands. By Horowitz's definition, this and many other equally well-planned exterminations of one tribal group by another do not constitute genocide, because the tribes lacked a state bureaucratic apparatus.

If people die *en masse* as a result of callous actions not specifically designed to kill them, does that count as genocide? Well-planned genocide includes that of Tasmanians by Australians, that of Armenians by Turks during the First World War, and (most notably) those committed by the Nazis during the Second World War. At the other extreme, when the Choctaw, Cherokee, and Creek Indians of south-eastern US states were forced to resettle west of the Mississippi River in the 1830s, it was not President Andrew Jackson's specific intent that many Indians should die en route, but he also did not take the measures that would have been necessary to keep them alive. Their numerous

SOME GENOCIDES, 1950-1990

	Deaths	Victims	Killers	Place	Date
1.	xx	Indians	Brazilians	Brazil	1957–68
2.	x	Aché Indians	Paraguayans	Paraguay	1970s
3.	xx	Argentine civilians	Argentine army	Argentina	1976–83
4.	xx	Moslems, Christians	Christians, Moslems	Lebanon	1975–90
5	x	Ibos	North Nigerians	Nigeria	1966
6	xx	opponents	dictator	Equatorial Guinea	1977–79
7	x	opponents	Emperor Bokassa	Central African Republic	1978–79
8	xxx	South Sudanese	North Sudanese	Sudan	1955–72
9.	xxx	Ugandans	Idi Amin	Uganda	1971–79
10.	xx	Tutsi	Hutu	Rwanda	1962–63
11.	xxx	Hutu	Tutsi	Burundi	1972–73
12.	x	Arabs	Blacks	Zanzibar	1964
13.	x	Tamils, Sinhalese	Sinhalese, Tamils	Sri Lanka	1985
14.	xxxx	Bengalis	Pakistan army	Bangladesh	1971
15	xxxx	Cambodians	Khmer Rouge	Cambodia	1975–79
16.	xxx	communists & Chinese	Indonesians	Indonesia	1965–67
17	xx	Timorese	Indonesians	East Timor	1975–76

x = less than 10,000; xx = 10,000 or more; xxx = 100,000 or more; xxxx = 1,000,000 or more

deaths were instead merely an inevitable result of forced marches in winter with little or no food or clothing.

An unusually candid statement about the role of intent in genocide emerged when the Paraguayan government was charged with complicity in the disappearance of the Guayaki Indians, who had been enslaved, tortured, deprived of food and medicine, and massacred. Paraguay's defence minister replied quite simply that there had been no intent to destroy the Guayaki: 'Although there are victims and victimizers, there is not the third element necessary to establish the crime of genocide – that is, "intent". Therefore, as there is no intent, one cannot speak of "genocide".' Brazil's Permanent Representative to the UN similarly rebutted charges of Brazilian genocide against Amazonian Indians: '. . . there was lacking the special malice or motivation necessary to characterize the occurrence of genocide. The crimes in question were committed for exclusively economic reasons, the perpetrators having acted solely to take possession of the lands of their victims.'

Some mass killings, such as those of Jews and gypsies by Nazis, were unprovoked; the slaughter was not in retaliation for previous murders committed by the slaughtered. In many other cases, however, a mass killing culminates a series of murders and countermurders. When a provocation is followed by massive retaliation out of all proportion to the provocation, how do we decide when 'mere' retaliation becomes genocide? At the Algerian town of Sétif in May 1945, celebrations of the end of the Second World War developed into a race riot in which Algerians killed 103 French. The savage French response consisted of planes destroying forty-four villages, a cruiser bombarding coastal towns, civilian commandos organizing reprisal massacres, and troops killing indiscriminately. The Algerian dead numbered 1,500 according to the French, 50,000 according to the Algerians. The interpretations of this event differ as do the estimates of the dead: to the French, it was suppression of a revolt; to the Algerians, it was a genocidal massacre.

*

Instances of genocide prove as hard to pigeonhole in their motivation as in their definition. While several motives may operate simultaneously, it is convenient to divide them into four types. In the first two types there is a real conflict of interest over land or power, whether or not the conflict is also disguised in ideology. In the other two types such conflict is minimal, and the motivation is more purely ideological or psychological.

Perhaps the commonest motive for genocide arises when a militarily stronger people attempt to occupy the land of a weaker people, who resist. Among the innumerable straightforward cases of this sort are not only the killing of Tasmanians and Australian Aborigines by white Australians, but also the killings of American Indians by white Americans, of Araucanian Indians by Argentinians, and of Bushmen and Hottentots by the Boer settlers of South Africa.

Another common motive involves a lengthy power struggle within a pluralistic society, leading to one group seeking a final solution by killing the other. Cases involving two different ethnic groups are the killing of Tutsi in Rwanda by Hutu in 1962–63, of Hutu in Burundi by Tutsi in 1972–73, of Serbs by Croats in Yugoslavia during the Second World War, of Croats by Serbs at the end of that war, and of Arabs in Zanzibar by blacks in 1964. However, the killer and killed may belong to the same ethnic group and may differ only in political views. Such was the case in history's largest known genocide, claiming an estimated twenty million victims in the decade 1929–39 and sixty-six million between 1917 and 1959 – that committed by the Russian government against its political opponents, many of whom were ethnic Russians. Political killings lagging far behind this record are the Khmer Rouge purge of several million fellow Cambodians during the 1970s, and Indonesia's killing of hundreds of thousands of communists in 1965–67.

In these two motives for genocide, the victims could be viewed as a significant obstacle to the killers' control of land or power. At the opposite extreme are scapegoat killings of a helpless minority blamed for frustrations of their killers. Jews were killed by fourteenth-century Christians as scapegoats for the bubonic plague, by early twentieth-century Russians as scapegoats for Russia's political problems, by Ukrainians after the First World War as scapegoats for the Bolshevist threat, and by the Nazis during the Second World War as scapegoats for Germany's defeat in the First World War. When the US Seventh Cavalry machine-gunned several hundred Sioux Indians at Wounded Knee in 1890, the soldiers were taking belated revenge for the Sioux's annihilating counterattack on Custer's Seventh Cavalry force at the Battle of the Little Big Horn fourteen years previously. In 1943–44, at the height of Russia's suffering from the Nazi invasion, Stalin ordered the killing or deportation of six ethnic minorities who served as scapegoats: the Balkars, Chechens, Crimean Tatars, Ingush, Kalmyks, and Karachai.

Racial and religious persecutions have served as the remaining class of motives. While I do not claim to understand the Nazi mentality, the Nazis' extermination of Gypsies may have stemmed from relatively 'pure' racial motivation, while scapegoating joined religious and racial motives in the extermination of Jews. The list of religious massacres is almost infinitely long. It includes the First Crusaders' massacre of all Moslems and Jews in Jerusalem when that city was finally captured in 1099, and the St Bartholomew's Day massacre of French Protestants by Catholics in 1572. Of course, racial and religious motives have contributed heavily to genocide provoked by land struggles, power struggles, and scapegoating.

*

Even if one allows for these disagreements over definitions and motives, plenty of cases of genocide remain. Let us now see how far back in and before our history as a species the record of genocide extends.

Is it true, as often claimed, that man is unique among animals in killing members of his own species? For example, the distinguished biologist Konrad Lorenz, in his book *On Aggression*, argued that animals' aggressive instincts are held in check by instinctive inhibitions against murder. But in human history this equilibrium supposedly became upset by the invention of weapons, and our inherited inhibitions were no longer strong enough to restrain our newly acquired powers of killing. This view of man as the unique killer and evolutionary misfit has been accepted by Arthur Koestler and many other popular writers.

Actually, studies in recent decades have documented murder in many, though certainly not all, animal species. Massacre of a neighbouring individual or troop may be beneficial to an animal, if it can thereby take over the neighbour's territory, food, or females. But attacks also involve some risk to the attacker. Many animal species lack the means to kill their fellows, and of those species with the means, some refrain from using them. It may sound utterly repugnant to do a cost and benefit analysis of murder, but such analyses nevertheless help one understand why murder appears to characterize only some animal species.

In non-social species, murders are necessarily just of one individual by another. However, in social carnivorous species, like lions, wolves, hyenas, and ants, murder may take the form of coordinated attacks by members of one troop on members of a neighbouring troop – that is, mass killings or 'wars'. The form of war varies among species. Males may spare and mate with neighbouring females, kill the infants, and drive off (langur monkeys) or even kill (lions) neighbouring males; or both males and females may be killed (wolves). As one example, here is Hans Kruuk's account of a battle between two hyena clans in Tanzania's Ngorongoro Crater:

> About a dozen of the Scratching Rock hyenas, though, grabbed one of the Mungi males and bit him wherever they could – especially in the belly, the feet, and the ears. The victim was completely covered by his attackers, who proceeded to maul him for about ten minutes. . . . The Mungi male was literally pulled apart, and when I later studied the injuries more closely, it appeared that his ears were bitten off and so were his feet and testicles, he was paralyzed by a spinal injury, had large gashes in the hind legs and belly, and subcutaneous haemor-rhages all over.

Of particular interest in understanding our genocidal origins is the behaviour of two of our three closest relatives, gorillas and common chimpanzees. Two decades ago, any biologist would have assumed that our ability to wield tools and to lay concerted group plans made us far more murderous than apes – if indeed apes were murderous at all. Recent discoveries about apes suggest, however, that a gorilla or common chimp stands at least as good a chance of being murdered as the average human. Among gorillas, for instance, males fight each other for ownership of harems of females, and the victor may kill the loser's infants as well as the loser himself. Such fighting is a major cause of death for infant and adult male gorillas. The typical gorilla mother loses at least one infant to infanticidal males in the course of her life. Conversely, thirty-eight per cent of infant gorilla deaths are due to infanticide.

Especially instructive, because it could be documented in detail, was the extermination of one of the common chimpanzee bands that Jane Goodall studied, carried out between 1974 and 1977 by another band. At the end of 1973 the two bands were fairly evenly matched: the Kasakela band to the north, with eight mature males and occupying fifteen square kilometers; and the Kahama band to the south, with six mature males and occupying ten square kilometers. The first fatal incident occurred in January 1974, when six of the Kasakela adult males, one adolescent male, and one adult female left behind the young Kasakela chimps, travelled south, then moved silently and more quickly south when they heard chimp calls from that direction, until they surprised a Kahama male referred to as Godi. One Kasakela male pulled the fleeing Godi to the ground, sat on his head, and pinned out his legs while the others spent ten minutes hitting and biting him. Finally, one attacker threw a large rock at Godi, and the attackers then left. Although able to stand up, Godi was badly wounded, bleeding, and had puncture marks. He was never seen again and presumably died of his injuries.

The next month, three Kasakela males and one female again travelled south and attacked the Kahama male Dé, who was already weak from a previous attack or illness. The attackers pulled Dé out of a tree, stamped on him, bit and hit him, and tore off pieces of his skin. A Kahama oestrus female with Dé was forced to return northwards with the attackers. Two months later Dé was seen still alive but emaciated, with his spine and pelvis protruding, some fingernails and part of a toe torn off, and his scrotum shrunk to one-fifth of normal size. He was not seen thereafter.

In February 1975 five adult and one adolescent Kasakela males tracked down and attacked Goliath, an old Kahama male. For eighteen minutes they hit, bit, and kicked him, stamped on him, lifted and dropped him, dragged him over the ground, and twisted his leg. At the end of the attack Goliath was unable to sit up and was not seen again.

While the above attacks were aimed at Kahama males, in September 1975 the Kahama female Madam Bee was fatally injured after at least four non-fatal attacks over the course of the preceding year. The attack was carried out by four Kasakela adult males, while one adolescent male and four Kasakela females (including Madam Bee's kidnapped daughter) watched. The attackers hit, slapped, and dragged Madam Bee, stamped and pounded on her, threw her to the ground, picked her up and slammed her down, and rolled her downhill. She died five days later.

In May 1977 five Kasakela males killed the Kahama male Charlie, but details of the fight were not observed. In November 1977 six Kasakela males caught the Kahama male Sniff and hit, bit, and pulled him, dragged him by the legs, and broke his left leg. He was still alive the next day but was not seen again.

Of the remaining Kahama chimps, two adult males and two adult females disappeared from unknown causes, while two young females transferred to the Kasakela band, which proceeded to occupy the former Kahama territory. However, in 1979 the next band to the south, the larger Kalande band with at least nine adult males, began to encroach on Kasakela territory and may have accounted for several vanished or wounded Kasakela chimps. Similar intergroup assaults have been observed in the sole other long-term field study of common chimps, but not in long-term studies of pygmy chimps.

If one judges these murderous common chimps by the standards of human killers, it is hard not to be struck by their inefficiency. Even though groups of three to six attackers assaulted a single victim, quickly rendered him or her defenceless, and continued the assault for ten to twenty minutes or more, the victim was always still alive at the end of that time. However, the attackers did succeed in immobilizing the victim and often causing eventual death. The pattern was that the victim initially crouched and may have tried to protect his head but then gave up any attempt at defence, and the attack continued beyond the point where the victim ceased moving. In this respect the inter-band attacks differ from the milder fights that often occur within a band. Chimps' inefficiency as killers reflects their lack of weapons, but it remains surprising that they have not learned to kill by strangling, although that would be within their capabilities.

Not only is each individual killing inefficient by our standards, but so is the whole course of chimp genocide. It took three years and ten months from the first killing of a Kahama chimp to the band's end, and all killings were of individuals, never of several Kahama chimps at once. In contrast, Australia's settlers often succeeded in eliminating a band of Aborigines in a single dawn attack. Partly, this inefficiency again reflects chimps' lack of weapons. Since all chimps are equally unarmed, killings can succeed only by several attackers overpowering a single victim, whereas Australia's settlers had the advantage of guns over unarmed Aborigines and could shoot many at once. Partly, too, genocidal chimps are much inferior to humans in brainpower and hence in strategic planning. Chimps apparently cannot plan a night attack or a coordinated ambush by a split assault team.

However genocidal chimps do seem to evince intent and unsophisticated planning. The Kahama killings resulted from Kasakela groups proceeding directly, quickly, silently, and nervously towards or into Kahama territory, sitting in trees and listening for nearly an hour, and finally running to Kahama chimps that they detected. Chimps also share xenophobia with us; they clearly recognize members of other bands as different from, and treat them very differently from, members of their own band.

In short, of all our human hallmarks – art, spoken language, drugs, and the others – the one that has been derived most straightforwardly from animal precursors is genocide. Common chimps already carried out planned killings, extermination of neighbouring bands, wars of territorial conquest, and abduction of young nubile females. If chimps were given spears and some instruction in their use, their killings would undoubtedly begin to approach ours in efficiency. Chimpanzee behaviour suggests that a major reason for our human hallmark of group living was defence against other human groups, especially once we acquired weapons and a large enough brain to plan ambushes. If this reasoning is correct, then anthropologists' traditional emphasis on 'man the hunter' as a driving force of human evolution might be valid after all – with the difference that we ourselves, not mammoths, were our own prey and the predator that forced us into group living.

*

Of the two patterns of genocide commonest among humans, both have animal precedents: killing both men and women fits the common chimpanzee and wolf pattern, while killing men and sparing women fits the gorilla and lion pattern. Unprecedented even among animals, however, is a procedure adopted from 1976 to 1983 by the Argentine military, in the course of killing over 10,000 political opponents and their families, the *desaparecidos*. Victims included the usual men, non-pregnant women, and children down to the age of three or four years, who were often tortured before being killed. But Argentina's soldiers made a unique contribution to animal behaviour by specializing in killing pregnant women, who were arrested, kept alive until they delivered, and then shot in the head, so that the newborn infant could be adopted by childless military parents.

If we are not unique among animals in our own propensity for murder, might our propensities nevertheless be a pathological fruit of modern civilization? Modern writers, disgusted by destruction of 'primitive' societies by 'advanced' societies, tend to idealize the former as noble savages who supposedly are peace-loving, or who commit only isolated murders rather than massacres. Erich Fromm believed the warfare of hunter-gatherer societies to be 'characteristically unbloody'. Certainly some pre-literate peoples (Pygmies, Eskimos) seem less warlike than some others (New Guineans, Great Plains and Amazonian Indians). Even the warlike peoples – so it is claimed – practise war in a ritualized fashion and stop when only a few adversaries have been killed. But this idealization does not match my experience of the New Guinea highlanders, who are often cited as practising limited or ritualized war. While most fighting in New Guinea consisted of skirmishes leaving no or few dead, groups sometimes did succeed in massacring their neighbours. Like other peoples, New Guineans tried to drive off or kill their neighbours on occasions when they found it advantageous, safe, or a matter of survival to do so.

When we consider early literate civilizations, written records testify to the frequency of genocide. The wars of the Greeks and Trojans, of Rome and Carthage, and of the Assyrians and Babylonians and Persians proceeded to a common end: the slaughter of the defeated irrespective of sex, or else the killing of the men and enslavement of the women. We all know the biblical account of how the walls of Jericho came tumbling down at the sound of Joshua's trumpets. Less often quoted is the sequel. Joshua obeyed the Lord's command to slaughter the inhabitants of Jericho as well as of Ai, Makkedeh, Libnah, Hebron, Debir, and many other cities. This was considered so ordinary that the Book of Joshua devotes only a phrase to each slaughter, as if to say, of course he killed all the inhabitants, what else would you expect? The sole account requiring elaboration is of the slaughter at Jericho itself, where Joshua did something really unusual; he spared the lives of one family (because they had helped his messengers).

We find similar episodes in accounts of the wars of the Crusaders, Pacific islanders, and many other groups. Obviously, I am not saying that slaughter of the defeated irrespective of sex has always followed crushing defeat in war. Yet either that outcome, or else milder versions like the killing of men and enslavement of women, happened often enough that they must be considered more than a rare aberration in our view of human nature. Since 1950 there have been nearly twenty episodes of genocide, including two claiming over a million victims each

(Bangladesh in 1971, Cambodia in the late 1970s) and four more with over a hundred thousand victims each (the Sudan and Indonesia in the 1960s, Burundi and Uganda in the 1970s) (see map on page 258).

Thus, genocide has been part of our human and prehuman heritage for millions of years. In light of this long history, what about our impression that twentieth-century genocide is unique? There is little doubt that Stalin and Hitler set new records for number of victims, because they enjoyed three advantages over killers of earlier centuries – denser populations of victims, improved communications for rounding up victims, and improved technology for mass killing. As another example of how technology can expedite genocide, the Solomon Islanders of Roviana Lagoon in the Southwest Pacific were famous for their head-hunting raids that depopulated neighbouring islands. However, as my Roviana friends explained to me, these raids did not blossom until steel axes reached the Solomon Islands in the Nineteenth Century. Beheading a man with a stone axe is difficult, and the axe blade quickly loses its sharp edge and is tedious to resharpen.

A much more controversial question is whether technology also makes genocide psychologically easier today, as Konrad Lorenz has argued. His reasoning goes as follows. As humans evolved from apes, we depended increasingly for our food on killing animals. However, we also lived in societies of more and more individuals, between whom cooperation was essential. Such societies could not maintain themselves unless we developed strong inhibitions about killing fellow humans. Throughout most of our evolutionary history, our weapons operated only at close quarters, so it was enough that we be inhibited from looking another person in the face and killing him/her. Modern push-button weapons bypassed these inhibitions by enabling us to kill without even seeing our victims' faces. Technology thus created the psychological prerequisites for the white-collar genocides of Auschwitz and Treblinka, of Hiroshima and Dresden.

I am uncertain whether this psychological argument really contributed significantly to the modern ease of genocide. The past *frequency* of genocide seems to have been at least as high as today's, though practical considerations limited the number of victims. To understand genocide further, we must leave dates and numbers and inquire about the ethics of killing.

*

That our urge to kill is restrained by our ethics almost all the time is obvious. The puzzle is: what unleashes it?

Today, while we may divide the world's people into 'us' and 'them', we know that there are thousands of types of 'them', all differing from each other as well as from us in language, appearance, and habits. To waste words on pointing this out seems silly: we all know it from books and television, and most of us also know it from first-hand experience of travel. It is hard to transfer ourselves back into the frame of mind prevailing throughout much of human history, described in Chapter Thirteen. Like chimpanzees, gorillas, and social carnivores, we lived in band territories. The known world was much smaller and simpler than it is today; there were only a few known types of 'them', one's immediate neighbours.

For example, in New Guinea until recently, each tribe maintained a shifting pattern of warfare and alliance with each of its neighbours. A person might enter the next valley on a friendly visit (never quite without danger) or on a war raid, but the chances of being able to traverse a sequence of several valleys in friendship were negligible. The powerful rules about treatment of one's fellow 'us' did not apply to 'them', those dimly understood, neighbouring enemies. As I walked between New Guinea valleys, people who themselves practised cannibalism and were only a decade out of the Stone Age routinely warned me about the unspeakably primitive, vile, and cannibalistic habits of the people whom I would encounter in the next valley. Even Al Capone's gangs in twentieth-century Chicago made a policy of hiring out-of-town killers, so that the assassin could feel that he was killing one of 'them' rather than of 'us'.

The writings of classical Greece reveal an extension of this tribal territorialism. The known world was larger and more diverse, but 'us' Greeks were still distinguished from 'them' barbarians. Our word 'barbarian' is derived from the Greek *barbaroi*, which simply means non-Greek foreigners. Egyptians and Persians, whose level of civilisation was like that of the Greeks, were nevertheless *barbaroi*. The ideal of conduct was not to treat all men equally, but instead to reward one's friends and to punish one's enemies. When the Athenian author Xenophon wanted to express the highest praise for his admired leader Cyrus, Xenophon related how Cyrus always repaid his friends' good turns more generously, and how Cyrus retaliated on his enemies' misdeeds more severely (for example, by gouging out their eyes or cutting off their hands).

Humans, like the Mungi and Scratching Rock hyena clans, practised a dual standard of behaviour, with strong inhibitions about killing one of 'us', but a green light to kill 'them' when it was safe to do so. Genocide was acceptable under this dichotomy, whether one considers the dichotomy as an inherited animal instinct or as a uniquely human ethical code. We all still acquire in childhood our own arbitrary dichotomous criteria for respecting or scorning other humans. I recall a scene at Goroka airport in the New Guinea highlands, when my Tudawhe field assistants were standing awkwardly in torn shirts and bare feet, and an unshaven, unwashed white man with a strong Australian accent and hat crumpled over his eyes approached. Even before he had begun to sneer at the Tudawhes as 'black bums, they won't be fit to run this country for a century', I had begun to think to myself, 'Dumb Aussie redneck, why doesn't he go home to his goddamn sheep dip.' There it was, a blueprint for genocide: I scorning the Australian, and he scorning the Tudawhes, based on collective characteristics taken in at a glance.

With time, this ancient dichotomizing has become increasingly unacceptable as a basis for an ethical code. Instead, there has been some tendency towards paying at least lip-service to a universal code – that is, one stipulating similar rules for treating different peoples. Genocide conflicts directly with a universal code.

Despite this ethical conflict, numerous modern perpetrators of genocide have managed to take unabashed pride in their accomplishments. When Argentina's General Julio Argentino Roca opened the pampas for white settlement by ruthlessly exterminating the Araucanian Indians, a delighted and grateful Argentinian nation elected him president in 1880. How do today's practitioners of genocide wriggle out of the conflict between their actions and a universal code of ethics? They resort to one of three types of rationalizations, all of which are variations on a simple psychological theme, 'Blame the victim!'

Firstly, most believers in a universal code still consider self-defence justified. This is a usefully elastic rationalization, because 'they' can invariably be provoked into some behaviour adequate to justify self-defence. For instance, the Tasmanians delivered an excuse to genocidal white colonists by killing an estimated total of 183 colonists over thirty-four years, while being provoked by a far greater number of mutilations, kidnappings, rapes, and murders. Even Hitler claimed self-defence in starting the Second World War, and he went to the trouble of faking a Polish attack on a German border post.

Possessing the 'right' religion or race or political belief, or claiming to represent progress or a higher level of civilization, is a second traditional justification for inflicting anything, including genocide, on those possessing the wrong principle. When I was a student in Munich in 1962, unrepentant Nazis still explained to me matter-of-factly that the Germans had had to invade Russia, because the Russian people had adopted Communism. My fifteen field assistants in New Guinea's Fakfak Mountains all looked pretty similar to me, but eventually they began explaining to me which of them were Moslems and which were

Christians, and why the former (or the latter) were irredeemably lower humans. There is an almost universal hierarchy of scorn, according to which literate peoples with advanced metallurgy (for instance, white colonialists in Africa) look down on herders (such as Tutsi, Hottentots), who look down on farmers (such as Hutu), who look down on nomads or hunter-gatherers (such as Pygmies, Bushmen).

Finally, our ethical codes regard animals and humans differently. Hence modern perpetrators of genocide routinely compare their victims to animals in order to justify the killings. Nazis considered Jews to be subhuman lice; the French settlers of Algeria referred to local Moslems as *ratons* (rats); 'civilized' Paraguayans described the Aché hunter-gatherers as rabid rats; Boers called Africans *bobbejaan* (baboons); and educated northern Nigerians viewed Ibos as subhuman vermin. The English language is rich in animal names used as pejoratives: you pig (ape, bitch, cur, dog, ox, rat, swine).

All three types of ethical rationalizations were employed by Australian colonists to justify exterminating Tasmanians. However, my fellow Americans and I can obtain a better insight into the rationalization process by focusing on the case that we have been trained from childhood to rationalize: our not-quite-complete extermination of American Indians. A set of attitudes that we absorb goes roughly as follows.

To begin with, we do not discuss the Indian tragedy much – not nearly as much as the genocide of the Second World War in Europe, for instance. Our great national tragedy is instead viewed as the Civil War. Insofar as we stop to think about white versus Indian conflict, we consider it as belonging to the distant past, and we describe it in military language, such as the Pequod War, Great Swamp Fight, Battle of Wounded Knee, Conquest of the West, and so on. Indians, in our view, were warlike and violent even towards other Indian tribes, masters of ambush and treachery. They were famous for their barbarity, notably for the distinctively Indian practices of torturing captives and scalping enemies. They were few in number and lived as nomadic hunters, especially bison hunters. The Indian population of the US as of 1492 is traditionally estimated at one million. This figure is so trivial, compared to the present US population of 250 million, that the inevitability of whites occupying this virtually empty continent becomes immediately apparent. Many Indians died from smallpox and other diseases. The aforementioned attitudes guided the Indian policy of the most admired US presidents and leaders from George Washington onwards (see quotations at the end of this chapter).

These rationalizations rest on a transformation of historical facts. Military language implies declared warfare waged by adult male combatants. Actually, common white tactics were sneak attacks (often by civilians) on villages or encampments to kill Indians of any age and either sex. Within the first century of white settlement, governments were paying scalp bounties to semi-professional killers of Indians. Contemporary European societies were at least as warlike and violent as Indian societies, when one considers the European frequency of rebellions, class wars, drunken violence, legalized violence against criminals, and total war including destruction of food and property. Torture was exquisitely refined in Europe: think of drawing and quartering, burning at the stake, and the rack. While the pre-contact Indian population of North America is the subject of widely varying opinions, plausible recent estimates are about eighteen million, a population not reached by white settlers of the US till around 1840. Although some Indians in the US were semi-nomadic hunters without agriculture, most were settled farmers living in villages. Disease may well have been the biggest killer of Indians, but some of the epidemics were intentionally transmitted by whites, and the epidemics still left plenty of Indians to kill by more direct means. It was only in 1916 that the last 'wild' Indian in the US (the Yahi Indian known as Ishi) died, and frank and unapologetic memoirs by the white killers of his tribe were still being published as recently as 1923.

In short, Americans romanticize the white/Indian conflict as battles of grown men on horseback, fought by US cavalry and cowboys against fierce nomadic bison-hunters able to offer strong resistance. The conflict is more accurately described as one race of civilian peasant farmers exterminating another. We Americans remember with outrage our own losses at the Alamo (*circa* 200 dead), on the battleship U.S.S. *Maine* (260 dead), and at Pearl Harbor (about 2,200 dead), the incidents that galvanized our support for the Mexican War, Spanish-American War,

Opposite page: ISHI, the last surviving Indian of the Yahi tribe of northern California. The photograph on the opposite page shows him, starving and terrified, on 29 August 1911, the day that he emerged from forty-one years of hiding in a remote canyon. Most of his tribe was massacred by white settlers between 1853 and 1870. In 1870 the sixteen survivors of the final massacre went into concealment in the Mount Lassen wilderness and continued to live as hunter-gatherers. In November 1908, when the survivors had dwindled to four, surveyors stumbled upon their camp and took all their tools, clothes, and winter food supplies, with the result that three of the Yahis (Ishi's mother, his sister, and an old man) died. Ishi remained alone for three more years until he could stand it no longer and walked out to white civilization, expecting to be lynched there. In fact, he was employed by the University of California Museum at San Francisco and died of tuberculosis in 1916. The photograph is from the archives of the Lowie Museum of Anthropology, University of California, Berkeley.

and the Second World War respectively. Yet these numbers of dead are dwarfed by the forgotten losses that we inflicted on the Indians. Introspection shows us how, in rewriting our great national tragedy, we like so many modern peoples reconciled genocide with a universal code of ethics. The solution was to plead self-defence and overriding principle, and to view the victims as savage animals.

*

Our rewriting of American history stems from the aspect of genocide that is of greatest practical importance in preventing it – its psychological effects on killers, victims, and third parties. The most puzzling question involves the effect, or rather the apparent non-effect, on third parties. On first thought, one might expect that no horror could grip public attention as much as the intentional, collective, and savage killing of many people. In reality, genocide rarely grips the public's attention in other countries, and even more rarely are interrupted by foreign intervention. Who among us paid much attention to the slaughter of Zanzibar's Arabs in 1964, or of Paraguay's Aché Indians in the 1970s?

Contrast our lack of response to these and all the other instances of genocide in recent decades with our strong reaction to the sole two cases of modern genocide that remain vivid in our imagination, that of the Nazis against the Jews and (much less vivid for most people) that of the Turks against the Armenians. These cases differ in three crucial respects from the genocide we ignore: the victims were whites, with whom other whites identify; the perpetrators were our war enemies whom we were encouraged to hate as evil (especially the Nazis); and there are articulate survivors in the US, who go to much effort to force us to remember. Thus, it takes a rather special constellation of circumstances to get third parties to focus on genocide.

The strange passivity of third parties is exemplified by that of governments, whose actions reflect collective human psychology. While the United Nations in 1948 adopted a Convention on Genocide that declared it a crime, the UN has never taken serious steps to prevent, halt, or punish it, despite complaints lodged before the UN against on-going genocide in Bangladesh, Burundi, Cambodia, Paraguay, and Uganda. To a complaint lodged against Uganda at the height of Idi Amin's terror, the UN Secretary-General responded only by asking Amin himself to investigate. The United States is not even among the nations that ratified the UN Convention on Genocide.

Is our puzzling lack of response because we did not know, or could not find out, about on-going genocide? Certainly not: many cases of genocide of the 1960s and 1970s received detailed publicity at the time, including those in Bangladesh, Brazil, Burundi, Cambodia, East Timor, Equatorial Guinea, Indonesia, Lebanon, Paraguay, Rwanda, Sudan, Uganda, and Zanzibar. (The casualties in Bangladesh and Cambodia each topped a million.) For example, in 1968 the Brazilian government filed criminal charges against 134 of the 700 employees of its Indian Protection Service for their acts in exterminating Amazonian Indian tribes. Among the acts detailed in the 5,115-page Figueiredo Report by Brazil's attorney general, and announced at a press conference by Brazil's minister of the interior, were the following: killing of Indians by dynamite, machine-guns, arsenic-laced sugar, and intentionally introduced smallpox, influenza, tuberculosis, and measles; kidnapping of Indian children as slaves; and the hiring of professional killers of Indians by land development countries. Accounts of the Figueiredo report appeared in the American and British press, but failed to stimulate much reaction.

One might thus conclude that most people simply do not care about injustice done to other people, or regard it as none of their business. This is undoubtedly part of the explanation, but not all of it. Many people care passionately about some injustices, such as apartheid in South Africa; why not also about genocide? This question was addressed poignantly, to the Organization of African States, by Hutu victims of the Tutsi in Burundi, where somewhere between 80,000 and 200,000 Hutu were killed in 1972.

Tutsi apartheid is established more ferociously than the apartheid of Vorster, more inhumanly than Portuguese colonialism. Outside of Hitler's Nazi movement, there is nothing to compete with it in world history. And the peoples of Africa say nothing. African heads of state receive the executioner Micombero [President of Burundi, a Tutsi] and clasp his hand in fraternal greeting. Sirs, heads of state, if you wish to help the African peoples of Namibia, Zimbabwe, Angola, Mozambique, and Guinea–Bissau to liberate themselves from their white oppressors, you have no right to let Africans murder other Africans. . . . Are you waiting until the entire Hutu ethnic group of Burundi is exterminated before raising your voices?

To understand this lack of reaction of third parties, we need to appreciate the reaction of surviving victims. Psychiatrists who have studied witnesses of genocide, such as Auschwitz survivors, describe the effects on them as 'psychological numbing'. Most of us have experienced the intense and lasting pain that comes when a loved friend or relative dies a natural death, out of sight. It is virtually impossible for us to imagine the multiplied intensity of pain when one is forced to watch at close hand many loved friends and relatives being killed with extreme savagery. For the survivors, there is a shattering of the implicit belief system under which such savagery was forbidden; a sense of stigma that one must indeed be worthless to have been singled out for such cruelty; and a sense of guilt at surviving, when one's companions died. Just as intense physical pain numbs us, so does intense psychological pain – there is no other way to survive and remain sane. For myself, these reactions were personified in a relative who survived two years in Auschwitz, and who remained practically unable to cry for decades afterwards.

As for the reactions of the killers, those killers whose ethical code distinguishes between 'us' and 'them' may be able to feel pride, but those reared under a universal ethical code may share the numbing of their victims, exacerbated by the guilt of perpetration. Hundreds of thousands of Americans who fought in Vietnam suffered this numbing. Even the descendants of practitioners of genocide – descendants who have no individual responsibility – may feel a collective guilt, the mirror image of the collective labelling of victims that defines genocide. To reduce the pain of guilt, the descendants often rewrite history; witness the response of modern Americans, or that of Ms Cobern and many other modern Australians.

We can now begin to understand better the lack of reaction of third parties to genocide. Genocide inflicts crippling and lasting psychological damage on the victims and killers who experience it first-hand. But it also may leave deep scars on those who hear about it only second-hand, such as the children of Auschwitz survivors, or the psychotherapists who treat the survivors and Vietnam veterans. Therapists who have trained professionally to be able to listen to human misery often cannot bear to hear the sickening recollections of those involved in genocide. If paid professionals cannot stand it, who can blame the lay public for refusing to listen?

Consider the reactions of Robert Jay Lifton, an American psychiatrist who had already had much experience with survivors of extreme situations before he interviewed survivors of the Hiroshima A-bomb:

> . . . now, instead of dealing with 'the atomic bomb problem', I was confronted with the brutal details of actual experiences of human beings who sat before me. I found that the completion of each of these early interviews left me profoundly shocked and emotionally spent. But very soon – within a few days, in fact – I noticed that my reactions were changing. I was listening to descriptions of the same horrors, but their effect upon me had lessened. The experience was an unforgettable demonstration of the 'psychic closing off' we shall see to be characteristic of all aspects of atomic bomb exposure. . . .

*

What genocidal acts can we expect from *Homo sapiens* in the future? There are plenty of obvious reasons for pessimism. The world abounds with trouble spots that seem ripe for genocide: South Africa, Northern Ireland, Sri Lanka, New Caledonia, and the Middle East, to name just a few. Totalitarian governments bent on genocide seem unstoppable. Modern weaponry permits one to kill ever larger numbers of victims, to be a killer while wearing a coat and tie, and even to effect a universal genocide of the human race.

At the same time, I see grounds for cautious optimism that the future need not be as murderous as the past. In many countries today, people of different races or religions or ethnic groups live together, with varying degrees of social justice but at least without open mass murder – for instance, Switzerland, Belgium, Papua New Guinea, Fiji, even the post-Ishi USA. Some attempts at genocide have been successfully interrupted, reduced, or prevented by the efforts or anticipated reactions of third parties. Even the Nazi extermination of Jews, which we view as the most efficient and unstoppable of genocides, was thwarted in Denmark, Bulgaria, and every other occupied state where the head of the dominant church publicly denounced deportation of Jews before or as soon as it began. A further hopeful sign is that modern travel, television, and photography enable us to see other people living 10,000 miles away as human, like us. Much as we damn twentieth-century technology, it is blurring the distinction between 'us' and 'them' that makes genocide possible. While genocide was considered socially acceptable or even admirable in the pre-first-contact world, the modern spread of international culture and knowledge of distant peoples have been making it increasingly hard to justify.

Still, the risk of genocide will be with us as long as we cannot bear to understand it, and as long as we delude ourselves with the belief that only rare perverts could commit it. Granted, it is hard not to go numb while reading about genocide. It is hard to imagine how we, and other nice ordinary people that we know, could bring ourselves to look helpless people in the face while killing them. I came closest to being able to imagine it when a friend whom I had long known told me of a genocidal massacre at which he had been a killer.

Kariniga is a gentle Tudawhe tribesman who worked with me in New Guinea. We shared life-threatening situations, fears, and triumphs, and I like and admire him. One evening after I had known Kariniga for five years, he told me of an episode from his youth. There had been a long history of conflict between the Tudawhes and a neighbouring village of Daribi tribesmen. Tudawhes and Daribis seem quite similar to me, but

Kariniga had come to view Daribis as inexpressibly vile. In a series of ambushes the Daribis finally succeeded in picking off many Tudawhes, including Kariniga's father, until the surviving Tudawhes became desperate. All the remaining Tudawhe men surrounded the Daribi village at night and set fire to the huts at dawn. As the sleepy Daribis stumbled down the steps of their burning huts, they were speared. Some Daribis succeeded in escaping to hide in the forest, where Tudawhes tracked down and killed most of them during the following weeks. The establishment of Australian government control ended the hunt before Kariniga could catch his father's killer.

Since that evening, I have often found myself shuddering as I recalled details of it – the glow in Kariniga's eyes as he told me of the dawn massacre; those intensely satisfying moments when he finally drove his spear into some of his people's murderers; and his tears of rage and frustration at the escape of his father's killer, whom he still hoped to kill some day with poison. That evening, I thought I understood how at least one nice person had brought himself to kill. The potential for genocide that circumstances thrust on Kariniga lies within all of us. As the growth of world population sharpens conflicts between and within societies, humans will have more urge to kill each other, and more effective weapons with which to do it. To listen to first-person accounts of genocide is unbearably painful. But if we continue to turn away and to not understand it, when will it be our own turn to become the killers, or the victims?

Appendix

INDIAN POLICIES OF
SOME FAMOUS AMERICANS

PRESIDENT GEORGE WASHINGTON. 'The immediate objectives are the total destruction and devastation of their settlements. It will be essential to ruin their crops in the ground and prevent their planting more.'

BENJAMIN FRANKLIN. 'If it be the Design of Providence to Extirpate these Savages in order to make room for Cultivators of the Earth, it seems not improbable that Rum may be the appointed means.'

PRESIDENT THOMAS JEFFERSON. 'This unfortunate race, whom we had been taking so much pains to save and to civilize, have by their unexpected desertion and ferocious barbarities justified extermination and now await our decision on their fate.'

PRESIDENT JOHN QUINCY ADAMS. 'What is the right of the huntsman to the forest of a thousand miles over which he has accidentally ranged in quest of prey?'

PRESIDENT JAMES MONROE. 'The hunter or savage state requires a greater extent of territory to sustain it, than is compatible with the progress and just claims of civilized life . . . and must yield to it.'

PRESIDENT ANDREW JACKSON. 'They have neither the intelligence, the industry, the moral habits, nor the desire of improvement which are essential to any favorable change in their condition. Established in the midst of another and a superior race, and without appreciating the causes of their inferiority or seeking to control them, they must necessarily yield to the force of circumstances and ere long disappear.'

CHIEF JUSTICE JOHN MARSHALL. 'The tribes of Indians inhabiting the country were savages, whose occupation was war, and whose subsistence was drawn from the forest. . . . That law which regulates, and ought to regulate in general, the relations between the conqueror and conquered was incapable of application to a people under such circumstances. Discovery [of America by Europeans] gave an exclusive right to extinguish the Indian title of occupancy, either by purchase or by conquest.'

PRESIDENT WILLIAM HENRY HARRISON. 'Is one of the fairest portions of the globe to remain in a state of nature, the haunt of a few wretched savages, when it seems destined by the Creator to give support to a large population and to be the seat of civilization?'

PRESIDENT THEODORE ROOSEVELT. 'The settler and pioneer have at bottom had justice on their side; this great continent could not have been kept as nothing but a game preserve for squalid savages.'

GENERAL PHILIP SHERIDAN. 'The only good Indians I ever saw were dead.'

PART FIVE

REVERSING OUR
PROGRESS OVERNIGHT

OUR SPECIES IS NOW AT THE PINNACLE OF ITS NUMBERS, ITS GEOGRAPHIC extent, its power, and the fraction of the Earth's productivity that it commands. That is the good news. The bad news is that we are also in the process of reversing all that progress much more rapidly than we created it. Our power threatens our own existence. We do not know whether we shall suddenly blow ourselves up before we would otherwise expire in a slow stew caused by global warming, pollution, habitat destruction, more mouths to feed, less food to feed those mouths, and extermination of other species that form our resource base. Are these dangers really new ones that arose since the Industrial Revolution, as widely assumed?

It is a common belief that species in a state of nature live in balance with each other and with their environment. Predators do not exterminate their prey, nor do herbivores overgraze their plants. According to this view, humans are the unique misfit. If this were true, Nature would hold no lessons for us.

There is something to this view, insofar as species do not go extinct under natural conditions as rapidly as we are exterminating them now, except under rare circumstances. Such a rare event was the mass fatality sixty-five million years ago, possibly due to an asteroidal impact, that finished the dinosaurs. Since evolutionary multiplications of species are very slow, natural extinctions obviously must also be slow, otherwise we would have been left with no species long ago. Expressed alternatively, the vulnerable species get eliminated quickly, and what we see persisting in Nature are the robust combinations of species.

That broad conclusion still leaves us with many instructive examples of species exterminating other species. Almost all known cases prove to combine two elements. Firstly, the cases involve species reaching environments where they did not occur before, and where they encounter prey populations that are naive about the threat of those invading predators. By the time that the ecological dust settles and a new equilibrium is reached, some of the new-found prey may have been exterminated. Secondly, the perpetrators of such exterminations prove to be so-called switching predators, which are not specialized to eat only a single prey species but can feed on many different ones. Although the predator exterminates some prey, it survives by switching to others.

Such exterminations often occur when humans intentionally or accidentally transfer a species from one part of the globe to another. Rats, cats, goats, pigs, ants, and even snakes are among these transferred killers. For instance, during the Second World War a tree snake native to the Solomon Islands was accidentally transported on ships or planes to the previously snake-free Pacific island of Guam. This predator has already exterminated or brought to the brink of extinction most of Guam's native forest bird species, which had had no opportunity to evolve behavioural defences against snakes. Yet the snake is in no danger itself despite having virtually eliminated its bird prey, because it can switch to bats, rats, lizards, and other victims. As another example, cats and foxes introduced into Australia by humans have been eating their way through Australia's small native marsupials and rats without endangering themselves, because there remain abundant rabbits and other prey species on which to feed.

We humans furnish the prime example of a switching predator. We eat everything from snails and seaweed to whales, mushrooms, and strawberries. We can overharvest some species to the point of extinction, and then just switch to other food. A wave of extinctions has ensued every time that humans have reached a previously unoccupied part of the globe. The dodo, whose name has become synonymous with extinction, formerly lived on the island of Mauritius, half of whose land and freshwater bird species became extinct following the island's discovery in 1507. Dodos in particular were big, edible, flightless, and easily caught by hungry sailors. Hawaiian bird species similarly died out *en masse* following Hawaii's discovery by Polynesians 1,500 years ago, as did America's large mammal species after ancestral Indians arrived 11,000 years ago. Extinction waves have also accompanied major improvements of hunting technology in lands long occupied by humans. For example, wild populations of the Arabian oryx, a beautiful antelope of the Near East, survived one million years of human hunting, only to succumb to high-powered rifles in 1972.

Thus, there are numerous animal precedents for our propensity to exterminate individual prey species but to sustain ourselves by switching to others. Is there any precedent for an animal population destroying its entire resource base and eating its way into extinction? This outcome is uncommon, because animal numbers are regulated by many factors that tend automatically to lower birth rates or increase death rates when the animal is numerous, and vice versa when it is rare. For example, mortality due to external factors like predators, diseases, parasites, and starvation tends to increase at high population densities. Responses of the animal itself to high densities also contribute, such as infanticide, postponed breeding, and increased aggression. These responses and external factors generally reduce the animal's population and relieve its pressure on its resources before they can be exhausted.

Nevertheless, some animal populations actually have eaten themselves into extinction. One example involves the progeny of twenty-nine reindeer that were introduced in 1944 to St Matthew Island in the Bering Sea. By 1957 they had multiplied nearly fifty-fold to 1,350, by the year 1963 another four-fold to 6,000. But reindeer depend for food on slow-growing lichens, which on St Matthew had no chance to recover from reindeer grazing, since the animals had nowhere to migrate. When a harsh winter struck in 1963–64, all the animals except forty-one females and one sterile male starved to death, leaving a doomed population on an island littered with thousands of skeletons. A similar example was the introduction of rabbits to Lisianski Island west of Hawaii in the first decade of this century. Within a decade the rabbits had eaten themselves into oblivion by consuming every plant on the island except two morning glories and a tobacco patch.

These and other similar examples of ecological suicide all involve populations that suddenly became free of the usual factors regulating their numbers. Rabbits and reindeer are normally subject to predators, and reindeer on continents use migration as a safety valve to leave an area and allow its vegetation to recover. But Lisianski and St Matthew Islands lacked predators, and emigration was impossible, so that the animals bred and ate unchecked.

On reflection, it is clear that the entire human species has been equally successful in recently escaping from the former controls on our numbers. We eliminated predation on ourselves long ago; twentieth-century medicine has greatly reduced our mortality from infectious disease; and some of our leading behavioural techniques of population control, such as infanticide, chronic war, and sexual abstinence, have become socially unacceptable. Our population is now doubling about every thirty-five years. Granted, that is not as fast as the St Matthew reindeer, and Island Earth is bigger than St Matthew Island, and some of our resources are more elastic than lichens (though other resources, like oil, are less elastic). Yet the qualitative conclusion remains the same: no population can grow indefinitely.

Thus, our present ecological predicament has familiar animal precursors. Like many switching predators, we exterminate some prey species when we colonize a new environment or acquire new destructive power. Like some animal populations that suddenly escaped their former limits on growth, we risk destroying ourselves by destroying our resource base. What about the view that we were in a state of relative ecological equilibrium until the Industrial Revolution, and that only since then have we begun seriously to exterminate species and overexploit our environment? That Rousseau-esque fantasy will be taken up in the three chapters of Part Five.

Chapter Seventeen examines the widespread belief in a former Golden Age, when we supposedly lived as noble savages practising a conservation ethic and in harmony with Nature. In reality, mass extinctions have coincided with each major extension of human *lebensraum* during the last ten thousand years and possibly much longer. Our direct responsibility for the extinctions is clearest in the case of the most recent expansions, where the evidence is still fresh – Europeans' expansion over the globe since 1492, and the earlier colonization of oceanic islands by Polynesians and the Malagasy. Older expansions such as the first human occupation of the Americas and Australia were also accompanied by mass extinctions, though the trail of evidence has had much more time to fade and so conclusions about cause and effect are necessarily weaker.

It is not just the case that the Golden Age was blackened by mass extinctions. While no large human population has eaten itself out of existence, some populations on small islands have done so, and many large populations have damaged their resources to the point of economic collapse. The clearest examples come from isolated cultures, such as the collapses of Easter Island and Anasazi civilizations. But environmental factors also drove the major shifts in Western civilization, including the successive collapses of the Middle Eastern, then Greek, then Roman hegemonies. Self-destructive abuse of our environment, far from being a modern invention, has long been a prime mover of human history.

Chapter Eighteen looks more closely at the biggest, most dramatic, and most controversial of these 'Golden Age mass extinctions'. Around 11,000 years ago most of the large mammals of two entire continents, North America and South America, became extinct. Around the same time appears the first unequivocal evidence for human occupation of the Americas, by the ancestors of American Indians. It was the biggest expansion of human territory since *Homo erectus* spread out of Africa to colonize Europe and Asia a million years ago. The temporal coincidence between the first Americans and the last big American mammals, the lack of mass extinctions elsewhere in the world at that same time, and proofs that some of the now-extinct beasts were hunted have suggested what is termed the New World blitzkrieg hypothesis. According to this interpretation, as the first wave of human hunters multiplied and spread from Canada to Patagonia, they encountered big animals that had never seen humans before, and they exterminated as they marched. However this theory's critics are at least as numerous as its backers; Chapter Eighteen will try to make sense of the debate.

The concluding chapter will seek to put approximate numbers on the count of species that we have already driven into extinction. We shall start with the firmest numbers: the species whose extinctions occurred in modern times and were well documented, and for which the search for survivors has been so thorough as to leave no doubt that there are no survivors. Next come estimates of three less certain numbers: the modern species that have not been seen alive for some time and that became extinct before anyone was aware of it; the modern species that have not even been 'discovered' and received a name; and the species that humans exterminated before the rise of modern science. That background will let us appraise the main mechanisms by which we exterminate, and the number of species that we are likely to exterminate within my sons' lifetime – if we proceed at our current rate.

SEVENTEEN

THE GOLDEN AGE THAT NEVER WAS

We cling to belief in a Rousseau-esque fantasy that the past was a Golden Age of environmentalism, when people lived in harmony with Nature. In reality, human societies, including those of stone-age farmers and possibly of hunter-gatherers as well, have been undermining their own subsistence by exterminating species and damaging environments for thousands of years. We differ from our supposedly conservationist forebears only in our greater numbers, more potent technology for inflicting damage, and access to written histories from which we refuse to learn.

*

Every part of the earth is sacred to my people. Every shining pine needle, every sandy shore, every mist in the dark woods, every clearing and humming insect is holy in the memory and experience of my people. . . . The white man . . . is a stranger who comes in the night and takes from the land whatever he needs. The earth is not his brother but his enemy. . . . Continue to contaminate your bed, and you will one night suffocate in your own waste. [From a letter written in 1855 to President Franklin Pierce, by Chief Seattle of the Duwanish tribe of American Indians.]

*

Environmentalists sickened by the damage that industrial societies are wreaking on the world often look to the past as a Golden Age. When Europeans began to settle America, the air and rivers were pure, the landscape green, the Great Plains teeming with bison. Today we breathe smog, worry about toxic chemicals in our drinking water, pave over the landscape, and rarely see any large wild animal. Worse is surely to come. By the time that my young sons reach retirement age, half of the world's species will be extinct, the air radioactive, and the seas polluted with oil. Undoubtedly, two simple reasons go a long way towards explaining our worsening mess: modern technology has far more power to cause havoc than did the stone axes of the past, and far more people are alive now than ever before. But a third factor may also have contributed, a change in attitudes. Unlike modern city-dwellers, at least some pre-industrial peoples – like the Duwanish, whose chief I quoted – depend on and revere their local environment. Stories abound of how such peoples are in effect practising conservationists. As a New Guinea tribesman once explained to me, 'It's our custom that if a hunter one day kills a pigeon in one direction from the village, he waits a week before hunting for pigeons again, and then goes in the opposite direction.' We are only beginning to realize how sophisticated are the conservationist policies of so-called 'primitive' peoples. For instance, well-intentioned foreign experts have made deserts out of large areas of Africa. In those same areas, local herders had thrived for uncounted millennia, by making annual nomadic migrations which ensured that land never became overgrazed.

The nostalgic outlook shared until recently by most of my environmentalist colleagues and myself is part of a human tendency to view the past as a Golden Age in many other respects. A famous exponent of this outlook was the eighteenth-century French philosopher Jean–Jacques Rousseau, whose *Discourse on the Origin of Inequality* traced our degeneration from the Golden Age to the human misery that Rousseau saw around him. When eighteenth-century European explorers encountered pre-industrial peoples like Polynesians and American Indians, those peoples became idealized in European salons as 'noble savages' living in a continued Golden Age, untouched by such curses of civilization as religious intolerance, political tyranny, and social inequality.

Even now, the days of classical Greece and Rome are widely considered to be the Golden Age of western civilization. Ironically, the Greeks and Romans also saw themselves as degenerates from a past Golden Age. I can still recite half-consciously those lines of the Roman poet Ovid that I memorized in tenth-grade Latin, '*Aurea prima sata est aetas, quae vindice nullo . . .*' ('First came the Golden Age, when men were honest and righteous of their own free will . . .') Ovid went on to contrast those virtues with the rampant treachery and warfare of his own times. I have no doubt that any humans still alive in the radioactive soup of the Twenty-second Century will write equally nostalgically about our own era, which will then seem untroubled by comparison.

Given this widespread belief in a Golden Age, some recent discoveries by archaeologists and paleontologists have come as a shock. It is now clear that pre-industrial societies have been exterminating species, destroying habitats, and undermining their own existence for thousands of years. Some of the best documented examples involve Polynesians and American Indians, the very peoples most often cited as exemplars of environmentalism. Needless to say, this revisionist view is hotly contested, not only in the halls of academia but also among lay people in Hawaii, New Zealand, and other areas with large Polynesian or Indian minorities. Are the new 'discoveries' just one more piece of racist pseudo-science by which white settlers seek to justify dispossessing indigenous peoples? How could the discoveries be reconciled with all the evidence for conservationist practices by modern pre-industrial peoples? If the discoveries were true, could we use them as case histories to help us predict the fate that our own environmental policies may bring upon us? Could the recent findings explain some otherwise mysterious collapses of ancient civilizations, like those of Easter Island or the Maya Indians?

Before we can answer these controversial questions, we need to understand the new evidence belying the assumed past Golden Age of environmentalism. Let's first consider evidence for past waves of exterminations, then evidence for past destruction of habitats.

*

When British colonists began to settle New Zealand in the 1800s, they found no native land mammals except bats. That was not surprising, for New Zealand is a remote island lying much too far from the continents for flightless mammals to reach. However, the colonists' ploughs uncovered instead the bones and eggshells of large birds that were then already extinct but that the Maori (the earlier Polynesian settlers of New Zealand) remembered by the name *moa*. From complete skeletons, some of them evidently recent and still retaining skin and feathers, we have a good idea how moas must have looked alive: they were ostrich-like birds comprising a dozen species, and ranging from little ones 'only' 3 feet high and forty pounds in weight up to giants of 500 pounds and 10 feet tall. Their food habits can be inferred from preserved gizzards containing twigs and leaves of dozens of plant species, showing them to have been herbivores. They thus used to be New Zealand's equivalents of big mammalian herbivores like deer and antelope.

While the moas are New Zealand's most famous extinct birds, many others have been described from fossil bones, totalling at least twenty-eight species that disappeared before Europeans arrived. Quite a few besides the moas were big and flightless, including a big duck, a giant coot, and an enormous goose. These flightless birds were descended from normal birds that had flown to New Zealand and that had then evolved to lose their expensive wing muscles in a land free of mammalian predators. Others of the vanished birds, such as a pelican, a swan, a giant raven, and a colossal eagle, were perfectly capable of flight.

Weighing up to thirty pounds, the eagle was by far the biggest and most powerful bird of prey in the world when it was alive. It dwarfed even the largest hawk now in existence, tropical America's harpy eagle. The New Zealand eagle would have been the sole predator capable of attacking adult moas. Although some moas were nearly twenty times heavier than the eagle, it still could have killed them by taking advantage of the moas' erect two-legged posture, crippling them with an attack on the long legs, then killing them with an attack on the head and long neck, and finally remaining for many days to consume the carcass, just as lions take their time at consuming a giraffe. The eagle's habits may explain the many headless moa skeletons that have been found.

Up to this point I have discussed New Zealand's big extinct animals. But fossil-hunters have also discovered the bones of small scampering animals of the size of mice and rats. Scampering or crawling on the ground were at least three species of flightless or weak-flying songbirds, several frogs, giant snails, many giant cricket-like insects up to double the weight of a mouse, and strange mouse-like bats that rolled up their wings and ran. Some of these little animals were completely extinct by the time that Europeans arrived. Others still survived on small offshore islands near New Zealand, but their fossil bones show that they were formerly abundant on the New Zealand mainland. Collectively, all these now-extinct species that had evolved in isolation on New Zealand would have provided New Zealand with the ecological equivalents of the continents' flightless mammals that had never arrived: moas instead of deer, flightless geese and coot instead of rabbits, big crickets and little songbirds and bats instead of mice, and colossal eagles instead of leopards.

Fossils and biochemical evidence indicate that the moas' ancestors had reached New Zealand millions of years ago. When and why, after surviving for so long, did the moas finally become extinct? What disaster could have struck so many species as different as crickets, eagles, ducks, and moas? Specifically, were all these strange creatures still alive when the ancestors of the Maoris arrived around 1000 AD?

At the time that I first visited New Zealand in 1966, the received wisdom was that moas had died out because of a change in climate, and that any moa species surviving to greet the Maoris were on their figurative last legs. New Zealanders took it as dogma that Maoris were conservationists and did not exterminate the moas. There is still no doubt that Maoris, like other Polynesians, used stone tools, lived mainly by farming and fishing, and lacked the destructive power of modern industrial societies. At most, it was assumed, Maoris might have given the *coup de grace* to populations already on the verge of extinction. However, three sets of discoveries have demolished this conviction.

Firstly, much of New Zealand was covered with glaciers or cold tundra during the last Ice Age ending about 10,000 years ago. Since then, the New Zealand climate has become much more favourable, with warmer temperatures and the spread of magnificent forests. The last moas died with their gizzards full of food, and enjoying the best climate that they had seen for tens of thousands of years.

Secondly, radiocarbon-dated bird bones from dated Maori archaeological sites prove that all known moa species were still present in abundance when the first Maoris stepped ashore. So were the extinct goose, duck, swan, eagle, and other birds now known only from fossil bones. Within a few centuries, the moas and most of those other birds were extinct. It would have been an incredible coincidence if every individual of dozens of species that had occupied New Zealand for millions of years chose the precise geological moment of human arrival as the occasion to drop dead in synchrony.

Finally, more than a hundred large archaeological sites are known – some of them covering dozens of acres – where Maoris cut up prodigious numbers of moas, cooked them in earth ovens, and discarded the remains. They ate the meat, used the skins for clothing, fashioned bones into fishhooks and jewellery, and blew out the eggs for use as water containers. During the Nineteenth Century moa bones were carted away from these sites by the wagonload. The number of moa skeletons in known Maori moa-hunter sites is estimated to be between 100,000 and 500,000, about ten times the number of moas likely to have been alive in New Zealand at any instant. Maoris must have been slaughtering moas for many generations.

Hence it is now clear that Maoris exterminated moas, at least partly by killing them, partly by robbing their nests of eggs, and probably partly as well by clearing some of the forests in which moas lived. Anyone who has hiked in New Zealand's rugged mountains will initially be incredulous at this thought. Just picture those travel posters of New Zealand's fiordland, with its steep-walled gorges 10,000 feet deep, its 400 inches of annual rainfall, and its cold winters. Even today, full-time professional hunters armed with telescopic rifles and operating from helicopters cannot control the numbers of deer in those mountains. How could the few thousand Maoris living on New Zealand's South Island and Stewart Island, armed only with stone axes and clubs and operating on foot, have hunted down the last moas?

But there would have been a crucial difference between deer and moas. Deers have been selected for tens of thousands of generations to flee from human hunters, while moas had never seen humans until Maoris arrived. Like the naive animals of the Galapagos Islands today, moas were probably tame enough for a hunter to walk up to one and club it. Unlike deer, moas may have had such low reproductive rates that a few hunters visiting a valley only once every couple of years could kill moas faster than they could breed. That is precisely what is happening today to New Guinea's largest surviving native mammal, a tree kangaroo in the remote Bewani Mountains. In areas settled by people, tree kangaroos are nocturnal, incredibly shy, live in trees, and are far harder to hunt than moas would have been. Despite all that, and despite the very low human population of the Bewanis, the cumulative effects of occasional hunting parties – literally one visit per valley per several years – have sufficed to bring this kangaroo to the verge of extinction. Having seen it happen to tree kangaroos, I now have no difficulty understanding how it happened to moas.

Not only moas, but also all of New Zealand's other extinct bird species, were still alive when Maoris landed. Most were gone a few centuries later. The larger ones – the swan and pelican, the flightless goose and coot – were surely hunted for food. The giant eagle, however, may have been killed by Maoris in self-defence. What do *you* think happened when that eagle, specialized at crippling and killing two-legged prey between three and ten feet tall, saw its first six-foot-tall Maoris? Even today, Manchurian eagles trained for hunting occasionally kill their human handlers, but the Manchurian birds were mere dwarfs beside New Zealand's giant, which was pre-adapted to become a man-killer.

Surely, though, neither self-defence nor hunting for food explains the rapid disappearance of New Zealand's peculiar crickets, snails, wrens, and bats. Why were so many of those species exterminated, either throughout their range or else everywhere except on some offshore islands?

Deforestation may be part of the answer, but the major reason was the other hunters that Maoris intentionally or accidentally brought with them – rats! Just as moas that evolved in the absence of humans were defenceless against humans, so, too, small insular animals that evolved in the absence of rats were defenceless against rats. We know that the rat species spread by Europeans played a major role in modern exterminations of many bird species on Hawaii and other previously rat-free oceanic islands. For example, when rats finally reached Big South Cape Island off New Zealand in 1962, they exterminated or decimated the populations of eight bird species and a bat within three years. That is why so many New Zealand species are restricted today to rat-free islands, the sole places where they could survive when the tide of rats accompanying the Maoris swept over the New Zealand mainland.

When the Maoris landed, they found an intact New Zealand biota of creatures so strange that we would dismiss them as science-fiction fantasies if we did not have their fossilized bones to convince us of their former existence. The scene was as close as we will ever get to what we might see if we could reach another fertile planet on which life had evolved. Within a short time, much of that community had collapsed in a biological holocaust, and some of the remaining community collapsed in a second holocaust following the arrival of Europeans. The end result is that New Zealand today has about half of the bird species that greeted the Maoris, and many of the survivors are either now at risk of extinction or else confined to islands with few introduced mammalian pests. A few centuries of hunting had sufficed to end millions of years of moa history.

*

Not only on New Zealand but on all other remote Pacific islands where archaeologists have looked recently in Polynesia, bones of many now-extinct bird species have been found at sites of the first settlers, proving there that the bird extinctions and human colonizations were somehow related. From all the main islands of Hawaii, paleontologists Storrs Olson and Helen James of the Smithsonian Institution have identified fossil bird species which disappeared during the Polynesian settlement that began around 500 AD. The fossils include not only small honey-creepers related to species still present but also bizarre flightless geese and ibises with no living close relatives at all. While Hawaii is notorious for its bird extinctions following European settlement, this earlier extinction wave had been unknown until Olson and James began publishing their discoveries in 1982. The known extinctions of Hawaiian birds before Captain Cook's arrival now total the incredible number of at least fifty species, nearly one-tenth of the number of bird species breeding on mainland North America.

That is not to say that all these Hawaiian birds were hunted out of existence. Although geese probably were indeed exterminated by overhunting, like the moas, small songbirds are more likely to have been eliminated by rats that arrived with the first Hawaiians, or else by destruction of forests that Hawaiians cleared for agriculture. Similar discoveries of extinct birds at archaeological sites of early Polynesians have also been made on Tahiti, Fiji, Tonga, New Caledonia, the Marquesas Islands, Chatham Islands, Cook Islands, Solomon Islands, and Bismarck Archipelago.

An especially intriguing collision of birds and Polynesians took place on Henderson Island, an extremely remote speck of land lying in the tropical Pacific Ocean 125 miles east of Pitcairn Island, which is in turn famous for its own isolation. (Recall that Pitcairn is so remote that the mutineers who wrested the H.M.S. *Bounty* from Captain Bligh lived undetected on Pitcairn for eighteen years until the island was re-discovered.) Henderson consists of jungle-covered coral riddled with crevices and totally unsuitable for agriculture. Naturally, the island is now uninhabited and has been ever since Europeans first saw it in 1606. Henderson has often been cited as one of the world's most pristine habitats, totally unaffected by humans.

It was therefore a big surprise when Olson and fellow paleontologist David Steadman recently identified bones of two large species of pigeons, one smaller pigeon, and three seabirds that had become extinct on Henderson some time between 500 and 800 years ago. The same six species or close relatives had already been found in archaeological sites on several inhabited Polynesian islands, where it was clear how they could have been exterminated by people. The apparent contradiction of birds also being exterminated by humans on uninhabited, seemingly un-inhabitable Henderson was solved by the discovery there of former Polynesian sites with hundreds of cultural artifacts, proving that the island had actually been occupied by Polynesians for several centuries. At those same sites, along with the bones of the six bird species that were exterminated on Henderson, were the bones of other bird species that survived, plus many fish.

Those early Polynesian colonists of Henderson evidently subsisted mainly on pigeons, seabirds, and fish until they had decimated the bird populations, at which point they had destroyed their food supply and either starved or else abandoned the island. The Pacific contains at least eleven other 'mystery' islands, besides Henderson, which were un-inhabited on European discovery but showed archaeological evidence of former occupation by Polynesians. Some of these islands had been settled for hundreds of years before their human population finally died out or left. All were small or in other respects marginally suitable for agri-culture, leaving human settlers heavily dependent on birds and other animals for food. Given the widespread evidence for over-exploitation of wild animals by early Polynesians, not only Henderson but the other mystery islands as well may represent the graveyards of human populations that ruined their own resource base.

*

Lest I leave the impression that Polynesians were in any way unique as pre-industrial exterminators, let's now jump nearly halfway around the globe to the world's fourth largest island, Madagascar, lying in the Indian Ocean off the coast of Africa. When Portuguese explorers arrived around 1500 AD, they found Madagascar already occupied by people now called the Malagasy. On geographic grounds, you might have expected their language to be related to African languages spoken a mere 200 miles to the west, on the coast of Mozambique. Astonishingly, though, it actually proved to belong to a group of languages spoken on the Indonesian island of Borneo, on the opposite side of the Indian Ocean thousands of miles to the northeast. Physically, the Malagasy range in appearance from typical Indonesians to typical blacks of East Africa. These paradoxes are due to the Malagasy having arrived between 1,000 and 2,000 years ago, as a result of Indonesian traders voyaging around the Indian Ocean coastline to India and eventually to East Africa. In Madagascar they proceeded to build a society based on herding cattle and goats and pigs, farming, and fishing, and linked to the East African coast by Muslim traders.

As interesting as Madagascar's people are the wild animals that it has – and those that it lacks. Living in enormous abundance on the nearby African mainland are many species of large and conspicuous beasts that run on the ground and are active by day – the antelopes, ostriches, zebras, baboons, and lions that draw modern tourists to East Africa. None of these animals, and no animals remotely equivalent to them, have occurred on Madagascar in modern times. They were kept out by the 200 miles of sea separating Madagascar from Africa, just as the sea also kept Australia's marsupials from reaching New Zealand. Instead, Madagascar supports two dozen species of small, monkey-like primates called lemurs, weighing only up to twenty pounds and mostly active at night and living in trees. Various species of rodents, bats, insectivores, and relatives of mongooses also occur, yet the largest still only weighs about twenty-five pounds.

However, littering Madagascar's beaches are proofs of vanished giant birds, in the form of countless eggshells of the size of a soccer ball. Eventually, bones turned up not only of the birds that laid those eggs, but also of a remarkable suite of vanished large mammals and reptiles. The egg-makers were half-a-dozen species of flightless birds up to 10 feet tall and weighing up to 1,000 pounds, like moas and ostriches but more massively built and hence now termed elephant birds. The reptiles were two species of giant land tortoises with shells about a yard long, and formerly very common, as indicated by the abundance of their bones. More diverse than either of these large birds or reptiles were a dozen species of lemurs up to the size of a gorilla, and all larger than or at least as large as the largest surviving lemur species. To judge from the small size of the eye orbits in their skulls, all or most of the extinct lemurs were probably diurnal rather than nocturnal. Some of them evidently lived on the ground like baboons, while others climbed in trees like orangutans and koala bears.

As if all this were not enough, Madagascar also yielded the bones of an extinct 'pygmy' hippopotamus ('only' the size of a cow), an aardvark, and a big mongoose-related carnivore built like a short-legged puma. Taken together, these extinct large animals formerly gave Madagascar the functional equivalents of the surviving large beasts for which tourists still flock to African game parks – just as did New Zealand's moas and other strange birds. The tortoises, elephant birds, and pygmy hippo would have been the herbivores replacing antelope and zebras; the lemurs would have replaced the baboons and great apes; and the mongoose-related carnivore made do for a leopard or scaled-down lion.

What happened to all these big extinct mammals, reptiles, and birds? We can be confident that at least some of them were alive to delight the eyes of the first arriving Malagasy, who used elephant bird eggshells as water containers and discarded butchered bones of the pygmy hippo and some of the other species in their rubbish heaps. In addition, the bones of all the other extinct species are known from fossil sites only a few thousand years old. Since they must have evolved and survived for millions of years until then, it is unlikely that all those animals had the foresight to give up the ghost just in those last few moments before hungry humans showed up. In fact, a few may still have been holding out in remote parts of Madagascar when Europeans arrived, since the seventeenth-century French governor Flacourt was given descriptions of an animal suggestive of the gorilla-sized lemur. The elephant birds may have survived long enough to have become known to Arab traders in the Indian Ocean, and to have given rise to the account of the rok (a giant bird) in the tale of Sinbad the Sailor.

Certainly some and probably all of Madagascar's vanished giants were somehow exterminated by the activities of the early Malagasy. It is not hard to understand why the elephant birds became extinct, since their eggshells made such convenient two-gallon jerrycans. While the Malagasy were herders and fisherman rather than big-game hunters, the other big animals would have been easy prey, since they had never seen humans before. Probably, like New Zealand's moas, they were as tame as Antarctic penguins and other creatures that evolved in the absence of humans. A hungry Malagasy could have walked up to one of those tame beasts, clubbed it, and enjoyed a quick barbecue. That is presumably why the easy-to-see, easy-to-catch lemurs big enough to be worth the effort of butchering them – the large, diurnal, terrestrial species – all became extinct, while the small, nocturnal, tree-living ones all survived.

However, unintended by-products of Malagasy activities probably killed more big animals than did hunting. Fires lit to clear forest for pasture and to stimulate growth of new grass each year would have destroyed habitats on which the beasts depended. Grazing cattle and goats also transformed habitats, as well as competing directly with grazing tortoises and elephant birds for food. Introduced dogs and pigs would have preyed on ground-dwelling animals, their young, and their eggs. By the time that the Portuguese arrived, Madagascar's once-abundant elephant birds had all been reduced to eggshells covering the beaches, skeletons in the ground, and vague memories of roks.

*

Madagascar and Polynesia merely provide well-documented examples of the extinction waves that probably unfolded on all large oceanic islands colonized by people before the European expansion of the last 500 years. Like New Zealand and Madagascar, all such islands where life had evolved in the absence of humans used to have unique species of big animals that modern zoologists never saw alive. Mediterranean islands like Crete and Cyprus had pygmy hippos and giant tortoises (just as did Madagascar), as well as dwarf elephants and dwarf deer. The West Indies lost monkeys, ground sloths, a bear-sized rodent, and owls of several sizes: normal, giant, colossal, and titanic. It seems likely that these big birds, mammals, and tortoises too somehow succumbed to the first Mediterranean peoples or American Indians to reach their islands. Nor were birds the only victims. Mammals, lizards, frogs, snails, and even large insects disappeared as well, comprising thousands of species when one adds up all oceanic islands. Olson describes these insular extinctions as 'one of the swiftest and most profound biological catastrophes in the history of the world'. However, we will not be sure that humans were responsible until the bones of the last animals and the remains of the first people have been dated more exactly for other islands, as has already been done for Polynesia and Madagascar.

In addition to these pre-industrial extermination waves on islands, other species may have fallen victim to extermination waves on continents, in the more distant past. About 11,000 years ago, around the probable time that the first ancestors of American Indians reached the New World, most large species of mammals became extinct throughout all of North and South America. The disappearances involved species as varied as lions, horses, giant armadillos, mammoths, and saber-toothed cats. A long-standing debate has raged over whether these big mammals were done in by Indian hunters, or whether they just happened to succumb to climate changes around the same time. I shall explain in the next chapter why I personally think that hunters did it. However, it is much harder to pinpoint dates and causes of events that happened around 11,000 years ago than it is for recent events, like the collision of the Maoris and the moa within the past thousand years. Similarly, within the past 50,000 years Australia lost most of its big mammal species *and* was colonized by the ancestors of today's Aboriginal Australians, but we are still uncertain whether the second event caused the first. Therefore, although it is now reasonably certain that the first pre-industrial peoples to reach islands wrought havoc among island species, the jury is still out on the question of whether this also happened on continents.

*

From all this evidence that the Golden Age was tarnished by exterminations of species, let's now turn to evidence for destruction of habitats. Three dramatic examples involve famous archaeological puzzles: the giant stone statues of Easter Island, the abandoned *pueblos* of the American Southwest, and the ruins of Petra.

An aura of mystery has clung to Easter Island ever since it and its Polynesian inhabitants were 'discovered' by the Dutch explorer Jakob Roggeveen in 1722. Lying in the Pacific Ocean 2,300 miles west of Chile, Easter surpasses even Henderson as one of the world's most isolated scraps of land. Hundreds of statues, weighing up to eighty-five tons and up to 37 feet tall, were carved from volcanic quarries, somehow transported several miles, and raised to an upright position on platforms, by people without metal or wheels and with no power source other than human muscle. Even more statues remain unfinished in the quarries, or lie finished but abandoned between the quarries and platforms. The scene today is as if the carvers and movers had suddenly walked off the job, leaving an eerily silent landscape.

When Roggeveen arrived, many statues were still standing, though new ones were no longer being carved. By 1840 all the erected statues had been deliberately toppled by the Easter Islanders themselves. How were such huge statues transported and erected, why were they eventually toppled, and why had carving ceased?

The first of those questions was answered when living Easter Islanders showed Thor Heyerdahl how their ancestors had used logs as rollers to transport the statues and then as levers to erect them. The other questions were solved by subsequent archaeological and paleontological studies that revealed Easter's gruesome history. When Polynesians settled Easter around 400 AD, the island was covered by forest that they gradually proceeded to clear, in order to plant gardens and to obtain logs for canoes and for erecting statues. By around 1500 AD the human population had built up to about 7,000 (over 150 per square mile), about a thousand statues had been carved, and at least 324 of those statues had been erected. But – the forest had been destroyed so thoroughly that not a single tree survived.

An immediate result of this self-inflicted ecological disaster was that the islanders no longer had the logs needed to transport and erect statues, so that carving ceased. But deforestation also had two indirect consequences that brought starvation. These were soil erosion, causing lower crop yields, plus lack of timber to build canoes, resulting in less protein available from fishing. As a result, the population was now greater than Easter could support, and island society collapsed in a holocaust of internecine warfare and cannibalism. A warrior class took over; spearpoints manufactured in huge quantities came to litter the landscape; the defeated were eaten or enslaved; rival clans pulled down each other's statues; and people took to living in caves for self-protection. What had once been a lush island supporting one of the world's most remarkable civilizations deteriorated into the Easter Island of today: a barren grassland littered with fallen statues, and supporting less than one-third of its former population.

*

Our second case study of pre-industrial habitat destruction involves the collapse of one of the most advanced Indian civilizations of North America. When Spanish explorers reached the US Southwest, they found gigantic multi-storey dwellings (*pueblos*) standing uninhabited in the middle of treeless desert. For example, the 650-room dwelling at Chaco Canyon National Monument in New Mexico was five stories high, 670 feet long, and 315 feet wide, making it the largest building ever erected in North America until topped by steel skyscrapers in the late Nineteenth Century. Navajo Indians in the region knew of the vanished builders only as 'Anasazi', meaning 'the Ancient Ones'.

Archaeologists subsequently established that construction of the Chaco *pueblos* began shortly after 900 AD, and that occupation ceased in the Twelfth Century. Why did the Anasazi erect a city in a barren wasteland, of all unpromising places? Where did they obtain their firewood, or the 16-foot-long wooden beams (200,000 of them!), that supported the roofs? Why did they then abandon the city that they had built at such enormous effort?

The conventional view, analogous to the claim that Madagascar's elephant birds and New Zealand's moas died out from natural changes in climate, attributes the abandonment of Chaco Canyon to a drought. However, a different interpretation emerges from the work of paleobotanists Julio Betancourt, Thomas Van Devender, and their colleagues, who used an ingenious technique to decipher changes in Chaco vegetation through time. Their method relied on the little rodents called packrats, which gather plants and other materials into shelters ('middens') that they eventually abandon after fifty or a hundred years but that remain well preserved under desert conditions. The plants can be identified centuries later, and the midden can be dated by radiocarbon techniques. Thus, each midden is virtually a time capsule of the local vegetation.

By this method, Betancourt and Van Devender were able to re-construct the following course of events. At the time that the Chaco pueblos were erected, they were not surrounded by barren desert but by pinyon and juniper woodland, with ponderosa pine forests nearby. This discovery at once solves the mystery of where the firewood and timber came from, and disposes of the apparent paradox of an advanced civilization rising from barren desert. As occupation continued at Chaco, however, the woodland and forest were cleared until the environment became the treeless wasteland that it remains today. The Indians were then having to go over ten miles to get firewood, and over twenty-five miles to get pine logs. When the pine forests had been felled, they built an elaborate road system to haul spruce and fir logs from mountain slopes over fifty miles away, relying on nothing more than their own muscle power. In addition, the Anasazi had solved the problems of agriculture in a dry environment by building irrigation systems to concentrate available water into valley bottoms. As deforestation caused progressively increas-ing erosion and water runoff, and as irrigation channels gradually dug gullies into the ground, the water table may finally have dropped below the level of the Anasazi fields, making irrigation without pumps impossible. Thus, while drought may have made some contribution to the Anasazi abandonment of Chaco Canyon, a self-inflicted ecological disaster was also a major factor.

*

Our remaining example of pre-industrial habitat destruction illuminates the gradual geographic shift in the power centre of ancient western civilizations. Recall that the first centre of power and innovation was the Mideast, where so many crucial developments arose – agriculture, animal domestication, writing, imperial states, battle chariots, and others. Ascendancy shifted between Assyria, Babylon, Persia, and occasionally Egypt or Turkey, but remained in or near the Mideast. With the overthrow of the Persian Empire by Alexander the Great, ascendancy moved finally westward, at first to Greece, then to Rome, and later to western and northern Europe. Why did the Mideast, Greece, and Rome in turn lose their primacy? (The transient current importance of the

Mideast, resting as it does on the single resource of oil, merely emphasizes by contrast the region's modern weakness in other respects.) Why do modern superpowers include the US and Russia, Germany and England, Japan and China, but no longer Greece and Persia?

This geographic shift in power is too big and lasting a pattern to have arisen by accident. A plausible hypothesis attributes it to each ancient centre of civilization in turn ruining its resource base. The Mideast and Mediterranean were not always the degraded landscape that they appear today. In ancient times much of this area was a lush, fertile mosaic of wooded hills and fertile valleys. Thousands of years of deforestation, overgrazing, erosion, and valley siltation converted this heartland of Western civilization into the relatively dry, barren, infertile landscape that predominates today. Archaeological surveys of ancient Greece have revealed several cycles of population growth alternating with population crashes and local abandonment of human settlements. In the growth phases, terracing and dams initially protected the landscape until felling of forests, clearing of steep slopes for agriculture, overgrazing by too many livestock, and planting of crops at too short intervals overwhelmed the system. The result each time was massive erosion of the hills, flooding of the valleys, and the collapse of local human society. One such event coincided with (and may have caused) the otherwise mysterious collapse of Greece's glorious Mycenean civilization, after which Greece fell back for several centuries into a dark age of illiteracy.

The support for this view of ancient environmental destruction comes from sources such as contemporary accounts and archaeological evidence. Yet a few sequences of photographs would constitute more decisive tests than all that anecdotal evidence combined. If we had snapshots of the same Greek hillside taken at thousand-year intervals, we could identify the plants, measure the ground cover, and calculate the shift from forest to goat-proof shrubs. We could thereby put numbers on the extent of environmental degradation.

Enter middens to the rescue again. While the Mideast does not have packrats, it does have rabbit-sized, marmot-like animals called hyraxes that build middens in the same way as packrats. (Surprisingly, the closest living relatives of hyraxes may be elephants.) Three Arizona scientists – Patricia Fall, Cynthia Lindquist, and Steven Falconer – studied hyrax middens at Jordan's famous lost city of Petra, which typifies the paradox of ancient Western civilization. Petra is now especially familiar to movie-going aficionados of Steven Spielberg and George Lucas, whose film *Indiana Jones and the Last Crusade* shows Sean Connery and Harrison Ford searching for the Holy Grail in Petra's magnificent rock tombs and temples amidst the desert sand. Anyone who sees those scenes of Petra must wonder how such a wealthy city could have arisen and supported itself in such a bleak landscape. In fact, there was already a Neolithic village near the site of Petra before 7000 BC, and farming and herding appeared there soon after. Under the Nabataean kingdom, of which it was the capital, Petra thrived as a commercial centre controlling trade between Europe, Arabia, and the Orient. The city grew even larger and richer under Roman, then Byzantine, control. Yet it was subsequently abandoned and so completely forgotten that its ruins were not re-discovered until 1812. What caused Petra's collapse?

Each hyrax midden from Petra yielded remains of up to 100 plant species, and the habitat prevailing when each midden's owner was alive could be calibrated by comparing pollen proportions in the midden with those in modern habitats. From the middens, the following trajectory was reconstructed for the degradation of Petra's environment.

Petra lies in an area of dry Mediterranean climate not unlike that of the wooded mountains behind my home in Los Angeles. The original vegetation would have been a woodland dominated by oak and pistachio trees. By Roman and Byzantine times, most of the trees had been felled, and the surroundings had been degraded to an open steppe, as expressed in the fact that only eighteen per cent of midden pollen came from trees, the rest from low plants. (For comparison, trees contribute forty to eighty-five per cent of the pollen in modern Mediterranean forests, eighteen per cent in forest steppes.) By 900 AD, a few centuries after Byzantine control of the Petra area ended, two-thirds of the remaining trees had disappeared. Even shrubs, herbs, and grasses had declined, converting the environment into the desert that we see now. Surviving trees today have their lower branches pruned off by goats and are scattered on goat-proof cliffs or in groves protected from goats.

Juxtaposing these data from hyrax middens with archaeological and literary data yields the following interpretation. Deforestation from Neolithic to Imperial times was driven by the clearing of land for agriculture, browsing by sheep and goats, gathering of firewood, and wood needs for house construction. Even Neolithic houses not only were supported by massive timbers but also consumed up to thirteen tons of firewood per house to make the plaster for the walls and floor. The Imperial population explosion quickened the pace of forest destruction and overgrazing. Elaborate systems of channels, pipes, and cisterns were needed to collect and store water for the orchards and city.

After Byzantine authority collapsed, orchards were abandoned and the population crashed, but land degradation continued as the remaining inhabitants became dependent on intensive grazing. The insatiable goats began to eat their way through the shrubs, herbs, and grasses. The Ottoman government decimated surviving woodlands before the First World War, to obtain the wood needed for the Hejaz Railway. I and many other movie-goers thrilled at the sight of Arab guerrillas led by Lawrence of Arabia (*a.k.a.* Peter O'Toole) blowing up that railway in widescreen technicolour, without realizing that we were watching the last act in the destruction of Petra's forests.

Petra's ravaged landscape today is a metaphor for what happened to the rest of the cradle of Western civilization. The modern surrounds of Petra could no more feed a city that commanded the world's main trade routes than the modern surrounds of Persepolis could feed the capital of a superpower such as the Persian Empire once was. The ruins of those cities, and of Athens and Rome, are monuments to states that destroyed their means of survival. Nor are Western civilizations the only literate societies that committed ecological suicide. The collapse of Classic Maya civilization in Central America, and of Harappan civilization in India's Indus Valley, are other obvious candidates for eco-disasters due to an expanding human population overwhelming its environment. While courses in the history of civilization often dwell on kings and barbarian invasions, deforestation and erosion may in the long run have been more important shapers of human history.

*

These are some of the recent discoveries making the supposed past Golden Age of environmentalism look increasingly mythical. Let's now go back to the larger issues I raised at the outset. Firstly, how can these discoveries of past environmental damage be reconciled with accounts of conservationist practices by so many modern pre-industrial peoples? Obviously, not all species have been exterminated, and not all habitats have been destroyed, so the Golden Age could not have been all black.

I suggest the following answer to this paradox. It is still true that small, long-established, egalitarian societies tend to evolve conservationist practices, because they have had plenty of time to get to know their local environment and to perceive their own self-interest. Instead, damage is likely to occur when people suddenly colonize an unfamiliar environment (like the first Maoris and Easter Islanders); or when people advance along a new frontier (like the first Indians to reach America), so that they can just move beyond the frontier when they have damaged the region behind; or when people acquire a new technology whose destructive power they have not had time to appreciate (like modern New Guineans, now devastating pigeon populations with shotguns). Damage is also likely in centralized states that concentrate wealth in the hands of rulers, who are out of touch with their environment. Some species and habitats are more susceptible to damage than others – such as flightless birds that had never seen humans (like moas and elephant birds), or the dry, fragile, unforgiving environments in which both Western civilization and Anasazi civilization arose.

Secondly, are there any practical lessons that we can learn from these recent archaeological discoveries? Archaeology is often regarded as a socially irrelevant academic discipline that becomes a prime target for budget cuts whenever money gets tight. In fact, archaeological research is one of the best bargains available to government planners. All over the world, we are launching developments that have great potential for doing irreversible damage, and that are really just more powerful versions of ideas put into operation by past societies. We cannot afford the experiment of developing five counties in five different ways and seeing which four counties get ruined. Instead, it will cost us much less in the long run if we hire archaeologists to find out what happened the last time, than if we go on making the same mistakes again.

Here is just one example. The American Southwest has over 100,000 square miles of pinyon and juniper woodland that we are exploiting more and more for firewood. Unfortunately, the US Forest Service has little data available to help it calculate sustainable yields and recovery rates in that woodland. Yet the Anasazi already tried the experiment and miscalculated, with the result that the woodland still has not recovered in Chaco Canyon after over 800 years. Paying some archaeologists to reconstruct Anasazi firewood consumption would be cheaper than committing the same mistake and ruining 100,000 square miles of the US, as we may now be doing.

Finally, let's face the touchiest question. Today, environmentalists view people who exterminate species and destroy habitats as morally bad. Industrial societies have jumped at any excuse to denigrate pre-industrial peoples, in order to justify killing them and appropriating their land. Are the purported new finds about moas and Chaco Canyon vegetation just pseudo-scientific racism that in effect is saying, Maoris and Indians do not deserve fair treatment because they were bad?

What has to be remembered is that it has always been hard for humans to know the rate at which they can safely harvest biological resources indefinitely, without depleting them. A significant decline in resources may not be easy to distinguish from a normal year-to-year fluctuation. It is even harder to assess the rate at which new resources are being produced. By the time that the signs of decline are clear enough to convince everybody, it may be too late to save the species or habitat. Thus, pre-industrial peoples who could not sustain their resources were guilty not of moral sins, but of failures to solve a really difficult ecological problem. Those failures were tragic, because they caused a collapse in lifestyle for the people themselves.

Tragic failures become moral sins only if one should have known better from the outset. In that regard there are two big differences between us and eleventh-century Anasazi Indians – those of scientific understanding, and literacy. We know, and they did not know, how to draw graphs that plot sustainable resource population size as a function of resource harvesting rate. We can read about all the ecological disasters of the past; the Anasazi could not. Yet our generation continues to hunt whales and clear tropical rainforest, as if no one had ever hunted moas or cleared pinyon and juniper woodland. The past was still a Golden Age, of ignorance, while the present is an Iron Age of wilful blindness.

From this point of view it is beyond understanding to see modern societies repeating the past's suicidal ecological mismanagement, with much more powerful tools of destruction in the hands of far more people. It is as if we had not already run that particular film many times before in human history, and as if we did not know the inevitable outcome. Shelley's sonnet 'Ozymandias' evokes Persepolis, Tikal, and Easter Island equally well; perhaps it will some day evoke to others the ruins of our own civilization.

I met a traveller from an antique land
Who said: Two vast and trunkless legs of stone
Stand in the desert. Near them, on the sand,
Half sunk, a shattered visage lies, whose frown,
And wrinkled lip, and sneer of cold command,
Tell that its sculptor well those passions read
Which yet survive, stamped on these lifeless things,
The hand that mocked them and the heart that fed;
And on the pedestal these words appear:
'My name is Ozymandias, king of kings:
Look on my works, ye Mighty, and despair!'
Nothing beside remains. Round the decay
Of that colossal wreck, boundless and bare
The lone and level sands stretch far away.

EIGHTEEN

BLITZKRIEG AND THANKSGIVING IN THE NEW WORLD

The colonization of the Americas eleven thousand years ago by ancestors of today's American Indians was the greatest extension of human lebensraum since Homo erectus emerged from Africa. It may also have been the first of the blitzkriegs against Nature that have since marked every expansion of humans into previously unpeopled areas. Within a short time of human arrival – perhaps only a few centuries – most of the big mammals of North and South America were extinct.

*

The United States devote two national holidays, Columbus Day and Thanksgiving Day, to celebrating dramatic moments in the European 'discovery' of the New World. No holidays commemorate the much earlier actual discovery by Indians. Yet archaeological excavations suggest that, in drama, that earlier discovery dwarfs the adventures of Christopher Columbus and of the Plymouth Pilgrims. Within perhaps as little as a thousand years after finding a way through an Arctic ice sheet and crossing the present border between the US and Canada, Indians had swept down to the tip of Patagonia and populated two productive and unexplored continents. The Indians' march southwards was the greatest range expansion in the history of *Homo sapiens*. Nothing remotely like it can ever happen again on our planet.

The sweep southwards was marked by another drama. When Indian hunters arrived, they found the Americas teeming with big mammals that are now extinct: elephant-like mammoths and mastodonts, ground sloths weighing up to three tons, armadillo-like glyptodonts weighing up to one ton, bear-sized beavers, and saber-toothed cats, plus American lions, cheetahs, camels, horses, and many others. Had those beasts survived, today's tourists in Yellowstone National Park would be watching mammoths and lions along with the bears and bison. The question of what happened at that moment of hunters–meet–beasts is still highly controversial among archaeologists and paleontologists. According to the interpretation that seems most plausible to me, the outcome was a blitzkrieg in which the beasts were quickly exterminated – possibly within a mere ten years at any given site. If that view is correct, it would have been the most concentrated extinction of big animals since an asteroid collision (it is believed) knocked off the dinosaurs sixty-five million years ago. It would also have been the first of the series of blitzkriegs that marred our supposed Golden Age of environmental innocence (Chapter Seventeen), and that have remained a human hallmark ever since.

*

That dramatic confrontation came as the finale to a long epic in which humans, spreading out of their centre of origin in Africa, occupied all the other habitable continents. Our African ancestors expanded to Asia and Europe around a million years ago, and from Asia to Australia around 50,000 years ago, leaving North and South America as the last habitable continents still without *Homo sapiens*.

From Canada to Tierra del Fuego, American Indians today are physically more homogeneous than the inhabitants of any other continent, implying that they arrived too recently to have become very diverse genetically. Even before archaeology uncovered evidence of the first Indians, it was clear that they must have originated from Asia, because modern Indians look similar to Asiatic Mongoloids. Much recent evidence from genetics and anthropology has made that conclusion certain. A glance at a map shows that by far the easiest route from Asia to America is across the Bering Straits separating Siberia from Alaska. The last such land bridge existed (with a few brief interruptions) from about 25,000 to 10,000 years ago.

However, colonization of the New World required more than a land bridge – there had to be people living at the Siberian end of the bridge. Because of its harsh climate the Siberian Arctic, too, was not colonized until late in human history (Chapter Two). Those colonists must have come from the cold temperate zones of Asia or Eastern Europe, as exemplified by stone-age hunters who lived in what is now the Ukraine and who built their houses out of neatly stacked bones of mammoths. By at least 20,000 years ago there were mammoth hunters in the Siberian Arctic as well, and by around 12,000 years ago stone tools similar to those of the Siberian hunters appear in Alaska's archaeological record.

After traversing Siberia and the Bering Straits, the ice-age hunters were still separated by one more barrier from their future hunting grounds in the US: a broad ice cap like that covering Greenland today, but stretching coast-to-coast across Canada. At intervals during the ice ages a narrow, ice-free, north/south corridor opened through this ice cap, just east of the Rocky Mountains. One such corridor closed around 20,000 years ago, but there had apparently as yet been no human in Alaska waiting to cross it. However, when the corridor next opened around 12,000 years ago, the hunters must have been ready, for their tell-tale stone tools appear soon thereafter not only at the south end of the corridor near Edmonton (Alberta) but also elsewhere south of the ice cap. At that point, hunters met America's elephants and other great beasts, and the drama began.

Archaeologists term these pioneering ancestral Indians the Clovis people, since their stone tools were first recognized at an excavation near the town of Clovis, ten miles inside New Mexico from the Texas border. However, Clovis tools or ones similar to them have been found in all forty-eight contiguous states of the US, and from Edmonton in the north to Mexico. Vance Haynes, a University of Arizona archaeologist, has emphasized that the tools are much like those of the earlier Eastern European and Siberian mammoth hunters, with one conspicuous exception: the flattish, two-faced, stone spear-points were 'fluted' on each face as a result of a longitudinal groove having been chipped out to make it easier to bind the stone point to the shaft. It is not clear whether the fluted points were mounted on spears to throw by hand, on darts to hurl by a throwing stick, or on lances to thrust. Somehow, though, the points were propelled into big mammals with such force that the points sometimes snapped in half, or else penetrated bone. Archaeologists have dug up skeletons of mammoths and bison with Clovis points inside the rib cage, including a mammoth from southern Arizona containing a total of eight points. At excavated Clovis sites, mammoths are by far the commonest prey (to judge from their bones), but other victims include bison, mastodonts, tapirs, camels, horses, and bears.

Among the startling discoveries about Clovis people is the speed of their spread. All Clovis sites in the US dated by the most advanced radiocarbon techniques were occupied for only a few centuries, in the period just before 11,000 years ago. A human site even at the southern tip of Patagonia is dated at about 10,500 years ago. Thus, within about a millenium of emerging from the ice-free corridor at Edmonton, humans had spread from coast to coast and over the entire length of the New World.

Equally startling is the rapid transformation of Clovis culture. Around 11,000 years ago Clovis points are abruptly replaced by a smaller, more finely made model now known as Folsom points (after a site near Folsom, New Mexico, where they were first identified). The Folsom points are often found associated with bones of an extinct wide-horned bison, never with the mammoths preferred by Clovis hunters.

There may be a simple reason why Folsom hunters switched from mammoths to bison: there were no more mammoths left. There also were no more mastodonts, camels, horses, giant ground sloths, nor several dozen other types of big mammals. In all, North America lost an astonishing seventy-three per cent, and South America eighty per cent, of their genera of big mammals around this time. Many paleontologists do not blame this American extinction spasm on Clovis hunters, since there is no surviving evidence of mass slaughter – only the fossilized bones of a few butchered carcasses here and there. Instead, those paleontologists attribute the extinctions to changes of climate and habitats at the end of the ice ages, just around the time that Clovis hunters arrived. That reasoning puzzles me for several reasons. Ice-free habitats for mammals expanded rather than contracted as glaciers yielded to grass and forest; big American mammals had already survived the ends of at least twenty-two previous ice ages without such an extinction spasm; and there were far fewer extinctions in Europe and Asia when the glaciers of those continents melted around the same time.

If changing climate had been the cause, one might have expected opposite effects on species preferring hot and cold climates. Instead, radiocarbon-dated fossils from the Grand Canyon show that the Shasta ground sloth and Harrington's mountain goat, derived from areas of hot and cold climates respectively, both died out within a century or two of each other, around 11,100 years ago. The sloths were common until just before their sudden extinction. In their softball-sized dung balls, still well-preserved in some southwestern US caves, botanists have identified remains of plants on which the last sloths chomped: the Mormon tea and globe mallow, which still occur around those caves today. It is highly suspicious that both those well-fed sloths and the goats of the Grand Canyon disappeared just after Clovis hunters reached Arizona. Juries have convicted murderers on the grounds of less compelling circumstantial evidence. If climate really was what did in the sloths, we would have to credit those supposedly stupid beasts with unsuspected intelligence, since they all chose to drop dead simultaneously at just the right moment to deceive some twentieth-century scientists into blaming Clovis hunters.

A more plausible explanation of this 'coincidence' is that it really was a case of cause and effect. Paul Martin, a geoscientist at the University of Arizona, describes the dramatic outcome of hunter-meets-elephant as a 'blitzkrieg'. According to his view, the first hunters to emerge from the ice-free corridor at Edmonton thrived and multiplied, because they found an abundance of tame, easy-to-hunt big mammals. As the mammals were killed off in one area, the hunters and their offspring kept fanning out into new areas that still had abundant mammals, and kept exterminating the mammal populations at the front of their advance. By the time that the hunters' front finally reached the south tip of South America, most of the big mammal species of the New World had been exterminated.

*

Martin's theory has attracted lots of vigorous criticism, most of it centring on four doubts. Could a band of 100 hunters arriving at Edmonton breed fast enough to populate a hemisphere in a thousand years? Could they spread fast enough to cover the nearly 8,000 miles from Edmonton to Patagonia in that time? Were Clovis hunters really the first people in the New World? And could stone-age hunters really pursue hundreds of millions of big mammals so efficiently that not a single individual survived, while nevertheless leaving little fossil evidence of their hunts?

Take first the question of breeding rates. Populations of modern hunter-gatherers on even their best hunting grounds number only about one per square mile. Hence, once the whole western hemisphere had been settled, its population of hunter-gatherers would have been at most ten million, since the New World's area outside of Canada and other areas covered by glaciers in Clovis times is about ten million square miles. In modern instances where colonists have arrived at an uninhabited land (for instance, when the H.M.S. *Bounty* mutineers reached Pitcairn Island), their population growth has been as rapid as 3.4% per year. That growth rate, which corresponds to each couple having four surviving children and a mean generation time of twenty years, would multiply 100 hunters into ten million in only 340 years. Thus, Clovis hunters should easily have been able to multiply to ten million within a millenium.

Could the descendants of the Edmonton pioneers have reached the south tip of South America in a thousand years? The overland straight-line distance is slightly under 8,000 miles, so that they would have to average eight miles a year. That is a trivial task – any fit hunter or huntress could have fulfilled the year's quota in a day and not moved for another 364 days. The quarry from which a Clovis tool was made can often be identified by its local type of stone, and we know in that way that individual tools travelled up to 200 miles. Some of the nineteenth-century Zulu migrations in southern Africa are known to have covered nearly 3,000 miles in a mere fifty years.

Were Clovis hunters the first humans to spread south of the Canadian ice sheet? That is a harder question, and it is extremely controversial among archaeologists. Primacy claims for Clovis are inevitably based on negative evidence: there are no unequivocal human remains or artifacts with universally accepted pre-Clovis dates anywhere in the New World south of the former Canadian ice sheet. Mind you, there are dozens of *claims* of sites with pre-Clovis human evidence, but all or almost all of them are marred by serious questions about whether the material used for radiocarbon dating was contaminated by older carbon, or whether the dated material was really associated with the human remains, or whether the tools supposedly made by humans were just naturally shaped rocks. The two most nearly convincing of those claimed pre-Clovis sites are Meadowcroft Rock Shelter in Pennsylvania, dated to about 16,000 years ago, and the Monte Verde site in Chile, dated to at least 13,000 years ago. Monte Verde is described as having amazingly good preservation of many types of human artifacts, but those results have not yet been published in detail, so they cannot yet be properly evaluated. At Meadowcroft there has been an unresolved debate about whether the radiocarbon dates are in error, especially because the plant and animal species from the site are ones expected to have been living there only much more recently than 16,000 years ago.

In contrast, the evidence for Clovis people is undeniable, is to be found in all forty-eight contiguous states, and is accepted by all archaeologists. Evidence for the still earlier settlement of the other habitable continents by more primitive humans is also unequivocal and universally accepted. At one Clovis site after another, you can see a level with Clovis artifacts and bones of numerous large extinct mammal species; immediately above (that is, younger than) the Clovis level, a level with Folsom artifacts but with the bones of not a single large extinct mammal except for bison; and immediately below the Clovis level, levels spanning thousands of years before Clovis times, reflecting benign environmental conditions, and full of the bones of large extinct mammals, but with not a single human artifact. How could people possibly have settled the New World in pre-Clovis times and *not* left behind the usual trail of abundant evidence that convinces archaeologists, like stone tools, hearths, occupied caves, and occasionally skeletons, with unequivocal radiocarbon dates? How could there have been pre-Clovis people who left no trace of their presence at Clovis sites, despite such favourable living conditions? How could people have arrived at Pennsylvania or Chile from Alaska, as if by helicopter, without leaving good evidence of their presence in all the intervening territory? For these reasons, I find it more plausible that the dates given for Meadowcroft and Monte Verde are somehow wrong, than that they are correct. The Clovis-first interpretation makes good sense, but the pre-Clovis interpretation just does not make sense to me.

*

The other hotly contested argument over Martin's blitzkrieg theory concerns the supposed over-hunting and extermination of big mammals. It seems hard to imagine how stone-age hunters could kill a mammoth at all, let alone hunt all mammoths to extinction. Even if the hunters *could* slaughter mammoths, why would they *want* to? And where are all the skeletons now?

Certainly, when we stand under a mammoth skeleton in a museum, the thought of using a stone-tipped spear to attack such a gigantic tusked beast feels utterly suicidal. Yet modern Africans and Asians with equally simple weapons do succeed in killing elephants, often hunting as a group relying on ambush or fire, but sometimes stalking an elephant as a single hunter armed with a spear or poisoned arrow. These modern elephant hunters still rate as amateur dabblers, compared to the mammoth hunters of Clovis times, heirs to hundreds of thousands of years of hunting experience with stone tools. Museum artists like to depict late stone-age hunters as naked brutes risking their lives to hurl boulders at an enraged charging mammoth, with one or two hunters already lying trampled to death on the ground. That is absurd. If any hunters had died in a typical mammoth hunt, mammoths would have exterminated hunters, rather than vice versa. Instead, a more realistic picture is of warmly-clad professionals safely spearing a terrified mammoth ambushed in a narrow stream bed.

Recall also that the big mammals of the New World had probably never seen humans before Clovis hunters, if the hunters indeed were the first people to reach the New World. We know from Antarctica and the Galapagos how tame and unafraid are animals that evolved in the absence of humans. When I visited New Guinea's isolated Foja mountains, which lack any human population, I found the large tree kangaroos so tame that I could walk up to within a few yards of them. Probably the New World's large mammals were equally naive and were killed off before they could have time to evolve a fear of man.

Could Clovis hunters have killed mammoths fast enough to exterminate them? Assume again that an average square mile supports one hunter-gatherer and (by comparison with elephants in Africa today) one mammoth, and that one-quarter of the Clovis population consisted of adult male hunters who each killed a mammoth every two months. That means six mammoths killed per four square miles per year, so the mammoths would have to reproduce their numbers in less than a year to keep up with the killing. Yet modern elephants are slow breeders that take about twenty years to reproduce their numbers, and few other large mammal species breed fast enough to reproduce their numbers in less than three years. It could plausibly have taken Clovis hunters only a few years to exterminate the large mammals locally and to move on to the next area. Archaeologists trying to document the slaughter today are searching for needles in a fossil haystack: a few years' worth of butchered mammoth bones among the bones of all the mammoths that died naturally over hundreds of thousands of years. It is no wonder that so few mammoth carcasses with Clovis points among the ribs have been found.

Why would a Clovis hunter even want to kill a mammoth every two months, when a 5,000-pound mammoth yielding 2,500 pounds of meat would provide ten pounds of meat per day for two months for the hunter, his wife, and two children? Ten pounds may sound like gross gluttony, but it actually approaches the daily meat ration per person on the US frontier in the last century. That is assuming that Clovis hunters really ate all 2,500 pounds of mammoth meat. But to keep the meat for two months would require drying it: would you go to the work of drying a ton of meat, when you could instead just go and kill a fresh mammoth? As Vance Haynes noted, Clovis mammoth kills prove to be only partly butchered, suggesting very wasteful and selective utilization of meat by people living amidst an abundance of game. Some hunting probably was not for meat at all but for ivory, hides, or just machismo. Seals and whales have similarly been hunted in modern times for oil or fur, leaving the meat to rot. In New Guinea fishing villages I often see the discarded carcasses of large sharks, killed only for their fins to make the delicacy of shark's fin soup.

We are all too familiar with the blitzkriegs by which modern European hunters nearly exterminated bison, whales, seals, and many other large animals. Recent archaeological discoveries on islands have shown that such blitzkriegs were an outcome whenever earlier hunters reached a land with animals unused to humans (Chapter Seventeen). New Zealand's giant flightless birds, the moas, were all exterminated by Māori colonists within a few centuries. The Indonesians and Africans who colonized Madagascar 1,500 years ago exterminated other giant flightless birds (the elephant birds), along with a dozen species of primates (the lemurs) ranging up to the size of a gorilla. Polynesian colonists of Hawaii exterminated numerous species of big flightless geese. Since the collision between humans and large naive animals has always ended in an extermination spasm, how could it have been otherwise when Clovis hunters entered a naive New World?

*

This end, though, would hardly have been foreseen by the first hunters to arrive at Edmonton. It must have been a dramatic moment when, after entering the ice-free corridor from an overpopulated, overhunted Alaska, they emerged to see herds of tame mammoths, camels, and other beasts. In front of them stretched the Great Plains to the horizon. As they began to explore, they must soon have realized (unlike Christopher Columbus and the Plymouth Pilgrims) that there were no people at all in front of them, and that they had truly arrived first at a fertile land. Those Edmonton Pilgrims, too, had cause to celebrate a Thanksgiving Day.

NINETEEN

THE SECOND CLOUD

Human society is now at risk of destroying itself by stumbling into either a nuclear holocaust or an environmental holocaust. The latter could arise from mass extinction of species. This chapter estimates how many species we have already exterminated, how many more we are likely to exterminate within the next century, and how mass extinction would affect us.

*

Until our own generation, no one had grounds to worry whether the next human generation would survive or enjoy a planet worth living on. Ours is the first generation to be confronted with these questions about its children's future. We devote much of our lives to training our children to support themselves and to get along with other people. Increasingly, we are asking ourselves whether all those efforts of ours might be wasted.

These concerns arise because of two clouds hanging over us – clouds that would have similar consequences, but that we view very differently. One, the risk of a nuclear holocaust, first revealed itself in the cloud over Hiroshima. Everyone agrees that the risk is real, since there are huge stockpiles of nuclear weapons and since politicians throughout history have occasionally made dumb miscalculations. Everyone agrees that, if a nuclear holocaust does happen, it will be bad for us and might even kill us all. This risk shapes much of current world diplomacy. The only thing about which we disagree is how best to handle it – for instance, whether we should aim for complete or partial nuclear disarmament, nuclear balance, or nuclear superiority.

The other cloud is the risk of an environmental holocaust, of which one often discussed potential cause is the gradual extinction of most of the world's species. In contrast to the case with nuclear holocaust, there is almost complete disagreement about whether the risk of a mass extinction is real and about whether it would really do us much harm if it happened. For instance, one of the most frequently cited estimates is that humans have caused about one per cent of the world's bird species to become extinct within the last few centuries. At one extreme, many thoughtful people – especially economists and industrial leaders, but also some biologists and many laypeople – think that that loss of one per cent would have been inconsequential, even if it had really happened. In fact, such people reason that one per cent is a gross *over*estimate, that most species are superfluous to us, and that it would do us no harm to lose ten times more species. At the opposite extreme, many other thoughtful people – especially conservation biologists and a growing number of laypeople belonging to environmentalist movements – think that the one per cent figure is a gross *under*estimate, and that mass extinction would undermine the quality or possibility of human life. Obviously, it will make a big difference to our children which of these two extreme views is closer to the truth.

The risks of a nuclear holocaust and of an environmental holocaust constitute the two really pressing questions facing the human race today. Compared to these two clouds, our usual obsessions with cancer, AIDS, and diet pale into insignificance, because those problems do not threaten the survival of the human species. If the nuclear and environmental risks should not materialize, we shall have plenty of leisure time to solve bagatelles like cancer. If we fail to avert those two risks, solving cancer will not have helped us anyway.

How many species have humans really driven into extinction already? How many more are likely to become extinct within our children's lifetimes? If more do become extinct, so what? How much do wrens contribute to our gross national product? Are not all species destined to become extinct sooner or later? Is the claimed mass extinction crisis an hysterical fantasy, a real risk for the future, or a proven event that is already well underway?

We need to go through three steps if we are to arrive at realistic estimates of the numbers involved in the mass extinction debate. Firstly, let's see how many species have become extinct in modern times (that is, since 1600). Secondly, let's estimate how many other species had become extinct before 1600. As the third step, let's try to predict how many further species are likely to become extinct within the lifespans of ourselves, our children, and our grandchildren. Finally, let's ask what difference it all makes to us anyway.

*

The first step, that of calculating the number of species that have become extinct in modern times, seems easy when one initially thinks about it. Just take some group of plants or animals, count up in a catalogue the total number of species, mark off the ones known to have become extinct since

1600, and add them up. As a group on which to try this exercise, birds have the advantage that they are easy to see and identify, and hoards of bird-watchers watch them. As a result, more is known about them than about any other group of animals.

Approximately 9,000 species of birds exist today. Only one or two previously unknown species are still being discovered each year, so virtually all living birds have already been named. The leading agency concerned with the status of the world's birds – the International Council for Bird Preservation (ICBP) – lists 108 species of birds, plus many additional subspecies, as having become extinct since 1600. Virtually all these cases of extinction were caused in one way or another by humans – more of that later. One hundred and eight is about one per cent of that total number of bird species: 9,000. That is where the one per cent figure I mentioned earlier comes from.

Before we take that as the final word on the number of modern birds that have become extinct, let's understand how the number of 108 was arrived at. The ICBP decides to list a species as extinct only after that bird has been specifically looked for in areas where it was previously known to occur or might have turned up, and after it has not been found for many years. In many cases, birders have watched a population dwindle down to a few individuals and have followed the fates of those last individuals. For example, the most recent subspecies of bird to have become extinct in the US was the dusky seaside sparrow that lived in marshes near Titusville, Florida. As its population shrank due to destruction of the marshes where it lived, wildlife agencies put identification bands on the few remaining sparrows so that they could be individually recognized. When only six remained, they were brought into captivity in order to protect and breed them. Unfortunately, one after another died. The last individual, and with it the subspecies itself, died on 16 June 1987.

Thus, there is no doubt that the dusky seaside sparrow is extinct. Equally little doubt attaches to the many other subspecies and the 108 full species of birds listed as extinct. The full species listed as having vanished in North America since European settlement, and the years in which the last individual of each died, are the great auk (1844), spectacled cormorant (1852), Labrador duck (1875), Carolina parakeet (1914), and passenger pigeon (1914). The great auk also formerly occurred in Europe, but no other European bird species is listed as having become extinct since 1600, though some species have disappeared within Europe while surviving on other continents.

What about all those remaining bird species that did not fulfil the ICBP's rigorous criteria for extinction? Can we be certain that they still exist? For most North American and European birds the answer is 'yes'. Hundreds of thousands of fanatical bird-watchers monitor all bird species on these continents every year. The rarer the species, the more fanatical is the annual search for it. No North American or European bird species could possibly drift into extinction unnoticed. There is only one North American bird species whose current existence is uncertain, the Bachman's warbler, last definitely recorded in 1977, but the ICBP hasn't given up hope for it because of more recent unconfirmed records. (The ivory-billed woodpecker may also be extinct, but the North American population is 'only' a subspecies; a few individuals of the other subspecies of this woodpecker survive in Cuba.) Thus, the number of North American bird species that have suffered extinction since 1600 is surely not less than five nor more than six. Every species but Bachman's warbler can be assigned to one of two categories – those that are 'definitely extinct', or 'definitely in existence'. Similarly, the number of European bird species extinct since 1600 is surely one – not two, not zero, but one.

Consequently we have an exact, unequivocal answer to the question of how many North American and European bird species have become extinct since 1600. If we could be equally definite for other groups of species, our first step in assessing the mass extinction debate would be complete. Unfortunately, this cut-and-dried situation does not apply to other groups of plants and animals, nor does it apply elsewhere in the world – least of all in the tropics, where the overwhelming majority of species lives. Most tropical countries have few or no bird-watchers, and no annual monitoring of birds. Many tropical areas have never again been monitored since they were first explored biologically many years ago. The status of many tropical species is unknown, because no one has seen them again or specifically looked for them since they were discovered. For instance, among the New Guinea birds that I study, Brass's friarbird is known only from eighteen specimens shot at one lagoon on the Idenburg River between 22 March and 29 April 1939. No scientist has revisited that lagoon, so we know nothing about the current status of Brass's friarbird.

At least, we know where to look for that friarbird. Many other species were described from specimens collected by nineteenth-century expeditions that provided only vague indications of the collecting site, such as, 'South America'. Try resolving the status of some rare species when you have only that broad hint where to look! The songs, behaviour, and habitat preferences of such species are unknown. Hence we do not know where to seek them, nor how to identify them if we glimpsed or heard them.

The status of many tropical species cannot be classified either as 'definitely extinct' or 'definitely in existence', but just as 'unknown'. Instead, it becomes a matter of chance which species happens to attract the attention of some ornithologist, becomes the object of a specific search, and hence may be recognized as possibly extinct.

Here is an example. The Solomon Islands are another of my favourite bird-watching areas in the tropical Pacific Ocean, and will be recalled by older Americans and Japanese as the site of some of the fiercest fighting in the Second World War. (Remember Guadalcanal, Henderson Field, President Kennedy's PT boat, the Tokyo Express?) The ICBP lists one Solomon bird species, Meek's crowned pigeon, as extinct. Yet when I tabulated all recent observations of all 164 known Solomon bird species, I noticed that twelve of those 164 species had not been encountered since 1953. Some of those twelve species are surely extinct, because they were formerly abundant and conspicuous. Several Solomon islanders told me that those birds had been exterminated by cats.

Twelve species possibly extinct out of 164 still may not sound like much to worry about. However, the Solomons are in much better shape environmentally than most of the remaining tropical world, because they have relatively few people, few bird species, little economic development, and much natural forest. More typical of the tropics is Malaysia, which is rich in species and has had most of its lowland forest cut down. Biological explorers had identified 266 fish species dependent on fresh water in Malaysia's forest rivers. A recent search that lasted four years was able to find only 122 of those 266 species – less than half. The other 144 Malaysian freshwater fish species must either be extinct, rare, or very local. They reached that status before anyone noticed it.

Malaysia is typical of the tropics in the pressure it faces from humans. Fish are typical of all species other than birds, in that they attract only patchy scientific attention. The estimate that Malaysia has already lost (or nearly lost) half of its freshwater fish is therefore a reasonable ballpark figure for the status of plants, invertebrates, and vertebrates other than birds in much of the rest of the tropics.

That is one complication in trying to pinpoint the number of extinctions since 1600: the status of many or most named species is unknown. But there is a further complication. So far, we have been trying to assess the extinction only of those species that had already been discovered and described (named). Could any species have become extinct before they were even described?

Of course they could, since sampling procedures suggest that the actual number of the world's species is near thirty million, but less than two million species have been described. Two examples illustrate the certainty of other species becoming extinct before description. Botanist Alwyn Gentry surveyed the plants of an isolated ridge in Ecuador called Centinela, where he found thirty-eight new species confined to that ridge. Shortly afterwards, the ridge was logged and those plants were exterminated. On Grand Cayman Island in the Caribbean, zoologist Fred Thompson discovered two new species of land snails confined to forest on a limestone ridge that was completely cleared a few years later for a housing development.

The fact that Gentry and Thompson accidentally visited those ridges before rather than after they were cleared means that we have names for those extinct species. But most tropical areas that are being developed are not first surveyed by biologists. There must have been land snails on Centinela, and plants and snails on innumerable other tropical ridges, that we exterminated before we discovered them.

In short, the problem of determining the number of modern species that have become extinct seems at first to be simple and to lead to modest estimates – for example, only five or six extinct bird species in all of North America plus Europe. On reflection, though, we appreciate two reasons why published lists of species known to be extinct must be gross underestimates of the actual numbers involved. Firstly, by definition the published lists consider only named species, whereas the great majority of species (except in well-studied groups like birds) have not even been named. Secondly, outside North America and Europe and except for birds, the published lists consist only of those few named species which some biologist happened to get interested in for one or another reason and found to be extinct. Among all those remaining species of unknown status, many are likely to be extinct or nearly so – for example, about half in the case of Malaysian freshwater fishes.

*

Now let's move on to the second step in evaluating the mass extinction debate. Our estimates up to this point have concerned only those species exterminated since 1600 AD, when scientific classification of species was beginning. These exterminations have taken place because the world's human population has grown in numbers, reached previously un-inhabited areas, and invented increasingly destructive technologies. Did these factors spring up suddenly in 1600, after several million years of human history? Were there no exterminations before 1600?

Of course not. Until fifty thousand years ago, humans were confined to Africa plus the warmer areas of Europe and Asia. Between then and 1600 AD our species underwent a massive geographic expansion that took us to Australia and New Guinea around 50,000 years ago, Siberia around 20,000 years ago, most of North and South America around 11,000 years ago, and most of the world's remote oceanic islands only since 2000 BC. We also underwent a massive expansion in numbers, from perhaps a few million people 50,000 years ago to about half a billion in 1600. Our destructiveness also increased, with the development of improved hunting skills in the last 50,000 years (Chapter Two), polished stone tools and agriculture in the last 10,000 years (Chapters Ten and Fourteen), and metal tools in the last 6,000 years.

In every area of the world that paleontologists have studied and that humans first reached within the last 50,000 years, human arrival approximately coincided with massive instances of prehistoric species' extinction waves. For Madagascar, New Zealand, Polynesia, and the Americas I have described those instances in the preceding two chapters. After people reached Australia, that continent lost its giant kangaroos, its 'marsupial lion', and other giant marsupials. Around the time that Indians reached North America 11,000 years ago, it lost lions, cheetahs, native wild horses, mammoths, mastodonts, giant ground sloths, and several dozen other large mammals. Mediterranean islands like Crete and Cyprus lost dwarf elephants and pygmy hippos, while Madagascar lost giant lemurs and flightless elephant birds. New Zealand lost its giant flightless moas, and Hawaii its flightless geese and dozens of smaller birds, when the Polynesians arrived around 1000 and 500 AD, respectively.

Ever since scientists became aware of these prehistoric extinction waves associated with human arrival, they have argued over whether people were the cause or just happened to arrive while animals were succumbing to climate changes. In the case of the extinction waves on Polynesian islands, there is now no reasonable doubt that Polynesian arrival in one way or another caused them. Bird extinction waves and Polynesian arrival coincided within a few centuries at a time when no big climate change was happening, and bones of thousands of roasted moas have been found in Polynesian ovens. The coincidence of timing is equally convincing for Madagascar. But the causes of the earlier extinction waves, especially those in Australia and the Americas, are still being debated.

As I explained on America's extinction waves in Chapter Eighteen, the evidence seems to me overwhelming that humans also played a role in those prehistoric cases of extinction outside Polynesia and Madagascar. In each part of the world an extinction wave occurred after the first arrival of humans, but did not occur simultaneously in other areas undergoing similar climate swings, and did not occur in the same area whenever si milar climate swings had occurred previously.

Hence I doubt that climate did it. Instead, all of you who have visited Antarctica or the Galapagos Islands know how tame are the animals there, being unaccustomed to humans until recently. Photographers can still walk up to those naive animals as easily as hunters used to. I assume that the first arriving hunters similarly walked up to naive mammoths and moas elsewhere in the world, while rats that came with the first hunters walked up to naive little birds of Hawaii and other islands.

It is not just in those areas of the world previously unoccupied by humans that prehistoric humans probably exterminated species. Within the last 20,000 years species also became extinct in the areas long occupied by humans – in Eurasia, woolly rhinos, mammoths, and giant deer ('Irish elk') died out, and Africa lost its giant buffalo, giant hartebeest, and giant horse. These big beasts may also have been among the victims of prehistoric humans who had already been hunting them for a long time, but who now were able to hunt them with better weapons than ever before. Eurasia's and Africa's big mammals were not unused to humans, but they disappeared for the same two simple reasons that California's grizzly bear, and Britain's bears, wolves, and beavers, succumbed only in recent times, after thousands of years of human persecution. Those reasons were more people, and better weaponry.

Can we at least estimate how many species were involved in these prehistoric extinction waves? No one has ever tried to guess the number of plants, invertebrates, and lizards exterminated by prehistoric habitat destruction, but virtually all oceanic islands explored by paleontologists have yielded remains of recently extinct bird species. Extrapolation to those islands not yet paleontologically explored suggests that about 2,000 bird species – one fifth of all the birds that existed a few thousand years ago – were island species already exterminated prehistorically. That does not include birds that may have been exterminated prehistorically on the continents. Among genera of large mammals, about seventy-three, eighty, and eighty-six per cent respectively became extinct in North America, South America, and Australia at the time of or after human arrival.

*

The remaining step in evaluating the mass extinction debate is to predict the future. Is the peak of the extinction wave that we have caused already past, or is most still to come? There are a couple of ways to assess this question.

A simple way is to reason that tomorrow's extinct species will be drawn from today's endangered species. How many species that still exist have populations already reduced to dangerously low levels? The ICBP estimates that at least 1,666 bird species are either endangered or at imminent risk of extinction – almost twenty per cent of the world's surviving birds. I said 'at least 1,666', because this number is an underestimate for the same reason I mentioned that the ICBP's estimate of extinct species was an underestimate. Both numbers are based just on those species whose status caught a scientist's attention, rather than on a reappraisal of the status of all bird species.

The alternative way of predicting what is to come is to understand the mechanisms by which we exterminate species. Extinction of species caused by humans may continue accelerating until human population and technology reach a plateau, but neither shows any signs of plateauing. Our population, which grew ten fold from half a billion in 1600 to over five billion now, is still growing at close to two per cent per year. Every day brings new technological advances for changing the earth and its denizens. There are four main mechanisms by which our growing population exterminates species: by overhunting, species introductions, habitat destruction, and ripple effects. Let's see if these four mechanisms have plateaued.

Overhunting – killing animals faster than they can breed – is the main mechanism by which we have exterminated big animals, from mammoths to California grizzly bears. (The latter appears on the flag of California, the state in which I live, but many of my fellow Californians do not recall that we exterminated our state's symbol long ago.) Have we already killed off all big animals that we might kill off? Obviously not. While the low numbers of whales led to an international ban on whaling for commercial reasons, Japan thereupon announced its decision to triple the rate at which it kills whales 'for scientific reasons'. We have all seen photos of the accelerating slaughter of Africa's elephants and rhinos, for their ivory and horns respectively. At current rates of change, not just elephants and rhinos but most populations of most other large mammals of Africa and Southeast Asia will be extinct outside game parks and zoos in a decade or two.

The second mechanism by which we exterminate is through intentionally or accidentally introducing certain species to parts of the world where they did not previously occur. Familiar examples of introduced species now firmly established in the US are Norway rats, European starlings, boll weevils, and the fungi causing Dutch elm disease and chestnut blight. Europe too has acquired introduced species, of which the misnamed Norway rat is an example (it originated in Asia, not Norway). When species are introduced from one region to another, they often proceed to exterminate some of the new species they encounter, by eating them or causing diseases. The victims evolved in the absence of the introduced pests and never developed defences against them. American chestnut trees have already been virtually exterminated in this way by chestnut blight, an Asian fungus to which Asian chestnut trees are resistant. Similarly, goats and rats have exterminated many plants and birds on oceanic islands.

Have we already spread all possible pests all around the world?

Obviously not; there are many islands still free of goats and Norway rats, and many insects and diseases to try to keep out of many countries by quarantines. The US Department of Agriculture has been trying at great expense, but apparently without success, to forestall the arrival of killer bees and Mediterranean fruit flies. In fact, what will probably prove to be the biggest extinction wave caused by an introduced predator in modern times has just started in Africa's Lake Victoria, home to hundreds of species of remarkable fishes found nowhere else in the world. A large predatory fish called the Nile perch, intentionally introduced in a misguided effort to establish a new fishery, is now eating its way through the lake's unique fishes.

Habitat destruction is the third means by which we exterminate. Most species occur in just a certain type of habitat: marsh warblers live in marshes, while pine warblers live in pine forests. If one drains marshes or cuts forests, one eliminates the species dependent on those habitats just as certainly as if one were to shoot every individual of the species. For example, when all the forest on Cebu Island in the Philippines was logged, nine of the ten birds unique to Cebu became extinct.

In the case of habitat destruction, the worst is still to come because we are just starting in earnest to destory tropical rainforests, the world's most species-rich habitats. The rainforests' biological richness is legendary – over 1,500 beetle species living in a single rainforest tree species in Panama, for instance. Rainforests cover only six per cent of the Earth's surface but harbour about half of its species. Each area of rainforest has large numbers of species unique to that area. To mention only some exceptionally rich rainforests now being destroyed, the felling of Brazil's Atlantic forest and Malaysia's lowland forest is already almost complete, and those of Borneo and the Philippines will be mostly logged within the next two decades. By the middle of the next century, the only large tracts of tropical rainforest likely to be still surviving will be in parts of Zaire and the Amazon Basin.

Every species depends on other species for food and for providing its habitat. Thus, species are connected to each other like branching chains of dominoes. Just as toppling one domino in a chain will topple some others, so too the extermination of one species may lead to the loss of others, which may in turn push still others over the brink. This fourth mechanism of extinction may be described as a ripple effect. Nature consists of so many species, connected to each other in such complex ways, that it is virtually impossible to foresee where the ripple effects from the extinction of any particular species may lead.

For example, fifty years ago no one anticipated that the extinction of big predators (jaguars, pumas, and harpy eagles) on Panama's Barro Colorado Island would lead to the extinction there of little antbirds, and to massive changes in the tree species composition of the island's forest. Yet it did so, because the big predators used to eat medium-sized predators like peccaries, monkeys, and coatimundis, and medium-sized seed-eaters like agoutis and pacas. With the disappearance of the big predators, there was a population explosion of the medium-sized predators, which proceeded to eat up the antbirds and their eggs. The medium-sized seed-eaters also exploded in abundance and ate large seeds that had fallen on the ground, thereby suppressing the propagation of tree species producing large seeds and favouring instead the spread of competing tree species with small seeds. That shift in forest tree composition is expected in turn to cause an explosion of mice and rats feeding on small seeds, and then to an explosion in hawks, owls, and ocelots preying on those small rodents. Thus, the extinction of three uncommon species of big predators will have triggered a rippling series of changes in the whole plant and animal community, including the extinction of many other species.

Through these four mechanisms – overhunting, species introductions, habitat destruction, and ripple effects – probably over half of existing species will be extinct or endangered by the middle of the next century, when this year's crop of human babies reaches the age of sixty. Like many fathers today, I often wonder how I will describe to my twin sons, who are now three years old, the world that I grew up in and that they will never see. By the time they would have been old enough to come with me to New Guinea, one of the world's biological treasurehouses where I have worked for the past twenty-five years, most of New Guinea's eastern highlands will be deforested.

When one adds the extinction of species we have already caused to that which we are about to cause, it is clear that the current extinction wave is surpassing the asteroid collision that may have wiped out the dinosaurs. Mammals, plants, and many other types of species survived that collision nearly unscathed, while the current wave is affecting everything from leeches and lilies to lions. Thus, the claimed extinction crisis is neither a hysterical fantasy, nor just a serious risk for the future. Instead, it is an event that has already been accelerating for 50,000 years and will start to approach completion in our children's lifetimes.

*

Let's finally consider two arguments that accept the reality of the extinction crisis but dismiss its significance. Firstly, is extinction not a natural process anyway? If so, why make a big deal about the wave of extinction happening now?

The answer to this first argument is that the current extinction rate caused by humans is far higher than the natural rate. If the estimate that half the world's total of thirty million species will become extinct in the next century is correct, then species are now becoming extinct at a rate of about 150,000 per year, or seventeen per hour. The world's 9,000 bird species are becoming extinct at a rate of at least two per year, but bird species under natural conditions were disappearing at a rate of less than one per century, so the present rate is at least 200 times the normal rate. Dismissing the extinction crisis on the grounds that extinction is natural would be just like dismissing genocide on the grounds that death is the natural fate of all humans.

The second argument is a simple one: so what? We care about our children, not about beetles and snail darters; who cares if ten million beetle species become extinct? The answer to this argument is equally simple. Like all species, we depend on other species for our existence, in many ways. Some of the most obvious ways are that other species produce the oxygen we breathe, absorb the carbon dioxide we exhale, decompose our sewage, provide our food, maintain the fertility of our soil, and provide our wood and paper.

Then could we not preserve only those particular species that we need, and let other species become extinct? Of course not, because the species we need also depend on other species. Just as Panama's antbirds could not have anticipated their need for jaguars, the ecological chain of dominoes is much too complex for us to have figured out which dominoes we can dispense with. For instance, could anyone please answer these three questions. Which ten tree species produce most of the world's paper pulp? For each of those ten tree species, which are the ten bird species that eat most of its insect pests, the ten insect species that pollinate most of its flowers, and the ten animal species that spread most of its seeds? Which other species do these ten birds, insects, and animals depend on? You would have to be able to answer those three impossible questions if you were the president of a timber company trying to figure out which species you could afford to allow to become extinct.

If you are trying to evaluate some proposed development project that would bring in a million dollars but might exterminate a few species, it is still tempting to prefer the certain profit over the uncertain risk. Then consider the following analogy. Suppose someone offers you a million dollars in return for the privilege of painlessly cutting out two ounces of your valuable flesh. You figure that two ounces is only one-thousandth of your body weight, so you will still have nine hundred and ninety-nine thousandths of your body left, which is plenty. That is fine if the two ounces come from your spare body fat and if they will be removed by a skilled surgeon. But what if the surgeon just hacks two ounces from any conveniently accessible part of your body, or does not know which parts are essential? You might then find that the two ounces came from your urethra. If you plan to sell off most of your body, as we now plan to sell off most of our planet's natural habitats, you are certain eventually to lose your urethra.

*

To conclude, let's place matters in perspective by comparing the two clouds which, as I mentioned at the outset, are hanging over our future. A nuclear holocaust is certain to prove disastrous, but it is not happening now, and it may or may not happen in the future. An environmental holocaust is equally certain to prove disastrous, but it differs in that it is already well underway. It started tens of thousands of years ago, is now causing more damage than ever before, is in fact accelerating, and will climax within about a century if unchecked. The only uncertainties are whether the resulting disaster would strike our children or our grand-children, and whether we choose to adopt now the many obvious countermeasures.

EPILOGUE

NOTHING LEARNED,
AND EVERYTHING
FORGOTTEN?

LET'S NOW DRAW TOGETHER THE THEMES OF THIS BOOK, BY TRACING OUR rise over the last three million years, as well as our incipient reversal of all our progress more recently.

The first indications that our ancestors were in any respect unusual among animals were our extremely crude stone tools that began to appear in Africa by around two-and-a-half million years ago. The quantities of tools suggest that they were beginning to play a regular, significant role in our livelihood. Among our closest relatives, in contrast, the pygmy chimpanzee and gorilla do not use tools, while the common chimpanzee occasionally makes some rudimentary ones but hardly depends on them for its existence.

Nevertheless, those crude tools of ours did not trigger any quantum jump in our success as a species. For another million-and-a-half years, we remained confined to Africa. Around a million years ago we did manage to spread to warm areas of Europe and Asia, thereby becoming the most widespread of the three chimpanzee species but still much less widespread than lions. Our tools progressed only at an infinitely slow rate, from extremely crude to very crude. By a hundred thousand years ago, at least the human populations of Europe and western Asia, the Neanderthals, were regularly using fire, but in other respects we continued to rate as just another species of big mammal. We had developed not a trace of art, agriculture, or high technology. It is unknown whether we had developed language, drug addiction, or our strange modern sexual habits and life-cycle, but Neanderthals rarely lived beyond the age of forty and hence may not yet have evolved female menopause.

Clear evidence of a Great Leap Forward in our behaviour appears suddenly in Europe around 40,000 years ago, coincident with the arrival of anatomically modern *Homo sapiens* from Africa via the Near East. At that point, we began displaying art, technology based on specialized tools, cultural differences from place to place, and cultural innovation with time. This leap in behaviour had undoubtedly been developing outside Europe, but the development must have been rapid, since the anatomically modern *Homo sapiens* populations living in southern Africa 100,000 years ago were still just glorified chimpanzees, judging by the debris in their cave sites. Whatever caused the leap, it must have involved only a tiny fraction of our genes, because we still differ from chimps in only 1.6% of our genes, and most of that difference had already developed long before our leap in behaviour. The best guess I can make is that the leap was triggered by the perfection of our modern capacity for language.

Although we usually think of the Cro-Magnons as the first bearers of our noblest traits, they also bore the two traits that lie at the root of our current problems: our propensities to murder each other *en masse* and to destroy our environment. Even before Cro-Magnon times, fossil human skulls punctured by sharp objects and cracked to extract the brains bear witness to murder and cannibalism. The suddenness with which Neanderthals disappeared after Cro-Magnons arrived provides a hint that genocide had now become efficient. Our efficiency at destroying our own resource base is suggested by the extinction of almost all large Australian animals following our colonization of Australia 50,000 years ago, and of some large Eurasian and African mammals as our hunting technology improved. If the seeds of self-destruction have been so closely linked with the rise of advanced civilizations in other solar systems as well, it becomes easy to understand why we have not been visited by any flying saucers.

At the end of the last Ice Age around 10,000 years ago, the pace of our rise quickened. We occupied the Americas, coincident with a mass extinction of big mammals that we may have caused. Agriculture emerged soon thereafter. Some thousands of years later, the first written texts start to document the pace of our technical inventiveness. They also show that we were already addicted to drugs, and that genocide had become routine and admired. Habitat destruction began undermining many societies, and the first Polynesian and Malagasy settlers caused blitzkrieg-like mass exterminations of species. From 1492 AD onwards, the worldwide expansion of literate Europeans lets us trace our rise and fall in detail.

Within the last few decades we have developed the means to send radio signals to other stars, and also to blow ourselves up overnight. Even if we do not blunder into that swift end, our harnessing of much of the Earth's productivity, our exterminations of species, and our damage to our environment are accelerating at a rate that cannot be sustained for even another century. One might object that, if we look around us, we see no obvious sign that the climax of our history will come soon. In fact, the signs become obvious if one observes and then extrapolates. Starvation, pollution, and destructive technology are increasing; usable farmland, food stocks in the sea, other natural products, and environmental capacity to absorb wastes are decreasing. As more people with more power scramble for fewer resources, something has to give way.

So, what is likely to happen?

There are many grounds for pessimism. Even if every human now alive were to die tomorrow, the damage that we have already inflicted on our environment would ensure that its degradation will continue for decades. Innumerable species already belong to the 'living dead', with populations fallen to levels from which they cannot recover, even though not all individuals have died yet. Despite all our past self-destructive behaviour from which we could have learned, many people who should know better dispute the need for limiting our population and continue to assault our environment. Others join that assault for selfish profit or out of ignorance. Even more people are too caught up in the desperate struggle for survival to be able to enjoy the luxury of weighing the consequences of their actions. All these facts suggest that the juggernaut of destruction has already reached unstoppable momentum, that we too are among the living dead, and that our future is as bleak as that of the other two chimpanzees.

This pessimistic view is captured by a cynical sentence that Arthur Wichmann, a Dutch explorer and professor, penned in another context in 1912. Wichmann had devoted a decade of his life to writing a monumental three-volume treatise on the history of New Guinea's exploration. In 1,198 pages he evaluated every source of information about New Guinea that he could find, from the earliest reports filtering through Indonesia to the great expeditions of the Nineteenth and early Twentieth centuries. He grew disillusioned as he realized that successive explorers committed the same stupidities again and again: they showed the same unwarranted pride in overstated accomplishments, refused to acknowledge disastrous oversights, ignored the experience of previous explorers, repeated previous errors, and hence blundered into unnecessary suffering and death. Looking back on this long history, Wichmann predicted that future explorers would continue to repeat the same errors. The bitter last sentence that concluded Wichmann's last volume was, 'Nothing learned, and everything forgotten!'

Despite all the grounds I have mentioned for being equally cynical about humanity's future, my view is that our situation is not hopeless. We are the only ones creating our problems, so it is completely within our power to solve them. While our language and art and agriculture are not quite unique, we really are unique among animals in our capacity to learn from the experience of others of our species living in distant places or in the distant past. Among the hopeful signs, there are many realistic, often discussed policies by which we could avoid disaster, such as limiting human population growth, preserving natural habitats, and adopting other environmental safeguards. Many governments are already doing some of these obvious things in some cases.

For example, awareness of environmental problems is spreading, and environmental movements are gaining political clout. Developers do not win all the battles, nor do short-sighted economic arguments always prevail. Many countries have lowered their rate of population growth in recent decades. While genocide has not vanished, the spread of communications technology has at least the potential for reducing our traditional xenophobia, and for making it harder to regard distant peoples as subhumans unlike ourselves. I was seven years old when the A-bombs were dropped on Hiroshima and Nagasaki, so I remember well the sense of an imminent risk of nuclear holocaust that prevailed for several decades thereafter. But nearly half a century has now passed without any further military use of nuclear weapons. The risk of a nuclear holocaust now seems more remote than at any other time since 9 August 1945.

My own outlook is conditioned by my experiences since 1979 as consultant to the Indonesian government on setting up a nature reserve system in Indonesian New Guinea (called Irian Jaya province). On the face of it, Indonesia does not seem a promising place to hope for much success in preserving our shrinking natural habitats. Instead, Indonesia exemplifies the problems of tropical Third World countries in acute form. With over 180 million inhabitants, it is the world's fifth most populous country, as well as one of the poorer ones. The population is growing rapidly; nearly half of all Indonesians are under fifteen years old. Some provinces with an inordinately high population density are exporting their population surpluses to the less populated provinces (such as Irian Jaya). There are no armies of bird-watchers, no broad-based indigenous environmental movements. The government is not a democracy in the Western sense, and corruption is viewed as pervasive. Indonesia depends on logging of its virgin rainforests, second only to exploitation of oil and natural gas, as a source of its foreign exchange.

For all these reasons, one might not expect preservation of species and habitats to be a national priority pursued seriously in Indonesia. When I first went to Irian Jaya, I was frankly doubtful that an effective conservation programme would result. Fortunately, my Wichmann-like cynicism proved wrong. Thanks to the leadership of a core of Indonesians convinced of the value of conservation, Irian Jaya now has the beginnings of a nature reserve system comprising twenty per cent of the province's area. Nor do those reserves exist just on paper. As my work proceeded, I was pleasantly surprised to come across sawmills abandoned because they conflicted with nature reserves, park guards out on patrol, and management plans being drawn up. All these measures were adopted not out of idealism, but out of a cold-blooded, correct perception of Indonesia's national self-interest. If Indonesia can do it, so can other countries with similar obstacles to environmentalism, as well as much richer countries with broad-based environmental movements.

We do not need novel, still-to-be invented technologies to solve our problems. We just need more governments to do many more of the same obvious things that some governments are already doing in some cases. Nor is it true that the average citizen is powerless. There are many causes of extinction that citizen groups have helped throttle in recent years – for instance, commercial whaling, hunting big cats for fur coats, and importing chimpanzees caught in the wild, to mention just a few examples. In fact, this is one area where it is particularly easy for a modest donation by the average citizen to have a big impact, because all conservation organizations now have such modest budgets. For instance, the annual combined budget for *all* primate conservation projects that the World Wild Fund for Nature supports throughout the world is only a few hundred thousand dollars. An extra thousand dollars means an extra project on some endangered monkey, ape, or lemur that might otherwise have been ignored. On pages 352–4 I suggest some specific starting points for interested readers.

Hence while I do see us facing serious problems with an uncertain prognosis, I am cautiously optimistic. Even the cynical last sentence of Wichmann's book proved false: New Guinea explorers since Wichmann really have learned from the past and avoided the disastrous stupidities of their predecessors. A motto more appropriate for our future than Wichmann's motto comes from the memoirs of the statesman Otto von Bismarck. As he reflected on the world around him towards the end of his long life, he too had reason to be cynical. Possessing a keen intellect and working at the centre of European politics for decades, Bismarck had witnessed a history of unnecessarily repeated errors as gross as those pervading the early history of New Guinea exploration. Yet Bismarck still considered it worthwhile to write his memoirs, to draw lessons from history, and to dedicate his memoirs 'to [my] children and grandchildren, towards an understanding of the past, and for guidance for the future'.

This is also the spirit in which I dedicate this book to my young sons and their generation. If we will learn from our past that I have traced, our own future may yet prove brighter than that of the other two chimpanzees.

FURTHER READING

These suggestions are for readers interested in reading further. In addition to key books and papers, I have also tended to favour recent references that provide comprehensive listings of the earlier literature. Journal titles are followed by the volume number, followed by the first and last page number, and then the year of publication in parentheses.

Chapter 1: A Tale of Three Chimps

The literature on deducing relationships among humans and other primates by means of the DNA clock consists of technical articles in scientific journals. Sibley and Ahlquist present their studies in three papers: C.G. Sibley and J.E. Ahlquist, 'The phylogeny of the hominoid primates, as indicated by DNA-DNA hybridization', *Journal of Molecular Evolution* 20, pp. 2–15 (1984); 'DNA hybridization evidence of hominoid phylogeny: results from an expanded data set', *Journal of Molecular Evolution* 26, pp. 99–121 (1987); and C.G. Sibley, J.A. Comstock, and J.E. Ahlquist, 'DNA hybridization evidence of hominoid phylogeny: a reanalysis of the data', *Journal of Molecular Evolution* 30, pp. 202–36 (1990). Sibley's and Ahlquist's many studies of bird relationships by means of the same DNA methods are summarized in two articles: C.G. Sibley and J.E. Ahlquist, 'The phylogeny and classification of birds based on the data of DNA-DNA hybridization', in the book *Current Ornithology*, edited by R.F. Johnston, vol. 1, pp. 245–92 (Plenum, New York, 1983); and C.G. Sibley, J.E. Ahlquist, and B.L. Monroe, 'A classification of the living birds of the world based on DNA-DNA hybridization studies', *Auk* 105, pp. 409–23 (1988).

Similar conclusions about human and primate relationships were obtained by DNA comparisons using a different method (termed the tetraethylammonium chloride method, rather than the hydroxyapatite method used by Sibley and Ahlquist). The results were described by A. Caccone and J.R. Powell in 'DNA divergence among hominoids', *Evolution* 43, pp. 925–42 (1989). A paper by the same authors explains how percentage similarity among DNAs can be calculated from DNA mixed melting points: A. Caccone, R. DeSalle, and J.R. Powell,

'Calibration of the changing thermal stability of DNA duplexes and degree of base pair mismatch', *Journal of Molecular Evolution* 27, pp. 212–16 (1988).

The above papers compare the entire genetic material (DNA) of two species by means of mixed melting points in order to obtain a single measure of overall similarity. Alternatively, a much more laborious method yielding much more detailed information about a tiny fraction of each species' DNA consists of determining the actual sequence of molecular units comprising that portion of DNA. Four studies stemming from a single laboratory and applying that method to human and primate relationships are M.M. Miyamoto *et al*, 'Phylogenetic relations of humans and African apes from DNA sequence in the Ψ-globin region', *Science* 238, pp. 369–73 (1987); M.M. Miyamoto *et al*, 'Molecular systematics of higher primates: genealogical relations and classification', *Proceedings of the National Academy of Sciences* 85, pp. 7627–31 (1988); M. Goodman *et al*, 'Molecular phylogeny of the family of apes and humans', *Genome* 31, pp. 316–35 (1989); and M. Goodman *et al*, 'Primate evolution at the DNA level and a classification of hominoids', *Journal of Molecular Evolution* 30, pp. 260–66 (1990). The same principle is applied to relationships among Lake Victoria's cichlid fishes by A. Meyer *et al*, 'Monophyletic origin of Lake Victoria cichlid fishes suggested by mitochondrial DNA sequences', *Nature* 347, pp. 550–53 (1990).

Two papers that vigorously criticize the DNA clock in general, and Sibley's and Ahlquist's application of it to human/primate relationships in particular, are J. Marks, C.W. Schmidt, and V.M. Sarich, 'DNA hybridization as a guide to phylogeny: relationships of the Hominoidea', *Journal of Human Evolution* 17, pp. 769–86 (1988); and V.M. Sarich, C.W. Schmidt, and J. Marks, 'DNA hybridization as a guide to phylogeny: a critical analysis', *Cladistics* 5, pp. 3–32 (1989). In my view, the criticisms by Marks, Schmidt, and Sarich have been adequately answered. The good agreement between conclusions about human/primate relationships based on the DNA clock as measured by Sibley and Ahlquist, the DNA clock as measured by Caccone and Powell, and DNA sequencing further supports the correctness of these conclusions.

Other papers on the DNA clock are in two issues of the *Journal of Molecular Evolution*, nos. 3 and 5 in vol. 30 (1990), that also include some of the papers cited above.

Chapter 2: The Great Leap Forward

Among the many books providing detailed accounts of human evolution, the recent one that I found the most useful is by Richard Klein,

The Human Career (University of Chicago Press, Chicago, 1989). A beautifully illustrated and less technical account is by Roger Lewin, *In the Age of Mankind* (Smithsonian Books, Washington DC, 1988).

Two books presenting multi-authored technical accounts of recent human evolution are edited by Fred H. Smith and Frank Spencer, *The Origins of Modern Humans* (Liss, New York, 1984) and by Paul Mellars and Chris Stringer, *The Human Revolution: Behavioural and Biological Perspectives on the Origins of Modern Humans* (Edinburgh University Press, Edinburgh, 1989). Some recent articles on the dating and geography of human evolution are by C.B. Stringer and P. Andrews, 'Genetic and fossil evidence for the origin of modern humans', *Science* 239, pp. 1263–68 (1988); H. Valladas *et al*, 'Thermoluminescence dating of Mousterian "proto-Cro-Magnon" remains from Israel and the origin of modern man', *Nature* 331, pp. 614–16 (1988); C.B. Stringer *et al*, 'ESR dates for the hominid burial site of Es Skhul in Israel', *Nature* 338, pp. 756–58 (1989); J.L. Bischoff *et al*, 'Abrupt Mousterian-Aurignacian boundaries at *c.* 40 ka bp: accelerator ^{14}C dates from l'Arbreda Cave (Catalunya, Spain)', *Journal of Archaeological Science* 16, pp. 563–76 (1989); V. Cabrera-Valdes and J. Bischoff, 'Accelerator ^{14}C dates for Early Upper Paleolithic (Basal Aurignacian) at El Castillo Cave (Spain)', *Journal of Archaeological Science* 16, pp. 577–84 (1989); and E.L. Simons, 'Human origins', *Science* 245, pp. 1343–50 (1989).

Three books with many beautiful illustrations of Ice Age art are by Randall White, *Dark Caves, Bright Visions* (American Museum of Natural History, New York, 1986); Mario Ruspoli, *Lascaux: the Final Photographs* (Abrams, New York, 1987); and Paul G. Bahn and Jean Vertut, *Images of the Ice Age* (Facts on File, New York, 1988).

Matthew H. Nitecki and Doris V. Nitecki, *The Evolution of Human Hunting* (Plenum Press, New York, 1986), provide a series of chapters by various authors on that subject.

The question of whether Neanderthals really did bury their dead is debated in an article by R.H. Gargett, 'Grave shortcomings: the evidence for Neanderthal burial', and in accompanying responses, published in *Current Anthropology* 30, pp. 157–90 (1989).

Three sources that will provide an entrance into the literature on the linked questions of human vocal tract anatomy and whether Neanderthals could speak are Philip Lieberman's *The Biology and Evolution of Language* (Harvard University Press, Cambridge, 1984); E.S. Crelin's *The Human Vocal Tract* (Vantage Press, New York, 1987); and an article by Arensburg *et al*, 'A Middle Palaeolithic human hyoid bone', *Nature* 338, 758–60 (1989).

Chapter 3: The Evolution of Human Sexuality
Chapter 4: The Science of Adultery

For anyone interested in an evolutionary approach to behaviour in general (including reproductive behaviour), two books are a must: E.O. Wilson, *Sociobiology* (Harvard University Press, Cambridge, 1975), and John Alcock, *Animal Behavior*, 4th edition (Sinauer, Sunderland, 1989).

Outstanding books that discuss the evolution of sexual behaviour include Donald Symons, *The Evolution of Human Sexuality* (Oxford University Press, Oxford, 1979); R.D. Alexander, *Darwinism and Human Affairs* (University of Washington Press, Seattle, 1979); Napoleon A. Chagnon and William Irons, *Evolutionary Biology and Human Social Behavior* (Duxbury Press, North Scituate, Massachusetts, 1979); Tim Halliday, *Sexual Strategies* (University of Chicago Press, Chicago, 1980); Glenn Hausfater and Sarah Hrdy, *Infanticide* (Aldine, Hawthorne, New York, 1980); Sarah Hrdy, *The Woman that Never Evolved* (Harvard University Press, Cambridge, 1981); Nancy Tanner, *On Becoming Human* (Cambridge University Press, New York, 1981); Frances Dahlberg, *Woman the Gatherer* (Yale University Press, New Haven, 1981); Martin Daly and Margo Wilson, *Sex, Evolution, and Behavior* (Willard Grant Press, Boston, 1983); Bettyann Kevles, *Females of the Species* (Harvard University Press, Cambridge, 1986); and Hanny Lightfoot-Klein, *Prisoners of Ritual: an Odyssey into Female Genital Circumcision in Africa* (Harrington Park Press, Binghamton, 1989).

Books dealing specifically with primate reproductive biology include C.E. Graham, *Reproductive Biology of the Great Apes* (Academic Press, New York, 1981); B.B. Smuts *et al, Primate Societies* (University of Chicago Press, Chicago, 1986); Jane Goodall, *The Chimpanzees of Gombe* (Harvard University Press, Cambridge, 1986); Toshisada Nishida, *The Chimpanzees of the Mahale Mountains, Sexual and Life History Strategies* (University of Tokyo Press, 1990); and Takayoshi Kano, *The Last Ape: Pygmy Chimpanzee Behavior and Ecology* (Stanford University Press, Stanford, 1991).

Articles on the evolution of sexual physiology and behaviour include the following: R.V. Short, 'The evolution of human reproduction', *Proceedings of the Royal Society (London)*, series B 195, pp. 3–24 (1976); R.V. Short, 'Sexual selection and its component parts, somatic and genetical selection, as illustrated by man and the great apes', *Advances in the Study of Behavior* 9, pp. 131–58 (1979); N. Burley, 'The evolution of concealed ovulation', *American Naturalist* 114, pp. 835–58 (1979); A.H. Harcourt *et al*, 'Testis weight, body weight, and breeding system in primates', *Nature* 293, pp. 55–57 (1981); R.D. Martin and R.M. May, 'Outward signs of breeding', *Nature* 293, pp. 7–9 (1981); M. Daly and

.M.I. Wilson, 'Whom are newborn babies said to resemble?', *Ethology and Sociobiology* 3, pp. 69–78 (1982); M. Daly, M. Wilson, and S.J. Weghorst, 'Male sexual jealousy', *Ethology and Sociobiology* 3, 11–27 (1982); A.F. Dixson, 'Observations on the evolution and behavioral significance of "sexual skin" in female primates', *Advances in the Study of Behavior* 13, pp. 63–106 (1983); S.J. Andelman, 'Evolution of concealed ovulation in vervet monkeys (*Cercopithecus aethiops*)', *American Naturalist* 129, pp. 785–99 (1987); and P.H. Harvey and R.M. May, 'Out for the sperm count', *Nature* 337, pp. 508–9 (1989).

Chapter Four discussed several examples illustrating how birds combine extramarital sex with apparent monogamy. Detailed examples of such studies are presented in papers by D.W. Mock, 'Display repertoire shifts and extra-marital courtship in herons', *Behaviour* 69, pp. 57–71 (1979); P. Mineau and F. Cooke, 'Rape in the lesser snow goose', *Behaviour* 70, pp. 280–91 (1979); D.F. Werschel, 'Nesting ecology of the Little Blue Heron: promiscuous behavior', *Condor* 84, pp. 381–84 (1982); M.A. Fitch and G.W. Shuart, 'Requirements for a mixed reproductive strategy in avian species', *American Naturalist* 124, pp. 116–26 (1984); and R. Alatalo *et al*, 'Extra-pair copulations and mate guarding in the polyterritorial pied flycatcher, *Ficedula hypoleuca*', *Behaviour* 101, pp. 139–55 (1987).

Chapter 5: How We Pick Our Mates and Sex Partners

Not surprisingly, this topic has called forth much scientific study. Some papers exemplifying the literature on mate choice by humans are E. Walster *et al*, 'Importance of physical attractiveness in dating behavior', *Journal of Personality and Social Psychology* 4, pp. 508–16 (1966); J.N. Spuhler, 'Assortative mating with respect to physical characteristics', *Eugenics Quarterly* 15, pp. 128–40 (1968); E. Berscheid and K. Dion, 'Physical attractiveness and dating choice: a test of the matching hypothesis', *Journal of Experimental Social Psychology* 7, 173–89 (1971); S.G. Vandenberg, 'Assortative mating, or who marries whom?', *Behavior Genetics* 2, pp. 127–57 (1972); G.E. DeYoung and B. Fleischer, 'Motivational and personality trait relationships in mate selection', *Behavior Genetics* 6, pp. 1–6 (1976); E. Crognier, 'Assortative mating for physical features in an African population from Chad', *Journal of Human Evolution* 6, pp. 105–114 (1977); P.N. Bentler and M.D. Newcomb, 'Longitudinal study of marital success and failure', *Journal of Consulting and Clinical Psychology* 46, pp. 1053–70 (1978); R.C. Johnson *et al*, 'Secular change in degree of assortative mating for ability?', *Behavior Genetics* 10, pp. 1–8 (1980); W.E. Nance *et al*, 'A model for the analysis of mate selection in the marriages of twins', *Acta Geneticae Medicae Gemellologiae* 29, pp. 91–101 (1980); D. Thiessen and B. Gregg, 'Human assortative mating and genetic equilibrium: an evolutionary perspective', *Ethology and Sociobiology* 1, pp. 111–40 (1980); D.M. Buss, 'Human mate selection', *American Scientist* 73, pp. 47–51 (1985); A.C. Heath and L.J. Eaves, 'Resolving the effects of phenotype and social background on mate selection', *Behavior Genetics* 15, pp. 75–90 (1985); and A.C. Heath *et al*, 'No decline in assortative mating for educational level', *Behavior Genetics* 15, pp. 349–69 (1985). Also relevant is a book by B.I. Murstein, *Who Will Marry Whom? Theories and Research in Marital Choice* (Springer, New York, 1976).

The literature on mate choice by animals is at least as extensive as that for humans. A good starting point is a book edited by Patrick Bateson, *Mate Choice* (Cambridge University Press, Cambridge, 1983). Bateson's own studies on Japanese quail are summarized in Chapter Eleven of that book, and also in his papers 'Sexual imprinting and optimal outbreeding', *Nature* 273, pp. 659–60 (1978) and 'Preferences for cousins in Japanese quail', *Nature* 295, pp. 236–37 (1982). Studies of mice and rats that grow up to prefer the perfumes of their mothers or fathers are described by T.J. Fillion and E.M. Blass, 'Infantile experience with suckling odors determines adult sexual behavior in male rats', *Science* 231, pp. 729–31 (1986), and by B. D'Udine and E. Alleva, 'Early experience and sexual preferences in rodents', pp. 311–27 in the book cited above by Patrick Bateson.

Finally, some other relevant papers are cited under the further readings for Chapters Three, Four and Six.

Chapter 6: Sexual Selection, and the Origin of Human Races

Darwin's own classic account is still a good introduction to natural selection: Charles Darwin, *On the Origin of Species by Means of Natural Selection, or the Preservation of Favoured Races in the Struggle for Life* (John Murray, London, 1859). An outstanding modern account is that of Ernst Mayr, *Animal Species and Evolution* (Harvard University Press, Cambridge, 1963).

Three books by Carleton S. Coon describe human geographic variation, compare it to geographic variation in climate, and attempt to account for human variation in terms of natural selection. They are *The Origin of Races* (Knopf, New York, 1962), *The Living Races of Man* (Knopf, New York, 1965), and *Racial Adaptations* (Nelson-Hall, Chicago, 1982). Three other relevant books are by Stanley M. Garn, *Human Races*, 2nd edition (Thomas, Springfield, Illinois, 1965), especially its Chapter Five; K.F. Dyer, *The Biology of Racial Integration* (Scientechnica, Bristol, 1974), especially its Chapters Two and Three; and A.S. Boughey, *Man and the Environment*, 2nd edition (Macmillan, New York, 1975).

Interpretations of geographic variation in human skin colour in terms of natural selection are put forward by W.F. Loomis, 'Skin-pigment regulation of vitamin-D biosynthesis in man', *Science* 157, pp. 501–6 (1967); Vernon Riley, *Pigmentation* (Appleton-Century-Crofts, New York, 1972), especially its Chapter Two; R.F. Branda and J.W. Eaton, 'Skin color and nutrient photolysis: an evolutionary hypothesis', *Science* 201, pp. 625–26 (1978); P.J. Byard, 'Quantitative genetics of human skin color', *Yearbook of Physical Anthropology* 24, pp. 123–37 (1981); and W.J. Hamilton III, *Life's Color Code* (McGraw-Hill, New York, 1983). Human geographic variation in response to cold is described by G.M. Brown and J. Page, 'The effect of chronic exposure to cold on temperature and blood flow of the hand', *Journal of Applied Physiology* 5, pp. 221–27 (1952), and T. Adams and B.G. Covino, 'Racial variations to a standardized cold stress', *Journal Of Applied Physiology* 12, pp. 9–12 (1958).

Just as for natural selection, Darwin's own account remains a good introduction to sexual selection: Charles Darwin, *The Descent of Man, and Selection in Relation to Sex* (John Murray, London, 1871). The further readings listed under Chapter Five for mate selection by animals are also relevant to this chapter. Malte Andersson describes his experiments on how female widowbirds responded to males with artificially shortened or lengthened tails in an article 'Female choice selects for extreme tail length in a widowbird', *Nature* 299, pp. 818–20 (1982). Three papers describing mate choice by white, blue, or pink snow geese are by F. Cooke and C.M. McNally: 'Mate selection and colour preferences in Lesser Snow Geese', *Behaviour* 53, pp. 151–70 (1975); F. Cooke *et al*, 'Assortative mating in Lesser Snow Geese (*Anser caerulescens*)', *Behavior Genetics* 6, pp. 127–40 (1976); and F. Cooke and J.C. Davies, 'Assortative mating, mate choice, and reproductive fitness in Snow Geese', pp. 279–95 in *Mate Choice* by Patrick Bateson, already cited.

Chapter 7: Why Do We Grow Old and Die?

The classic paper in which George Williams presented an evolutionary theory of aging is 'Pleiotropy, natural selection, and the evolution of senescence', *Evolution* 11, pp. 398–411 (1957). Other papers that have employed evolutionary approaches are by G. Bell, 'Evolutionary and non-evolutionary theories of senescence', *American Naturalist* 124, pp.

600–3 (1984); E. Beutler, 'Planned obsolescence in humans and in other biosystems', *Perspectives in Biology and Medicine* 29, pp. 175–79 (1986); R.J. Goss, 'Why mammals don't regenerate – or do they?', *News in Physiological Sciences* 2, 112–15 (1987); L.D. Mueller, 'Evolution of accelerated senescence in laboratory populations of *Drosophila*', *Proceedings of the National Academy of Sciences* 84, pp. 1974–77 (1987); and T.B. Kirkwood, 'The nature and causes of ageing', pp. 193–206 in a book edited by D. Evered and J. Whelan, *Research and the Ageing Population* (John Wiley, Chichester, 1988).

Two books exemplifying the physiological (proximate-cause) approach to aging are by R.L. Walford, *The Immunologic Theory of Aging* (Munksgaard, Copenhagen, 1969), and MacFarlane Burnett, *Intrinsic Mutagenesis: A Genetic Approach to Ageing* (John Wiley, New York, 1974).

Some papers exemplifying the literature on biological repair and turnover are by R.W. Young, 'Biological renewal: applications to the eye', *Transactions of the Opthalmological Societies of the United Kingdom* 102, pp. 42–75 (1982); A. Bernstein *et al*, 'Genetic damage, mutation, and the evolution of sex', *Science* 229, pp. 1277–81 (1985); J.F. Dice, 'Molecular determinants of protein-half lives in eukaryotic cells', *Federation of American Societies for Experimental Biology Journal* 1, pp. 349–57 (1987); P.C. Hanawalt, 'On the role of DNA damage and repair processes in aging: evidence for and against', pp. 183–98 in a book edited by H.R. Warner *et al*, *Modern Biological Theories of Aging* (Raven Press, New York, 1987); and M. Radman and R. Wagner, 'The high fidelity of DNA duplication', *Scientific American*, pp. 40–46 (August 1988).

While all readers will be aware of the changes in their own bodies with age, three papers describing the cruel facts for three different systems are R.L. Doty *et al*, 'Smell identification ability: changes with age', *Science* 226, pp. 1441–43 (1984); J. Menken *et al*, 'Age and infertility', *Science* 233, pp. 1389–94 (1986); and R. Katzman, 'Normal aging and the brain', *News in Physiological Sciences* 3, pp. 197–200 (1988).

'The Adventure of the Creeping Man' will be found in Arthur Conan Doyle's *The Complete Sherlock Holmes* (Doubleday, New York, 1960). If you think that attempts at self-rejuvenation by hormonal injections were only a fantasy of Doyle's, read how it was actually attempted in David Hamilton, *The Monkey Gland Affair* (Chatto and Windus, London, 1986).

Chapter 8: Bridges to Human Language

How Monkeys See the World (University of Chicago Press, Chicago, 1990), by Dorothy Cheney and Robert Seyfarth, is not only a readable account of vervet vocal communications, but also a good introduction to studies of how animals in general communicate to each other and view the world.

Derek Bickerton has described his studies of creolization and his views on human language origins in two books and several papers. The books are *Roots of Language* (Karoma Press, Ann Arbor, 1981) and *Language and Species* (University of Chicago Press, Chicago, 1990). The papers include 'Creole languages', in *Scientific American* 249, no. 1, pp. 116–22 (1983); 'The language bioprogram hypothesis', in *Behavioral and Brain Sciences* 7, pp. 173–221 (1984); and 'Creole languages and the bioprogram', in *Linguistics: the Cambridge Survey* 2, pp. 267–84, edited by F.J. Newmeyer (Cambridge University Press, Cambridge, 1988). The second of those articles includes, and the third is immediately followed by, presentations by other authors whose views often diverge from Bickerton's.

Pidgin and Creole Languages, by Robert A. Hall, Jnr (Cornell University Press, Ithaca, 1966), is a less recent account of its subject. The best introduction to Neo-Melanesian is *The Jacaranda Diary and Grammar of Melanesian Pidgin* by F. Mihalic (Jacaranda Press, Milton, Queensland, 1971).

Among the many influential books on language by Noam Chomsky are *Language and Mind* (Harcourt Brace, New York, 1968) and *Knowledge of Language: Its Nature, Origin, and Use* (Praeger, New York, 1985).

References to some related fields that I mentioned only briefly in Chapter Eight will also be of interest. Susan Curtiss's book *Genie: a Psycholinguistic Study of a Modern-Day "Wild Child"* (Academic Press, New York, 1977) both relates a gut-wrenching human tragedy and is a detailed study of a child whose parents' pathologies isolated her from normal human language and contact until the age of thirteen. Recent accounts of efforts to teach language-like communication to captive apes include Carolyn Ristau's and Donald Robbins's paper 'Language and the great apes: a critical review', in *Advances in the Study of Behavior*, vol. XII, pp. 141–255, edited by J.S. Rosenblatt *et al* (Academic Press, New York, 1982); E.S. Savage-Rumbaugh, *Ape Language: from Conditioned Response to Symbol* (Columbia University Press, 1986); and 'Symbols: their communicative use, comprehension, and combination by bonobos (*Pan paniscus*)', by E.S. Savage-Rumbaugh *et al*, in *Advances in Infant Research* vol. VI, pp. 221–78, edited by Carolyn Rovee-Collier and Lewis Lipsitt (Ablex Publishing Corporation, Norwood, New Jersey, 1990). Some starting points in the large literature on early language learning by children include Melissa Bowerman's chapter 'Language Development' in the *Handbook of Cross-cultural Psychology: Developmental Psychology*, vol. IV, pp. 93–185, edited by Harvey Triandis and Alastair Heron (Allyn and Bacon, Boston, 1981); Eric Wanner and Lila Gleitman, *Language Acquisition: the State of the Art* (Cambridge University Press, Cambridge, 1982); Dan Slobin, *The Crosslinguistic Study of Language Acquisition*, vols I and II (Lawrence Erlbaum Associates, Hillsdale, New Jersey, 1985); and Frank S. Kessel, *The Development of Language and Language Researchers: Essays in Honor of Roger Brown* (Lawrence Erlbaum Associates, Hillsdale, New Jersey, 1988).

Chapter 9: Animal Origins of Art

The book that describes elephant art and illustrates it with photographs of the artist and of her drawings is by David Gucwa and James Ehmann, *To Whom It May Concern: An Investigation of the Art of Elephants* (Norton, New York, 1985). For a similar account of ape art, see Desmond Morris, *The Biology of Art* (Knopf, New York, 1962). Animal art is also treated by Thomas Sebeok, *The Play of Musement* (Indiana University Press, Bloomington, 1981).

There are two fine illustrated books on bowerbirds and birds of paradise, with pictures of their bowers: E.T. Gilliard, *Birds of Paradise and Bower Birds* (Natural History Press, Garden City, New York, 1969), and W.T. Cooper and J.M. Forshaw, *The Birds of Paradise and Bower Birds* (Collins, Sydney, 1977). For a more recent technical account, see my article 'Biology of birds of paradise and bowerbirds', *Annual Reviews of Ecology and Systematics* 17, pp. 17–37 (1986). I published two accounts of the bowerbird species with the fanciest bower, 'Bower building and decoration by the bowerbird *Amblyornis inornatus*', *Ethology* 7, pp. 177–204 (1987); and 'Experimental study of bower decoration by the bowerbird *Amblyornis inornatus*, using colored poker chips', *American Naturalist* 131, pp. 631–53 (1988). Gerald Borgia proved by experiments that female bowerbirds really do care about males' bower decorations, in his paper, 'Bower quality, number of decorations and mating success of male satin bowerbirds *(Ptilonorhynchus violaceus)*: an experimental analysis', *Animal Behaviour* 33, pp. 266–71 (1985). Birds of paradise with somewhat similar habits are described by S.G. and M.A. Pruett-Jones in 'The use of court objects by Lawes' Parotia', *Condor* 90, pp. 538–45 (1988).

Chapter 10: Agriculture's Two-Edged Sword

The health consequences of giving up hunting for farming receive detailed treatment in a book edited by Mark Cohen and George Armelagos, *Paleopathology at the Origins of Agriculture* (Academic Press, Orlando, 1984), and in *The Paleolithic Prescription* (Harper and Row, New York, 1988) by S. Boyd Eaton, Marjorie Shostak, and Melvin Konner.

The world's hunter-gatherers are summarized in a book edited by Richard B. Lee and Irven DeVore, *Man the Hunter* (Aldine, Chicago, 1968). References describing the work schedule of hunter-gatherers, and in some cases comparing it with that of farmers, include the same book, plus the book by Richard Lee *The !Kung San* (Cambridge University Press, Cambridge, 1979), and the following articles: K. Hawkes *et al*, 'Aché at the settlement: contrasts between farming and foraging', *Human Ecology* 15, pp. 133–61 (1987); K. Hawkes *et al*, 'Hardworking Hadza grandmothers', pp. 341–66 in *Comparative Socioecology of Mammals and Man*, edited by V. Standen and R. Foley (London, Blackwell, 1987); and K. Hill and A.M. Hurtado, 'Hunter-gatherers of the New World', *American Scientist* 77, pp. 437–43 (1989). The slow spread of ancient farmers across Europe is described by Albert J. Ammerman and L.L. Cavalli-Sforza, *The Neolithic Transition and the Genetics of Populations in Europe* (Princeton University Press, Princeton, 1984).

Chapter 11: Why Do We Smoke, Drink, and Use Dangerous Drugs?

Amotz Zahavi explains his handicap theory in two papers, 'Mate selection – a selection for a handicap', *Journal of Theoretical Biology* 53, pp. 205–14 (1975), and 'The cost of honesty (further remarks on the handicap principle)', *Journal of Theoretical Biology* 67, pp. 603–5 (1977). Two other well-known models of how animals evolve to choose their mates are the runaway selection model and the truth-in-advertising model. The former was developed in a book by R.A. Fisher, *The Genetical Theory of Natural Selection* (Clarendon Press, Oxford, 1930); the latter, in a paper by A. Kodric-Brown and J.H. Brown 'Truth in advertising: the kinds of traits favoured by sexual selection', *American Naturalist* 14, pp. 309–23 (1984). Melvin Konner develops another perspective on risky human behaviour patterns in a chapter 'Why the reckless survive' from his book with the same title (Viking, New York, 1990). For discussions of American Indian enemas, see Peter Furst's and Michael Coe's account of the discovery of Maya enema vases in their article 'Ritual enemas', *Natural History Magazine* 86, pp. 88–91 (March 1977); Johannes Wilbert's book *Tobacco and Shamanism in South America* (Yale University Press, New Haven, 1987); and Justin Kerr's *The Maya Vase Book*, 2 vols (Kerr Associates, New York, 1989 and 1990), illustrating Maya vases and analysing one enema vase in detail on pp. 349–61 of Vol. II. Also relevant are the many further readings on sexual selection and mate choice already listed under Chapters Five and Six.

Chapter 12: Alone in a Crowded Universe

Everything that you might want to know about woodpeckers in general, and about each particular species of them, is contained in a book by Lester L. Short, *Woodpeckers of the World* (Delaware Museum of Natural History, Greenville, Delaware, 1982). Pioneering calculations arguing for the existence of intelligent extraterrestrial life were carried out by I.S. Shklovskii and Carl Sagan, *Intelligent Life in the Universe* (Holden-Day, San Francisco, 1966).

Chapter 13: The Last First Contacts

Bob Connolly's and Robin Anderson's book *First Contact* (Viking Penguin, New York, 1987) describes first contact in the New Guinea highlands through the eyes of both the whites and the New Guineans who met there. The quotation on page 207 is taken from their book. Other gripping accounts of first contacts and of pre-contact conditions include Don Richardson's *Peace Child* (Regal Books, Ventura, 1974) for the Sawi people of southwest New Guinea, and Napoleon A. Chagnon's *Yanomamo, The Fierce People*, 3rd edition (Holt, Rinehart and Winston, New York, 1983) for the Yanomamo Indians of Venezuela and Brazil. A clear history of the exploration of New Guinea is by Gavin Souter, *New Guinea: The Last Unknown* (Angus and Robertson, London, 1963). The leaders of the Third Archbold Expedition describe their entrance into the Grand Valley of the Balim River in the report by Richard Archbold *et al*, 'Results of the Archbold Expeditions. No. 41. Summary of the 1938– 1939 New Guinea expedition', *Bulletin of the American Museum of Natural History* 79, pp. 197–288 (1942). Two accounts by earlier explorers who attempted to penetrate the mountains of New Guinea are by A.F.R. Wollaston, *Pygmies and Papuans* (Smith Elder, London, 1912), and A.S. Meek, *A Naturalist in Cannibal Land* (Fisher Unwin, London, 1913).

Chapter 14: Accidental Conquerors

Books that discuss plant as well as animal domestication in relation to the development of civilization include C.D. Darlington, *The Evolution of Man and Society* (Simon and Schuster, New York, 1969); Peter J. Ucko and G.W. Dimbleby, *The Domestication and Exploitation of Plants and Animals* (Aldine, Chicago, 1969); Erich Isaac, *Geography of Domestication* (Prentice-Hall, Englewood Cliffs, New Jersey, 1970); and David R. Harris and Gordon C. Hillman, *Foraging and Farming* (Unwin Hyman, London, 1989).

References on animal domestication include S. Bokonyi, *History of Domestic Mammals in Central and Eastern Europe* (Akademiai Kiado, Budapest, 1974); S.J.M. Davis and F.R. Valla, 'Evidence for domestication of the dog 12,000 years ago in the Natufian of Israel', *Nature* 276, pp. 608–10 (1978); Juliet Clutton-Brock, 'Man-made dogs', *Science* 197, pp. 1340–42 (1977), and *Domesticated Animals from Early Times* (British Museum of Natural History, London, 1981); Andrew Sherratt, 'Plough and pastoralism: aspects of the secondary products revolution', pp. 261–305 in a book edited by Ian Hodder *et al*, *Pattern of the Past* (Cambridge University Press, Cambridge, 1981); Stanley J. Olsen, *Origins of the Domestic Dog* (University of Arizona Press, Tucson, 1985); E.S. Wing, 'Domestication of Andean mammals', pp. 246–64 in *High Altitude Tropical Biogeography*, edited by F. Vuilleumier and M. Monasterio (Oxford University Press, New York, 1986); Simon J.M. Davis, *The Archaeology of Animals* (Yale University Press, New Haven, 1987); Dennis C. Turner and Patrick Bateson, *The Domestic Cat: The Biology of its Behaviour* (Cambridge University Press, Cambridge, 1988); and Wolf Herre and Manfred Rohrs, *Haustiere – zoologisch gesehen*, 2nd edition (Fischer, Stuttgart, 1990).

Domestication specifically of the horse, and its importance, are the subjects of books by Frank G. Row, *The Indian and the Horse* (University of Oklahoma Press, Norman, 1955); Robin Law, *The Horse in West African History* (Oxford University Press, Oxford, 1980); and Matthew J. Kust, *Man and Horse in History* (Plutarch Press, Alexandria, Virginia, 1983). The development of wheeled vehicles, including war chariots, is treated in books by M.A. Littauer and J.H. Crouwel, *Wheeled Vehicles and Ridden Animals in the Ancient Near East* (Brill, Leiden, 1979) and by Stuart Piggott, *The Earliest Wheeled Transport* (Thames and Hudson, London, 1983). Edward Shaughnessy describes the arrival of the horse and chariot in China in 'Historical perspectives on the introduction of the chariot into China', *Harvard Journal of Asiatic Studies* 48, pp. 189–237 (1988).

For general accounts of plant domestication, see Kent V. Flannery, 'The origins of agriculture', *Annual Review of Anthropology* 2, pp. 271–310 (1973); Charles B. Heiser, Jnr, *Seed to Civilization*, 2nd edition (Freeman, San Francisco, 1981), and *Of Plants and Peoples* (University of Oklahoma Press, Norton, 1985); David Rindos, *The Origins of Agriculture: an Evolutionary Perspective* (Academic Press, New York, 1984); and Hugh H. Iltis, 'Maize evolution and agricultural origins', pp. 195–213 in *Grass Systematics and Evolution*, edited by T.R. Soderstrom *et al* (Smithsonian Institution Press, Washington DC, 1987). This and other papers by Iltis are a stimulating source of ideas about the differing ease of cereal domestication in the Old and New World.

Plant domestication specifically in the Old World is treated by Jane

Renfrew, *Palaeoethnobotany* (Columbia University Press, New York, 1973), and by Daniel Zohary and Maria Hopf, *Domestication of Plants in the Old World* (Clarendon Press, Oxford, 1988). Corresponding accounts for the New World include Richard S. MacNeish, 'The food-gathering and incipient agricultural stage of prehistoric Middle America', pp. 413–26 in the *Handbook of Middle American Indians*, edited by Robert Wauchope and Robert C. West, Vol. I: *Natural Environment and Early Cultures* (University of Texas Press, Austin, 1964); P.C. Mangelsdorf *et al*, 'Origins of agriculture in Middle America', pp. 427–45 in the book by Wauchope and West; D. Ugent, 'The potato', *Science* 170, pp. 1161–66 (1970); C.B. Heiser, Jnr, 'Origins of some cultivated New World plants', *Annual Reviews of Ecology and Systematics* 10, pp. 309–26 (1979); H.H. Iltis, 'From teosinte to maize: the catastrophic sexual dismutation', *Science* 222, pp. 886–94 (1983); William F. Keegan, *Emergent Horticultural Economies of the Eastern Woodlands* (Southern Illinois University, Carbondale, 1987); and B.D. Smith, 'Origins of agriculture in eastern North America', *Science* 246, pp. 1566–71 (1989). Three pioneering books point out the asymmetrical intercontinental spread of diseases, pests, and weeds: William H. McNeill, *Plagues and Peoples* (Anchor Press, Garden City, New York, 1976); and Alfred W. Crosby, *The Columbian Exchange: Biological and Cultural Consequences of 1492* (Greenwood Press, Westport, 1972), and *Ecological Imperialism: The Biological Expansion of Europe, 900–1900* (Cambridge University Press, Cambridge, 1986).

Chapter 15: Horses, Hittites, and History

Two stimulating, knowledgeable recent books summarizing the Indo-European problem are by Colin Renfrew, *Archaeology and Language* (Jonathan Cape, London, 1987), and J.P. Mallory, *In Search of the Indo-Europeans* (Thames and Hudson, London, 1989). For the reasons explained in my chapter, I agree with Mallory's conclusions, and disagree with Renfrew's, concerning the approximate time and place of proto-Indo-European origins.

An older but still useful comprehensive multi-authored book is by George Cardona *et al*, *Indo-European and Indo-Europeans* (University of Pennsylvania Press, Philadelphia, 1970). A journal titled (what else?) *The Journal of Indo-European Studies* is the main outlet for technical publication in this field.

The view that both Mallory and I find persuasive is supported in the writings of Marija Gimbutas, who is the author of four books in this field: *The Balts* (Praeger, New York, 1963), *The Slavs* (Thames and Hudson,

London, 1971), *The Goddesses and Gods of Old Europe* (Thames and Hudson, London, 1982), and *The Language of the Goddess* (Harper and Row, New York, 1989). Gimbutas also described her work in chapters in the book by Cardona *et al* cited above, in the books by Polomé and by Bernhard and Kandler-Pálsson cited below, and in the *Journal of Indo-European Studies* 1, pp. 1–20 and 163–214 (1973); 5, pp. 277–338 (1977); 8, pp. 273–315 (1980); and 13, pp. 185–201 (1985).

Books or monographs dealing with early Indo-European peoples themselves are by Emile Benveniste, *Indo-European Language and Society* (Faber and Faber, London, 1973); Edgar Polomé, *The Indo-Europeans in the Fourth and Third Millenia* (Karoma, Ann Arbor, 1982); Wolfram Bernhard and Anneliese Kandler-Pálsson, *Ethnogenese europäischer Völker* (Fischer, Stuttgart, 1986); and Wolfram Nagel, 'Indogermanen und Alter Orient: Rückblick und Ausblick auf den Stand des Indogermanen-problems', *Mitteilungen der Deutschen Orient-Gesellschaft zu Berlin* 119, pp. 157–213 (1987). Books on the languages themselves include those by Henrik Birnbaum and Jaan Puhvel, *Ancient Indo-European Dialects* (University of California Press, Berkeley, 1966); W.B. Lockwood, *Indo-European Philology* (Hutchinson, London, 1969); Norman Bird, *The Distribution of Indo-European Root Morphemes* (Harrassowitz, Wiesbaden, 1982); and Philip Baldi, *An Introduction to the Indo-European Languages* (Southern Illinois University Press, Carbondale, 1983). Paul Friedrich's book *Proto-Indo-European Trees* (University of Chicago Press, Chicago, 1970) uses the evidence of tree names in an attempt to deduce the Indo-European homeland.

W.P. Lehmann and L. Zgusta provide and discuss a sample of reconstructed proto-Indo-European in their chapter 'Schleicher's tale after a century', pp. 455–66 in *Studies in Diachronic, Synchronic, and Typological Linguistics*, edited by Bela Brogyanyi (Benjamins, Amsterdam, 1979).

The references to the domestication and importance of horses cited under Chapter Fourteen are also relevant to the role of horses in the Indo-European expansion. Papers specifically on this subject are by David Anthony, 'The "Kurgan culture", Indo-European origins and the domestication of the horse: a reconsideration', *Current Anthropology* 27, pp. 291–313 (1986); and by David Anthony and Dorcas Brown, 'The origins of horseback riding', *Antiquity* 65, pp. 22–38 (1991).

Chapter 16: In Black and White

Three books providing general surveys of genocide are by Irving Horowitz, *Genocide: State Power and Mass Murder* (Transaction Books,

New Brunswick, 1976); Leo Kuper, *The Pity of it All* (Gerald Duck-worth, London, 1977); and Leo Kuper, *Genocide: Its Political Use in the 20th Century* (Yale University Press, New Haven, 1981). A gifted psychiatrist, Robert J. Lifton, has published studies of the psychological effects of genocide on its perpetrators and survivors, including *Death in Life: Survivors of Hiroshima* (Random House, New York, 1967) and *The Broken Connection* (Simon and Schuster, New York, 1979).

Books that describe the extermination of the Tasmanians and other native Australian groups include N.J.B. Plomley, *Friendly Mission: The Tasmanian Journals and Papers of George Augustus Robinson 1829–1834* (Tasmanian Historical Research Association, Hobart, 1966); C.D. Rowley, *The Destruction of Aboriginal Society*, Vol. I (Australian National University Press, Canberra, 1970); and Lyndall Ryan, *The Aboriginal Tasmanians* (University of Queensland Press, St. Lucia, 1981). Patricia Cobern's letter indignantly denying that Australian whites exterminated the Tasmanians has been reprinted as an appendix to the book by J. Peter White and James F. O'Connell, *A Prehistory of Australia, New Guinea, and Sahul* (Academic Press, New York, 1982).

Among the many books detailing the extermination of American Indians by white settlers are Wilcomb E. Washburn, 'The moral and legal justification for dispossessing the Indians', pp. 15–32 in *Seventeenth Century America*, edited by James Morton Smith (University of North Carolina Press, Chapel Hill, 1959); Alvin M. Josephy, Jnr, *The American Heritage Book of Indians* (Simon and Schuster, New York, 1961); Howard Peckham and Charles Gibson, *Attitudes of Colonial Powers Towards the American Indian* (University of Utah Press, Salt Lake City, 1969); Francis Jennings, *The Invasion of America: Indians, Colonialism, and the Cant of Conquest* (University of North Carolina Press, Chapel Hill, 1975); Wilcomb E. Washburn, *The Indian in America* (Harper and Row, New York, 1975); Arrell Morgan Gibson, *The American Indian, Prehistory to the Present* (Heath, Lexington, Massachusetts, 1980); and Wilbur H. Jacobs, *Dispossessing the American Indian* (University of Oklahoma Press, Norman, 1985). The extermination of the Yahi Indians, and the survival of Ishi, are the subjects of Theodora Kroeber's classic book *Ishi in Two Worlds: A Biography of the Last Wild Indian in North America* (University of California Press, Berkeley, 1961). The extermination of Brazil's Indians is treated by Sheldon Davis, *Victims of the Miracle* (Cambridge University Press, Cambridge, 1977).

Genocide under Stalin is described in books by Robert Conquest, including *The Harvest of Sorrow* (Oxford University Press, New York, 1986).

Accounts of murder and mass murder of animals by other animals of the same species are given by E.O. Wilson, *Sociobiology* (Harvard

University Press, Cambridge, 1975); Cynthia Moss, *Portraits in the Wild*, 2nd edition (University of Chicago Press, Chicago, 1982); and Jane Goodall, *The Chimpanzees of Gombe* (Harvard University Press, Cambridge, 1986).

Chapter 17: The Golden Age that Never Was

Extinction of animals in the Late Pleistocene and Early Recent era are described comprehensively in the book edited by Paul Martin and Richard Klein, *Quaternary Extinctions* (University of Arizonia Press, Tucson, 1984). For the history of deforestation, see John Perlin's book *A Forest Journey* (Norton, New York, 1989).

Comprehensive accounts of New Zealand's plants, animals, geology, and climate will be found in a book edited by G. Kuschel, *Biogeography and Ecology in New Zealand* (Junk, V.T. Hague, 1975). New Zealand examples of extinction are summarized in chapters 32–34 of the book by Martin and Klein, cited above. Moas are the subject of a supplement to the *New Zealand Journal of Ecology*, Vol. XII (1989); see especially the articles by Richard Holdaway on pp. 11–25, and by Ian Atkinson and R.M. Greenwood on pp. 67–96. Other key articles relevant to moas are by G. Caughley, 'The colonization of New Zealand by the Polynesians', *Journal of the Royal Society of New Zealand* 18, pp. 245–70 (1988), and by A. Anderson, 'Mechanics of overkill in the extinction of New Zealand moas', *Journal of Archaeological Science* 16, pp. 137–151 (1989).

Examples of extinction in Madagascar and Hawaii are described in Chapters 26 and 35 respectively of the book by Martin and Klein, cited above. The Henderson Island story is told by David Steadman and Storrs Olson, 'Bird remains from an archaeological site on Henderson Island, South Pacific: man-caused extinctions on an "uninhabited" island', *Proceedings of the National Academy of Sciences* 82, pp. 6191–95 (1985). See under suggested reading for Chapter Eighteen for accounts of species' extinction in the Americas.

The grisly end of Easter Island civilization is recounted by Patrick V. Kirch in his book *The Evolution of the Polynesian Chiefdoms* (Cambridge University Press, Cambridge, 1984). Easter's deforestation was reconstructed by J. Flenley, 'Stratigraphic evidence of environmental change on Easter Island', *Asian Perspectives* 22, pp. 33–40 (1979), and by J. Flenley and S. King, 'Late Quaternary pollen records from Easter Island', *Nature* 307, pp. 47–50 (1984).

Some accounts of the rise and fall of Anasazi settlement at Chaco Canyon are J.L. Betancourt and T.R. Van Devender, 'Holocene vegetation in Chaco Canyon, New Mexico', *Science* 214, pp. 656–58

(1981); M.L. Samuels and J.L. Betancourt, 'Modeling the long-term effects of fuelwood harvests on pinyon-juniper woodlands', *Environmental Management* 6, pp. 505–15 (1982); J.L. Betancourt *et al*, 'Prehistoric long-distance transport of construction beams, Chaco Canyon, New Mexico', *American Antiquity* 51, pp. 370–75 (1986); Kendrick Frazier, *People of Chaco: A Canyon and its Culture* (Norton, New York, 1986); and Alden C. Hayes *et al*, *Archaeological Surveys of Chaco Canyon* (University of New Mexico Press, Albuquerque, 1987).

Everything that anyone would want to know about *Packrat Middens* is described in the eponymous book by Julio Betancourt, Thomas Van Devender, and Paul Martin (University of Arizona Press, Tucson, 1990). In particular, Chapter Nineteen of that book analyses the hyrax middens from Petra.

The possible link between environmental damage and the decline of Greek civilization is explored by K.O. Pope and T.H. Van Andel in 'Late Quaternary civilization and soil formation in the southern Argolid: its history, causes and archaeological implications', *Journal of Archaeological Science* 11, pp. 281–306 (1984); T.H. van Andel *et al*, 'Five thousand years of land use and abuse in the southern Argolid', *Hesperia* 55, pp. 103–28 (1986); and C. Runnels and T.H. van Andel, 'The evolution of settlement in the southern Argolid, Greece: an economic explanation', *Hesperia* 56, pp. 303–34 (1987).

Books on the rise and fall of Maya civilization include those by T. Patrick Culbert, *The Classic Maya Collapse* (University of New Mexico Press, Albuquerque, 1973); Michael D. Coe, *The Maya*, 3rd edition (Thames and Hudson, London, 1984); Sylvanus G. Morley *et al*, *The Ancient Maya*, 4th edition (Stanford University Press, Stanford, 1983); and Charles Gallenkamp, *Maya: The Riddle and Rediscovery of A Lost Civilization*, 3rd revised edition (Viking Penguin, New York, 1985).

For a comparative account of collapses of civilizations, see *The Collapse of Ancient States and Civilizations*, edited by Norman Yoffee and George L. Cowgill (University of Arizona Press, Tucson, 1988).

Chapter 18: Blitzkrieg and Thanksgiving in the New World

Three books provide good starting points and many references to the large, contentious literature on human settlement and the extinction of large animals in the New World. They are the book by Paul Martin and Richard Klein cited under Chapter Seventeen; Brian Fagan, *The Great Journey* (Thames and Hudson, New York, 1987); and Ronald C. Carlisle (editor), *Americans Before Columbus: Ice-Age Origins* (Ethnology Monographs No. 12, Department of Anthropology, University of Pittsburgh, 1988).

The blitzkrieg hypothesis was outlined by Paul Martin in his article 'The Discovery of America', *Science* 179; pp. 969–74 (1973), and modelled mathematically by J.E. Mosimann and Martin in 'Simulating overkill by Paleoindians', *American Scientist* 63, pp. 304–13 (1975).

The series of articles that C. Vance Haynes, Jnr has published on Clovis culture and its origins include a chapter on pp. 345–53 of the book by Martin and Klein, cited under Chapter Seventeen, and the following selected articles: 'Fluted projectile points: their age and dispersion', *Science* 145, pp. 1408–13 (1961); 'The Clovis culture', *Canadian Journal of Anthropology* 1, pp. 115–21 1980); and 'Clovis origin update', *The Kiva* 52, pp. 83–93 (1987).

For the simultaneous extinction of the Shasta ground sloth and Harrington's mountain goat, see J.I. Mead *et al*, 'Extinction of Harrington's mountain goat', *Proceedings of the National Academy of Sciences* 83, pp. 836–39 (1986). Critiques of pre-Clovis claims are provided by Roger Owen in a chapter 'The Americas: the case against an Ice-Age human population', pp. 517-63 in *The Origins of Modern Humans*, edited by Fred H. Smith and Frank Spencer (Liss, New York, 1984); by Dena Dincauze, 'An archaeo-logical evaluation of the case for pre-Clovis occupations', in *Advances in World Archaeology* 3, pp. 275–323 (1984); and by Thomas Lynch, 'Glacial-age man in South America? A critical review', in *American Antiquity* 55, pp. 12–36 (1990). Arguments in support of a pre-Clovis date for human occupation levels at Meadowcroft Rockshelter are summarized by James Adovasio in 'Meadowcroft Rockshelter, 1973–1977: a synopsis', pp. 97–131 in J.E. Ericson *et al*, *Peopling of the New World* (Los Altos, California, 1982), and in 'Who are those guys?: some biased thoughts on the initial peopling of the New World', pp. 45–61 in *Americans Before Columbus: Ice-Age Origins*, edited by Ronald C. Carlisle, cited above. The first of several projected volumes with a detailed description of the Monte Verde site is by T.D. Dillehay, *Monte Verde: A Late Pleistocene Settlement in Chile*; Vol. I: *Palaeoenvironment and Site Contexts* (Smithsonian Institution Press, Washington DC, 1989).

Readers interested in keeping up on the story of the first Americans and the last mammoths will enjoy subscribing to a quarterly newspaper, *Mammoth Trumpet*, obtainable from the Center for the Study of the First Americans, 495 College Avenue, Orono, Maine 04473.

Chapter 19: The Second Cloud

Species-by-species accounts of extinct and endangered species are contained in the Red Data Books published by the International Union for Conservation of Nature and Natural Resources (abbreviated IUCN).

There are separate books for various groups of plants and animals, and separate books are also now appearing for different continents. Corresponding books for birds have been prepared by the International Council for Bird Preservation (abbreviated ICBP): Warren B. King, editor, *Endangered Birds of the World: The ICBP Red Data Book* (Smithsonian Institution Press, Washington DC, 1981); and N.J. Collar and P. Andrew, *Birds to Watch: The ICBP World Checklist of Threatened Birds* (ICBP, Cambridge, 1988).

A summary and analysis of modern and Ice-Age extinction waves and their mechanisms are provided by my article 'Historic extinctions: a Rosetta Stone for understanding prehistoric extinctions', pp. 824–62 in *Quaternary Extinctions* by Martin and Klein, cited under Chapter Seventeen. The problem of overlooked species extinctions is discussed in my article 'Extant unless proven extinct? Or extinct unless proven extant?' in *Conservation Biology* 1, pp. 77–79 (1987). Terry Erwin estimates the total number of living species in a paper 'Tropical forests: their richness in Coleoptera and other arthropod species', *The Coleopterists' Bulletin* 36, pp. 74–75 (1982).

Further readings on Pleistocene and Early Recent cases of extinction are given under Chapters Seventeen and Eighteen. In addition, Storrs Olson reviews the extinction of island birds in an article 'Extinction on islands: man as a catastrophe', pp. 50–53 of *Conservation for the Twenty-first Century*, edited by David Western and Mary Pearl (Oxford University Press, New York, 1989). Ian Atkinson's article on pp. 54–75 of the same book, 'Introduced animals and extinctions', summarizes the havoc wrought by rats and other pests.

Epilogue
Nothing Learned, and Everything Forgotten?

Many excellent books discuss the present and future of the extinction crisis and the other crises now facing humanity, their causes, and what to do about them. Among them are the following:

John J. Berger, *Restoring the Earth: How Americans are Working to Renew our Damaged Environment* (Knopf, New York, 1985); editor, *Environmental Restoration: Science and Strategies for Restoring the Earth* (Island Press, Washington DC, 1990).

John Cairns, Jnr, *Rehabilitating Damaged Ecosystems* (CRC Press, Boca Raton, Florida, 1988); with K.L. Dickson and E.E. Herricks, *Recovery and Restoration of Damaged Ecosystems* (University Press of Virginia, Charlottesville, 1977).

Anne and Paul Ehrlich, *Extinction* (Random House, New York, 1981); *Earth* (Franklin Watts, New York, 1987); *The Population Explosion* (Simon and Schuster, New York, 1990); *Healing Earth* (Addison Wesley, New York, 1991).

Paul Ehrlich *et al*, *The Cold and the Dark* (Norton, New York, 1984).

D. Furguson and N. Furguson, *Sacred Cows at the Public Trough* (Maverick Publications, Bend, Oregon, 1983).

Suzanne Head and Robert Heinzman, editors, *Lessons of the Rainforest* (Sierra Club Books, San Francisco, 1990).

Jeffrey A. McNeely, *Economics and Biological Diversity* (International Union for the Conservation of Nature, Gland, 1988); Jeffrey A. McNeely *et al*, *Conserving the World's Biological Diversity* (International Union for the Conservation of Nature, Gland, 1990).

Norman Myers, *Conversion of Tropical Moist Forests* (National Academy of Sciences, Washington DC, 1980); *Gaia: an Atlas of Planet Management* (Doubleday, Garden City, New York, 1984); *The Primary Source* (Norton, New York, 1985).

Michael Oppenheimer and Robert Boyle, *Dead Heat: the Race against the Greenhouse Effect* (Basic Books, New York, 1990).

Walter V. Reid and Kenton R. Miller, *Keeping Options Alive: the Scientific Basis for Conserving Biodiversity* (World Resources Institute, Washington DC, 1989).

Sharon L. Roan, *Ozone Crisis: the Fifteen-Year Evolution of a Sudden Global Emergency* (Wiley, New York, 1989).

Robin Russell Jones and Tom Wigley, editors, *Ozone Depletion: Health and Environmental Consequences* (Wiley, New York, 1989).

Steven H. Schneider, *Global Warming: Are We Entering the Greenhouse Century?*, second edition (Sierra Club Books, San Francisco, 1990).

Michael E. Soulé, editor, *Conservation Biology: the Science of Scarcity and Diversity* (Sinauer, Sunderland, Massachusetts, 1986).

John Terborgh, *Where Have All the Birds Gone?* (Princeton University Press, Princeton, 1990).

E.O. Wilson, *Biophilia* (Harvard University Press, Cambridge, Massachusetts, 1984); editor, *Biodiversity* (National Academy Press, Washington DC, 1988).

Finally, readers interested enough to want to pursue further readings may also want suggestions about what to do to reduce the risk that our children's generation will become extinct. As I explain in the text, the average citizen can do a good deal, both by being active politically and by giving even modest amounts of money to conservation organizations. Here are the names and addresses of a few of the best-known and largest such organizations, among the many that are worthy of support:

World Wide Fund for Nature, Panda House, Weyside Park, Godalming, Surrey, GU7 1XR, UK.

Greenpeace, 30–1 Islington Green, London, N1 8XE, UK.

International Council for Bird Preservation, 32 Cambridge Road, Girton, Cambridge, CB3 OPJ, UK.

International Union for Conservation of Nature and Natural Resources, Avenue du Mont-Blanc, CH-1196 Gland, Switzerland.

Friends of the Earth, 26–28 Underwood Street, London, N1 7JQ, UK.

Conservation Foundation, Lowther Lodge, 1 Kensington Gore, London, SW7 2AR, UK.

ACKNOWLEDGEMENTS

It is a pleasure for me to acknowledge the contributions of many people to this book. My parents and my teachers at Roxbury Latin School taught me how to pursue interests along many lines simultaneously. My debt to my many New Guinean friends will be obvious from the frequency with which I cite their experiences. I owe an equal debt to my many scientist friends and professional colleagues, who patiently explained the subtleties of their subjects and read my drafts. Earlier versions of most of the chapters appeared as articles in *Discover* magazine and in *Natural History* magazine. I have been fortunate to have as my editors Leon Jaroff, Fred Golden, Gil Rogin, Paul Hoffman, and Marc Zabludoff at *Discover*, Alan Ternes and Ellen Goldensohn at *Natural History*, Thomas Miller at HarperCollins Publishers, Neil Belton at Hutchinson Radius Publishers, and my wife Marie Cohen.

JARED M. DIAMOND
Los Angeles
January 1991

INDEX